BRAID GROUP, KNOT THEORY AND STATISTICAL MECHANICS

ADVANCED SERIES IN MATHEMATICAL PHYSICS

Editors-in-charge
V G Kac (*Massachusetts Institute of Technology*)
D H Phong (*Columbia University*)
S-T Yau (*Harvard University*)

Associate Editors
L Alvarez-Gaumé (*CERN*)
J P Bourguignon (*Ecole Polytechnique, Palaiseau*)
T Eguchi (*University of Tokyo*)
B Julia (*CNRS, Paris*)
F Wilczek (*University of California, Santa Barbara*)

Also in the Series

Volume 1: Mathematical Aspects of String Theory
edited by S-T Yau (*Harvard*)
Volume 2: Bombay Lectures on Highest Weight Representations of Infinite Dimensional Lie Algebras
by V G Kac (*MIT*) & A K Raina (*Tata*)
Volume 3: Kac-Moody and Virasoro Algebras: A Reprint Volume for Physicists
edited by P Goddard (*Cambridge*) & D Olive (*Imperial College*)
Volume 4: Harmonic Mappings, Twistors and σ-Models
edited by P. Gauduchon (*CNRS, Paris*)
Volume 5: Geometric Phases in Physics
edited by F. Wilczek and A. Shapere (*IAS, Princeton*)
Volume 7: Infinite Dimensional Lie Algebras and Groups
edited by V Kac (*MIT*)
Volume 8: Introduction to String Field Theory
by W. Siegel (*SUNY*)

Advanced Series in Mathematical Physics
Vol. 9

BRAID GROUP, KNOT THEORY AND STATISTICAL MECHANICS

Editors

C. N. YANG
M. L. GE

World Scientific
Singapore • New Jersey • London • Hong Kong

Published by

World Scientific Publishing Co. Pte. Ltd.,
P O Box 128, Farrer Road, Singapore 9128
USA office: 687 Hartwell Street, Teaneck, NJ 07666
UK office: 73 Lynton Mead, Totteridge, London N20 8DH

BRAID GROUP, KNOT THEORY AND STATISTICAL MECHANICS

ISBN 9971-50-828-1
9971-50-833-8 (pbk)

Printed in Singapore by JBW Printers & Binders Pte. Ltd.

FOREWORD

The theory of knots is a problem in topology which underwent in 1985 a kind of phase transition with the discovery by Jones of his new famous invariant polynomials. Striking as this discovery was, it was even more striking that his work was related to some well known problems in statistical mechanics. There was thus ushered in a new era of cross fertilization of ideas in physics and in mathematics. While the full extent of this welcome development is not yet fathomable, we thought it worthwhile to collect in this single volume articles by workers in the field who have very different approaches. We believe the collection will be useful to many people.

Chen Ning Yang
Mo-Lin Ge

CONTENTS

BRAID GROUP, KNOT THEORY AND STATISTICAL MECHANICS

NOTES ON SUBFACTORS AND STATISTICAL MECHANICS

V.F.R. JONES
Dept. of Mathematics,
University of California at Berkeley,
Berkeley, CA 94720, USA

0. Introduction

A lot has been made in the last few years of connections between knot theory, statistical mechanics, field theory and von Neumann algebras. Because of their more technical nature, the von Neumann algebras have tended to be neglected in surveys. This is not an accurate reflection of their fundamental role in the subject, both as a continuing inspiration and as the vehicle of the discovery of the original ties between statistical mechanics and knot theory. In this largely expository article, we attempt to redress this balance by talking almost entirely about von Neumann algebras and how they occur as algebras of transfer matrices in statistical mechanical models. We shall focus mostly on the Potts model and the model of Fateev and Zamolodchikov, with a brief exposition of how vertex models are related to type III factors.

1. Spin Models in Statistical Mechanics

A spin model may be defined on any graph G. One is given a set S, with Q elements, of spin states per site, and a state of the system is a function from the set of vertices V of G to S. For each edge e in G we suppose given an energy function $E_e : S \times S \longrightarrow \mathbb{R}$ such that $E_e(\sigma, \tau)$ is the energy of the "bond" represented by e if the ends of e are in states σ and τ. (If the edges of G are directed, E_e does not have to be symmetric — see [A–YBP] — but we will not consider this more general case). We then define $w_e : S \times S \longrightarrow \mathbb{R}^+$ by $w_e(\sigma, \tau) = \exp(-\beta E_e(\sigma, \tau))$, β being a constant. Given a state $\sigma : V \longrightarrow S$, its energy is then $\Sigma_e E_e(\sigma(\alpha), \sigma(\beta))$, α and β being the ends of e, and the corresponding Boltzmann weight is then $w(\sigma) = \Pi_e w_e(\sigma(\alpha),\ \sigma(\beta))$.

The *partition function* of the system is then defined to be

$$Z = \sum_{\sigma:V\to S} \prod_e w_e(\sigma(\alpha), \sigma(\beta)) \ .$$

If the graph G has a special form, one may make restrictions on w_e. The most commonly encountered situation in 2 dimensions is a graph which is part of a square lattice

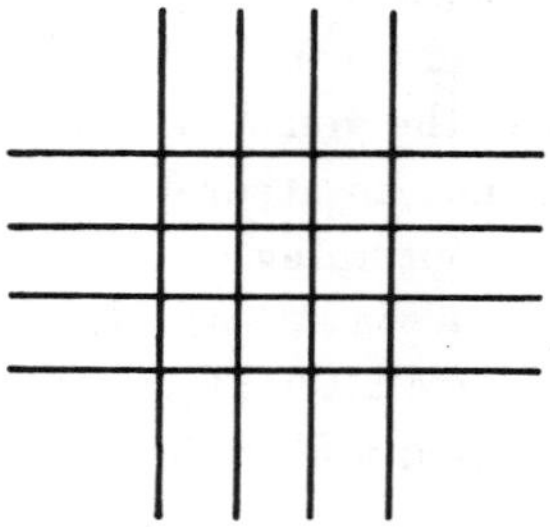

where the boundary conditions may require that identifications are made on the boundary (e.g., periodic boundary conditions when the graph is wrapped on a torus). The edges of the graph can now be divided horizontally and vertically, and the Boltzmann weights are only allowed to be of two kinds, one for each direction: w_{H} and w_{V}.

To calculate the partition function in the above situation, it is very convenient to use a *transfer matrix*, which builds the lattice up row by row. Thus we consider a vector space V with basis S and define the linear map $T : \otimes^n V \longrightarrow \otimes^n V$ by

$$T(\sigma_1 \otimes \sigma_2 \ldots \otimes \sigma_n) = \sum_{\tau_1,\tau_2,\ldots,\tau_n} \left(\prod_{i=1}^{n-1} w_{\mathrm{H}}(\sigma_i, \sigma_{i+1}) \right) \times \left(\prod_{i=1}^{n} w_{\mathrm{V}}(\sigma_i, \tau_i) \right) \tau_1 \otimes \tau_2 \ldots \otimes \tau_n \ .$$

It is easy to see then that the $(\sigma_1, \ldots, \sigma_n; \tau_1, \ldots \tau_n)$ entry of T^m is then the partition function for the system:

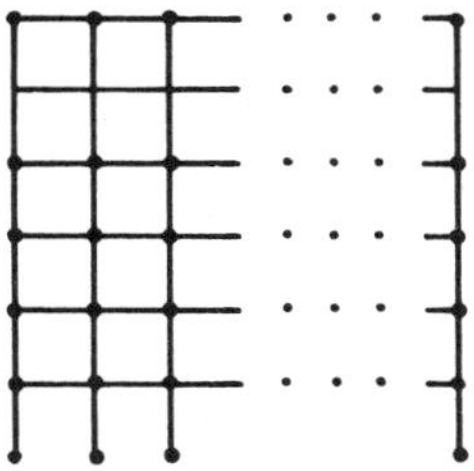

with m rows and n columns, where the spins at the top and bottom are fixed to be $(\tau_1, \tau_2, \ldots, \tau_n)$ and $(\sigma_1, \sigma_2, \ldots, \sigma_n)$, respectively. We see that the row-to-row transfer matrix T further splits up as a product

$$T = R_1 R_3 \ldots R_{2n-1} R_2 R_4 R_6 \ldots R_{2n-2} \,,$$

where

$$\begin{aligned} R_{2i-1}(\sigma_1 \otimes \ldots \otimes \sigma_n) &= \sum_\tau w_{\mathrm{V}}(\sigma_i, \tau) \sigma_1 \otimes \ldots \otimes \tau \otimes \ldots \otimes \sigma_n \\ R_{2i}(\sigma_1 \otimes \ldots \otimes \sigma_n) &= w_{\mathrm{H}}(\sigma_i, \sigma_{i+1}) \sigma_1 \otimes \ldots \otimes \sigma_n \,. \end{aligned}$$

The individual R_i's can generate some very interesting algebras and this was the reason for the connection between von Neumann algebras, statistical mechanics and knot theory referred to in the introduction. The reason for the interesting algebra from the statistical mechanics point of view is the following: Suppose all Boltzmann weights depend on a parameter λ. Suppose further that for each pair λ, λ' in parameter space, there is a third λ'' such that $R_i(\lambda)R_{i+1}(\lambda')R_i(\lambda'') = R_{i+1}(\lambda'')R_i(\lambda')R_{i+1}(\lambda)$ (the so-called star triangle relation), then an elegant argument (on p. 93 of [B]) shows that the diagonal-to-diagonal transfer matrices for the system wrapped on a torus commute for all parameters in the family. Note that the star triangle relation only needs to be checked for $i = 1$ and 2.

We give two examples of the algebraic relations we have in mind.

Example 1.1. The Potts model

This is the model defined by $w_{\mathrm{H}}(\sigma, \tau) = \exp(K_1 \delta_{\sigma,\tau})$ and $w_{\mathrm{V}}(\sigma, \tau) = \exp(K_2 \delta_{\sigma,\tau})$. Then it is clear that if we define e_{2i-1} by $e_{2i-1}(\sigma_1 \otimes \ldots \otimes \sigma_n) = \frac{1}{Q}\Sigma_\tau \sigma_1 \otimes \ldots \otimes \tau \otimes \ldots \sigma_n$, then $R_{2i-1} = Q e_{2i-1} + (e^{K_2} - 1)\mathbb{1}$, $\mathbb{1}$ being the identity matrix, and if $e_{2i}(\sigma_1 \otimes \ldots \otimes \sigma_n) = \delta_{\sigma_i, \sigma_{i+1}}(\sigma_1 \otimes \ldots \otimes \sigma_n)$ then

$R_{2i} = (e^{K_1} - 1)e_{2i} + \mathbb{1}$. It is easy to verify the following relations

$$\text{(TL)} \qquad \begin{aligned} e_i^2 &= e_i \\ e_i e_{i\pm 1} e_i &= \frac{1}{Q} e_i \\ e_i e_j &= e_j e_i \text{ if } |i-j| \geq 2 \,. \end{aligned}$$

The usefulness of these relations was first noticed by Temperley and Lieb ([TL]), who used the easily proven fact that if we define $p = e_1 e_3 \ldots e_{2n-1}$, then $p^2 = p$ and $pxp = \varphi(x)p$ for any x in the algebra generated by $e_1, e_2, \ldots e_{2n-1}$. The proof shows that the linear functional φ is completely determined by the algebraic relations (TL). The linear functional φ is also essentially the sum of all the matrix entries so that the partition function for the Potts model is also determined by (TL). We will see that these relations occur again in our discussion of von Neumann algebras. This remarkable coincidence was first pointed out by D. Evans in 1984. The above star-triangle equation is true for the Potts model but an analysis of the argument shows that the transfer matrices will commute only provided K_1 and K_2 satisfy a relation. This relation is believed to be the same as criticality for the model.

Example 1.2. The model of Fateev and Zamolodchikov

The Potts model has an obvious symmetry group given by the full symmetric group S_Q, since the Boltzmann weight $w(\sigma, \tau)$ depends only on whether $\sigma = \tau$. It is equally clear that the Potts model is the only model with this property. If one reduces the size of the symmetry group one can expect to find other models. The most natural starting case would be $\mathbb{Z}/Q\mathbb{Z}$ so that the Boltzmann weights would have to satisfy the property that $w(a,b)$ depends only on $a - b \pmod{Q}$. Write $w_{\mathrm{H}}(a,b) = x_{\mathrm{H}}(a-b)$ and $w_{\mathrm{V}}(a,b) = x_{\mathrm{V}}(a-b)$. Such a model, with a parameter λ, has been found by Fateev and Zamolodchikov in [FZ]. The functions x_{H} and x_{V} are:

$$x_{\mathrm{H}}(n,\lambda) = \prod_{k=0}^{n-1} \frac{\sin\left(\pi \frac{k}{Q} + \frac{\lambda}{2Q}\right)}{\sin\left(\pi(k+1)/Q - \frac{\lambda}{2Q}\right)} \,.$$

The function $x_{\mathrm{V}}(n,\lambda)$ is the finite Fourier transform of the function $x_{\mathrm{H}}(n,\lambda)$.

To take advantage of the $\mathbb{Z}/Q\mathbb{Z}$ symmetry, it is convenient to analyze the transfer matrices in terms of operators,

$$\begin{aligned} &U_1, U_2, \ldots U_n, \text{ where } U_i U_{i+1} = e^{2\pi i/Q} U_{i+1} U_i, \\ &U_i U_j = U_j U_i \text{ if } |i-j| \geq 2 \text{ and } U_i^Q = 1 \,. \end{aligned}$$

These relations are represented on $\otimes^n V$ by

$$U_{2i-1}(\sigma_1 \otimes \ldots \otimes \sigma_n) = \sigma_1 \otimes \ldots \otimes (\sigma_i + 1) \otimes \ldots \otimes \sigma_n \ ,$$
$$U_{2i}(\sigma_1 \otimes \ldots \otimes \sigma_n) = \omega^{\sigma_i - \sigma_{i+1}}(\sigma_1 \otimes \ldots \otimes \sigma_n) \ .$$

The horizontal transfer matrices for the above model are then $\Sigma_{k=0}^{Q-1} x_{\mathrm{H}}(k,\lambda)\ U_{2i-1}^k$ and the vertical ones are (at least up to a constant) $\Sigma_{k=0}^{Q-1} x_{\mathrm{H}}(k,\lambda) U_{2i}^k$.

2. von Neumann Algebras

If $\mathcal{H}$ is a complex Hilbert space with inner product $\langle\ ,\rangle$, a *von Neumann algebra* M is an algebra of bounded linear operators on $\mathcal{H}$ such that

a) $id \in M$
b) If $a \in M$, then $a^* \in M$.
c) If a_n is a set of elements of M with $\langle a_n\xi, \eta\rangle \longrightarrow \langle a\xi, \eta\rangle$ for some a, then $a \in M$.

The simplest example of such an M is the algebra of all bounded operators $\mathcal{B}(\mathcal{H})$. This example has "atoms", i.e., projections p on to one-dimensional subspaces, so that $pMp = \mathbb{C}p$. In finite dimensions any von Neumann algebra is abstractly (i.e. as a $*$-algebra) isomorphic to a direct sum of finitely many copies of $\mathcal{B}(\mathcal{H})$. Thus it is determined as an algebra by integers $n_1, n_2, \ldots n_k$ which can be conveniently represented as the row of integers $n_1, n_2, \ldots n_k$. For instance

$$2 \quad 3 \quad 1$$

stands for the algebra $M_2(\mathbb{C}) \oplus M_3(\mathbb{C}) \oplus \mathbb{C}$. What we have just said completely describes finite-dimensional von Neumann algebras up to abstract isomorphism. To understand how they can act on Hilbert spaces, one need only consider the case of a simple direct summand $M_n(\mathbb{C})$. Elementary arguments show that, since the identity operator on $\mathcal{H}$ must be in $M = M_n(\mathbb{C})$, the dimension of $\mathcal{H}$ is a multiple mn of n and that there is a tensor product splitting $\mathcal{H} \simeq \mathbb{C}^n \otimes \mathcal{H}_2$ with the action of $a \in M$ being as $a \otimes id_{\mathcal{H}_2}$. The integer m is then called the *multiplicity* of the algebra M on $\mathcal{H}$.

Already, in the finite-dimensional case, we see the importance of the direct summands in the decomposition. They are characterized as algebras with trivial centre and this is taken as a definition in the general case: a *factor* is a von Neumann algebra whose centre is just the scalar multiples of the identity. Our only infinite-dimensional example $\mathcal{B}(\mathcal{H})$ is certainly a factor. Any factor

M with "atoms" (i.e. elements p with $p^* = p^2 = p$ and $pMp = \mathbb{C}p$) is called *type I* and is abstractly isomorphic to $\mathcal{B}(\mathcal{H})$ for some $\mathcal{H}$, not necessarily the $\mathcal{H}$ on which it acts. As in finite dimensions, type I factors acting on Hilbert spaces induce tensor product factorizations.

Not all factors are of type I. This was the great discovery of Murray and von Neumann in [MvN]. There are infinite-dimensional factors that in some ways are more like $M_n(\mathbb{C})$ than $\mathcal{B}(\mathcal{H})$ (dim $\mathcal{H} = \infty$). These are the *type II_1* factors which admit a trace tr: $M \longrightarrow \mathbb{C}$ which is a (unique) linear function such that $\mathrm{tr}(ab) = \mathrm{tr}(ba)$, $\mathrm{tr}(1) = 1$. We shall construct some below. One can show that if one restricts the trace to the projections of a II_1 factor, the allowed values are precisely the unit interval $[0, 1]$, where one would have obtained $\{0, \frac{1}{n}, \frac{2}{n}, \ldots, 1\}$ for $M_n(\mathbb{C})$. This is *continuous dimensionality.*

Beyond II_1 factors, one may construct type II_∞ ones as tensor products $\mathrm{II}_1 \otimes \mathcal{B}(\mathcal{H})$. There are still other factors, called type III, which admit no projections p such that the algebra pMp has a trace as above. The type III factors do admit a complicated splitting as "crossed products" of type II objects by groups. For an excellent survey of the global structure of factors, see [Co].

We now want to take up the construction of a type II factor. The method we shall give is quite limited and somewhat complicated but it will introduce ideas that will be crucial later. We will start with finite-dimensional von Neumann algebras and build them up to II_1 factors. For this, it behooves us to study inclusions of finite-dimensional von Neumann algebras. These are described by "Bratteli diagrams" as amply illustrated in the following example:

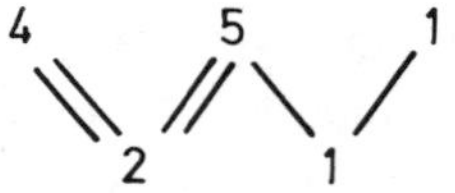

This diagram means two algebras $N \subseteq M$, where $N = M_2(\mathbb{C}) \oplus \mathbb{C}$, $M = M_4(\mathbb{C}) \oplus M_5(\mathbb{C}) \oplus \mathbb{C}$ and an element $A \oplus x$ of N is the element

$$\begin{pmatrix} A & 0 \\ 0 & A \end{pmatrix} \oplus \begin{pmatrix} A & 0 & 0 \\ 0 & A & 0 \\ 0 & 0 & x \end{pmatrix} \oplus x \text{ of } M .$$

One can stack such diagrams on top of each other to obtain infinite-dimensional abstract algebras as limits of finite-dimensional ones. Thus

$$\begin{array}{c} 1 \\ \| \\ 2 \\ \| \\ 4 \\ \| \\ 8 \\ \| \\ \text{etc.} \end{array}$$

is the Bratteli diagram for an algebra A which may alternatively be described as the infinite tensor product $\otimes_{i=1}^{\infty} M_2(\mathbb{C})$. A more interesting case is the following

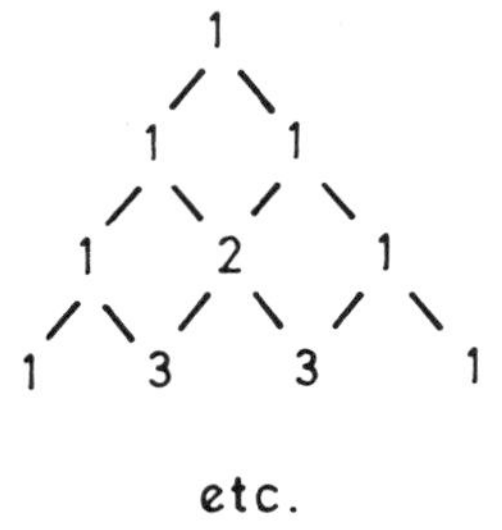

This is Pascal's triangle and this algebra may be obtained from the previous one by looking at the fixed points for the action of the circle group $\mathbb{T}$, an element $e^{i\theta}$ of which acts by

$$x \longrightarrow \left[\bigotimes_{i=1}^{\infty} \begin{pmatrix} e^{i\theta} & 0 \\ 0 & e^{-i\theta} \end{pmatrix}\right] x \left[\bigotimes_{i=1}^{\infty} \begin{pmatrix} e^{-i\theta} & 0 \\ 0 & e^{i\theta} \end{pmatrix}\right].$$

The algebras A obtained by this process do not act naturally on a Hilbert space. To make them into von Neumann algebras, one uses the so-called GNS construction ([Ar]). (Note that for this construction, A need not be of the form described above). One begins with a linear functional $\varphi : A \longrightarrow \mathbb{C}$ with the properties $\varphi(a^*a) \geq 0$ and $\varphi(1) = 1$ (a "state"). The set of all a such that $\varphi(a^*a) = 0$ is then a subspace I and A/I inherits a pre-Hilbert structure: $\langle a, b\rangle = \varphi(b^*a)$. The algebra A acts on A/I, by left multiplication, as bounded operators which therefore extend to the Hilbert space completion $\mathcal{H}_\varphi$ of A/I. The relevant von Neumann algebra M_φ is the closure of A in the topology implicit in the definition of a von Neumann algebra. M_φ is to be thought of as the *von Neumann completion of A*.

The simplest situation is when φ is a trace, i.e., $\varphi(ab) = \varphi(ba)$. Then φ extends to a trace on M_φ which thus, if a factor, will be a II_1 factor. Now traces are easy to keep track of on finite-dimensional von Neumann algebras. Moreover, it is easy to show that if the algebra A admits precisely one trace tr, then the von Neumann algebra M_{tr} is a II_1 factor. It is very easy to give criteria on a Bratteli diagram of A for this to happen. Basically, if the width of the diagram is finite and it is not too lopsided, there is a unique trace. For instance, the algebra

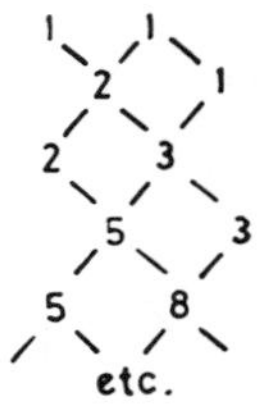

admits a unique trace, whereas the algebra

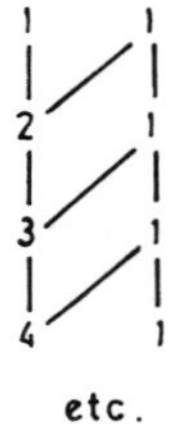

does not. (Note that a given trace on a Bratteli diagram may define a II_1 factor without the diagram satisfying these conditions — there is a one-parameter family of such traces on our Pascal's triangle example.)

Thus we have a whole host of II_1 factors. Just choose any Bratteli diagram satisfying the criteria and it has a unique trace, which gives a II_1 factor as the von Neumann algebra completion of A. It is a remarkable result of Murray and von Neumann that *all such II_1 factors are isomorphic*, and the resulting factor is called the hyperfinite II_1 factor, R. By a deep result of Connes [Co2], R is a "minimal" II_1 factor. Any II_1 subfactor is again isomorphic to R, and R is contained in any II_1 factor. It can also be constructed by many other means. It is naturally the Clifford algebra of an infinite-dimensional Hilbert space.

Before passing to subfactors, we must say a few words about the "multiplicity" of a II_1 factor on a Hilbert space. This is the analogue of the multiplicity defined by the tensor product splitting that comes with the action of a type I factor on a Hilbert space. It is most conveniently thought of by considering the II_1 factor M as an abstract object capable of acting on many Hilbert spaces $\mathcal{H}$ which become "Hilbert spaces over M". In this analogy, M behaves like a field and the Hilbert space as a vector space over that field. The multiplicity then plays the role of the dimension of the vector space and we shall write it $\dim_M(\mathcal{H})$. The striking fact is that for II_1 factors, all positive real values of $\dim_M(\mathcal{H})$ (and ∞) are allowed, where we would have found only integers for a type I factor (and no obvious coherent way of measuring dimension for a type III factor). If M were constructed from A by the GNS construction with respect to a trace tr (with $\mathrm{tr}(a^*a) > 0$ for $a \neq 0$) the space $\mathcal{H}_{\mathrm{tr}}$ would be a "free module of rank 1", thus $\dim_M(\mathcal{H}_{\mathrm{tr}}) = 1$. (This Hilbert space is often denoted $L^2(M, \mathrm{tr})$ as it is also obtained by applying the GNS construction to the algebra M itself, and if M were $L^\infty(X, \mu)$ with trace $\int f d\mu$, one would obtain $L^2(X, \mu)$.)

To see Hilbert spaces of all real dimensions over M, first take $L^2(M, \mathrm{tr})$ and look at the space $\mathcal{H} = L^2(M, \mathrm{tr})p$ where p is a projection in M. (Since tr is a trace, M also acts on the right on the GNS Hilbert space). Not surprisingly, $\dim_M(\mathcal{H}) = \mathrm{tr}(p)$ and by continuous dimensionality, this can be any real between 0 and 1. Values larger than 1 can be obtained simply by taking direct sums of the Hilbert spaces constructed above.

3. Subfactors

Pursuing the analogy which would have it that a II_1 factor is a field, the study of subfactors is an extension of Galois theory and we can expect to see finite groups come into play. Thus the situation we want to consider is two II_1 factors $N \subseteq M$, with the same identity. If N and M were fields, the most immediate entity to attach would be the integer giving the degree of the extension, the dimension of M over N. But following our previous discussion, we can define this number using our "$\dim_N$" ideas by the simple expedient of applying the GNS construction to complete M to the Hilbert space $L^2(M)$. Thus we define

$$[M : N] = \dim_N(L^2(M)), \text{ called the } \textit{index of } N \textit{ in } M.$$

We shall restrict our attention in this article to the case $[M : N] < \infty$.

The simplest example of all for a finite index subfactor is obtained by tensoring by a matrix algebra, thus $M = N \otimes M_n(\mathbb{C})$, N considered as a subfactor of M via the obvious diagonal embedding. A moment's thought shows that the index in this example must be n^2. Despite its triviality, many constructions to follow are easily worked out for this example and it is instructive to do so.

The next most interesting example is to take a finite group G and suppose that it acts by automorphisms on M. If all the automorphisms are outer (i.e., not of the form $x \longrightarrow uxu^{-1}$) except the identity, it is well known that M^G is again a II_1 factor. It is no surprise that $[M : M^G] = |G|$. This example is very important in the theory, especially as it is known ([JI]) that any finite group can act on R in one and only one way by outer automorphisms. It is also important that this example is irreducible in the sense that no element of M commutes with all of M^G except the identity. This makes it quite different from the previous $N \otimes M_n(\mathbb{C})$ example.

Note that we have only seen integers so far as the indices of subfactors, whereas according to the continuous dimension idea, the index is just a real number ≥ 1. The now obvious question of deciding what real numbers occur as index values is completely answered by the following:

Theorem (J2)

a) If $N \subseteq M$ are II_1 factors and $[M : N] < 4$, there is an integer $n \geq 3$ with $[M : N] = 4\cos^2 \pi/n$.

b) For each number $r = 4\cos^2 \pi/n$ as in a), and for each real number $r \geq 4$, there is a subfactor $R_r \subseteq R$ with $[R : R_r] = r$.

The original proof of this theorem involved a basic construction which has since proved to be quite powerful and useful in many situations (see [We1]). The key ingredient is the conditional expectation $E_N : M \longrightarrow N$ (see [U]), which is the restriction to M of the orthogonal projection e_N from $L^2(M, \mathrm{tr})$ to $L^2(N, \mathrm{tr})$. Identifying M with its image under the GNS construction on $L^2(M, \mathrm{tr})$, it is easy to show that e_N satisfies the following properties:

(3.1) $e_N^2 = e_N^* = e_N$

(3.2) If $x \in M$, then $xe_N = e_N x$ iff $x \in N$.

(3.3) For $x \in M$, $e_N x e_N = E_N(x) e_N$ so $e_N M e_N = N e_N$.

The situation presented by these equations is quite interesting and one may enquire as to the algebra generated by M and e_N in the cases a) where M and N are as constructed above b) where M, N and e_N are all in some larger algebra, and E_N in some map.

In case a), we write the algebra as $\langle M, e_N \rangle$ and call this the *basic construc-*

tion. In case b), the algebra is not as easy to pin down but can, in **general**, be determined by an analysis of traces. It invariably contains $\langle M, e_N \rangle$ as a two-sided ideal. For a precise statement, see theorem 1.1 of [We2].

The basic construction is even useful, and quite enjoyable, in finite dimensions. The result is that if $N \subseteq M$ is given by a Bratteli diagram, then $M \subseteq \langle M, e_N \rangle$ is given by the mirror image of that diagram (independent of the trace used, provided $\mathrm{tr}(a^*a) > 0$ for $a \neq 0$). For instance, if $N \subseteq M$ is

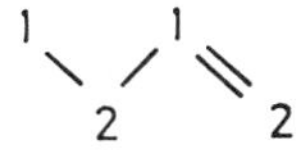

then $M \subseteq \langle M, e_N \rangle$ is

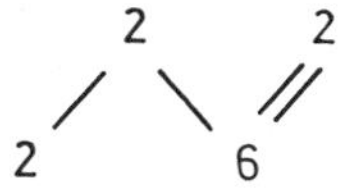

We have digressed a little, so let us return to the II_1 factor case. The crucial result is:

Lemma 3.1. If $[M : N] < \infty$, then $\langle M, e_N \rangle$ is a II_1 factor and
(i) $[\langle M, e_N \rangle : M] = [M : N]$
(ii) $\mathrm{tr}(x e_N) = [M : N]^{-1}\, \mathrm{tr}(x)$ if $x \in M$.
(Note that we do not need to specify which trace we mean since II_1 factors admit only one normalized trace.)

Property (i) of the lemma meakes it tempting to iterate the basic construction. One obtains a tower $M_i \subseteq M_{i+1}$ of II_1 factors defined by $M_1 = N, M_2 = M, M_{i+1} = \langle M_i, e_{M_{i-1}} \rangle$.

Writing $e_{M_i} = e_i$, properties (3.1), (3.2) and (3.3) make it easy to show:

$$\text{(TL)} \qquad \begin{aligned} & e_i^2 = e_i \\ & e_i e_{i\pm 1} e_i = \tau e_i \qquad\qquad (\tau = [M : N]^{-1}) \\ & e_i e_j = e_j e_i \quad |i - j| \geq 2 \ . \end{aligned}$$

Note that these are *precisely* the relations used by Temperley and Lieb for the Potts model! Part a) of the theorem is now proved by using positivity of the restriction of $\mathrm{tr}(aa^*)$ to the algebra generated by $e_1, e_2, \ldots$, and (ii) of lemma 3.1.

Part b) of the theorem is proved for $[M : N] < 4$ by iterating the basic construction for a pair of finite-dimensional von Neumann algebras and then using the resulting von Neumann algebra generated by $\{e_1, e_2, \ldots\}$ as R and R_r the von Neumann algebra generated by $\{e_2, e_3, \ldots\}$.

4. The Tower of Relative Commutants

Although only the e_i's were used in the proof of the theorem of Sec. 3, there is a much richer structure available and attention was turned to it immediately after the proof of the theorem. Its existence follows from the next almost trivial result.

Lemma 4.1. If $N \subseteq M$ has finite index then $N' \cap M = \{x \in M | xy = yx$ for all y in $N\}$ is a finite-dimensional von Neumann algebra.

Thus if we consider the tower

$$N \subseteq M \subseteq M_3 \subseteq M_4 \subseteq M_4 \subseteq \ldots$$

previously constructed, then all the algebras $N' \cap M_i$ and $M' \cap M_i$ are finite-dimensional. They also contain the e_i's which is precious information. We thus have two Bratteli diagrams, for $N' \cap M_i$ and $M' \cap M_i$ which are also connected by "horizontal" inclusions. The simplest nontrivial example to analyze is that in which M carries an outer action of a finite abelian group G and $N = M^G$, the fixed point algebra. Standard duality arguments then give the following Bratteli diagram for both towers $N' \cap M_i$, $M' \cap M_i$

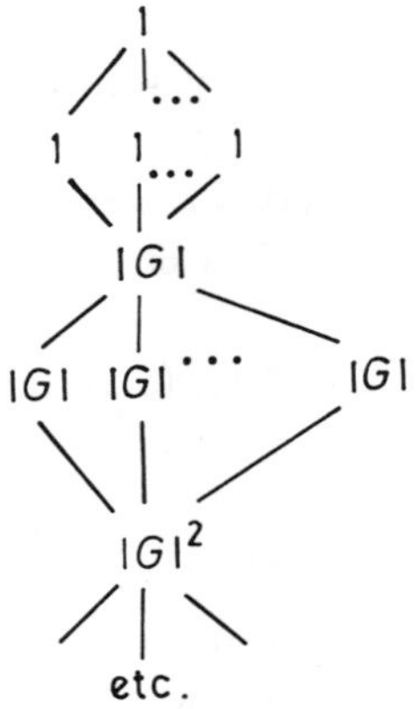

The inclusion diagram between the towers is the obvious one. If we had chosen a non-abelian group, the two towers would be different — the $M' \cap M_i$ tower would be as above but the second level of the $N' \cap M_i$ tower would be given by the group algebra.

The most penetrating analysis of the two towers has been made by Ocneanu ([O1]) who has announced a characterization of the possible combinatorial structures. The presence of the e_i's in the towers means they always have the property that the mirror image of the previous inclusion is always part of the next inclusion. Thus we may throw away all the reflected stuff and look only at what Ocneanu calls the *principal graph.*

For instance, the following is an example of a tower

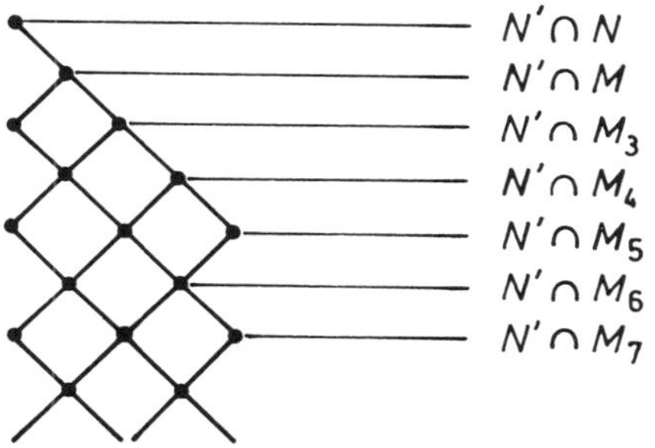

etc.

Throwing away the reflections, we just see

One may recognize the graph A_5 of the Coxeter-Dynkin theory. It was known to the author in 1982 that if $[M : N] < 4$, the principal graph must be an A, D, E Coxeter graph (for a full account, see [GHJ]). For instance the diagram D_4 can be realized by choosing $G = \mathbb{Z}/3\mathbb{Z}$ in the previous example. According to Ocneanu's announcements, the graphs D_{2n+1} and E_7 cannot be obtained.

The principal graph may be either finite or infinite and the subfactor is said to be of finite or infinite depth, accordingly. (The subfactors R_r for $r \geq 4$ are of infinite depth.)

The most impressive result about the towers from a von Neumann algebra point of view is the following, proved by Popa in [Po] and announced by Ocneanu in [O2].

If $R_0 \subseteq R$ is a finite depth subfactor, then the inclusion $R_0 \subseteq R$ is isomorphic to the inclusion coming from the appropriately completed towers of relative commutants.

This result reduces the classification of finite depth subfactors to the combinatorial elucidation of all possible towers. Ocneanu has developed another approach to this problem using bimodules rather than algebras, the equivalence of the two approaches hinging on the fact that $\langle M, e_N \rangle$ is $M \otimes_N M$ as an $M - M$ bimodule. (See [O2].)

From the statistical mechanics point of view, the interest of the towers lies more in the possibility of defining solvable models using their combinatorial structure. This led Pasquier to define his A-D-E models ([Pa]) generalizing the Andrews-Baxter-Forrester models [(ABF)]. These are IRF models, not spin models, so we just refer the interested reader to [Pa] and references therein.

This thoroughly motivates the search for examples of subfactors. Although Ocneanu's announced machinery is in principle computable, most of the beautiful examples of subfactors have been found by other means. In [We2], Wenzl constructed many examples using Hecke algebras. The combinatorial ingredients are the Young diagrams, and corresponding statistical mechanical models occur in the work of Date, Miwa and Jimbo (see Jimbo's article in this volume). Wenzl has also found many examples using the Birman-Wenzl algebra ([BW]) which was discovered through the knot theory connection. There are also many sporadic examples (see [GHJ]) and some new ones of small index (< 5) have been announced recently by Ocneanu, and Haagerup. Presumably, they all have solvable models associated with them.

5. Some Examples of Bratteli Diagrams

We have seen many examples of algebras obtained as limits of finite-dimensional von Neumann algebras, both in statistical mechanics and subfactors. In this section, we given the Bratteli diagrams for some of them. In all cases except the Hecke algebra, the structure can be calculated by an elaboration of the basic construction technique in finite dimensions.

1) The Q-state Potts model

The Temperley-Lieb algebra generated by the elementary transfer matrices R_i has the following Bratteli diagram.

$Q = 1$:

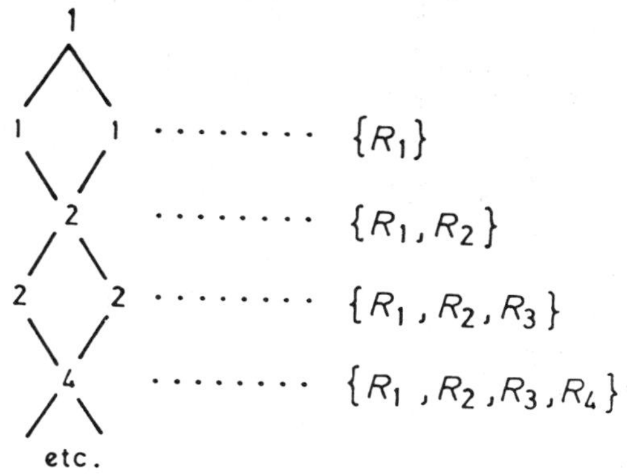

$Q = 3$:

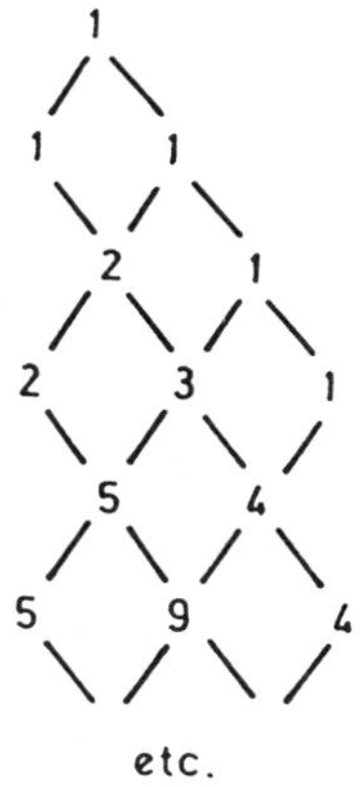

(R_i's as above)

$Q \geqslant 4$:

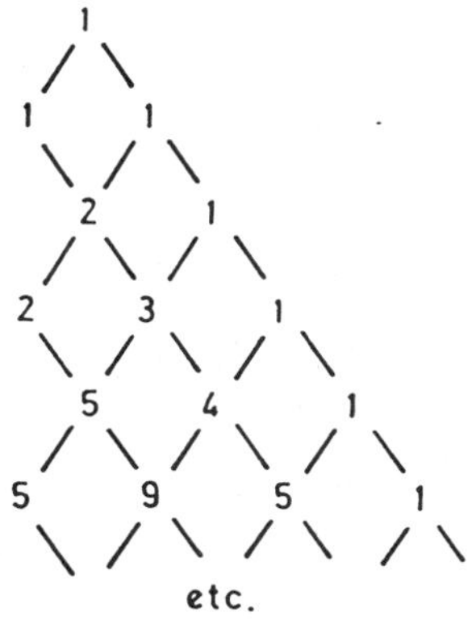

(R_i's as above)

2) The Fateev-Zamolodchikov model

a) The algebra generated by the U_i's of Sec. 1 (common to any $\mathbb{Z}/Q\mathbb{Z}$ symmetric model).

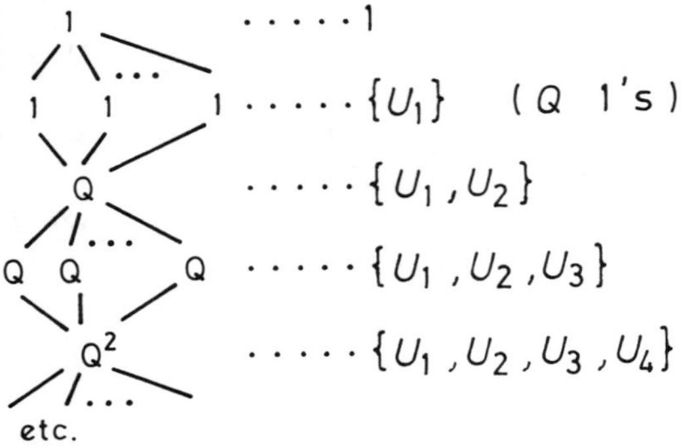

(Note – this is the same as the tower diagram for $G = \mathbb{Z}/Q\mathbb{Z}$ in Sec. 4.

b) The algebra generated by the R_i's ($R_i = \Sigma x_{\mathrm{H}}(k, \lambda)$)($U_{2i-1}$)

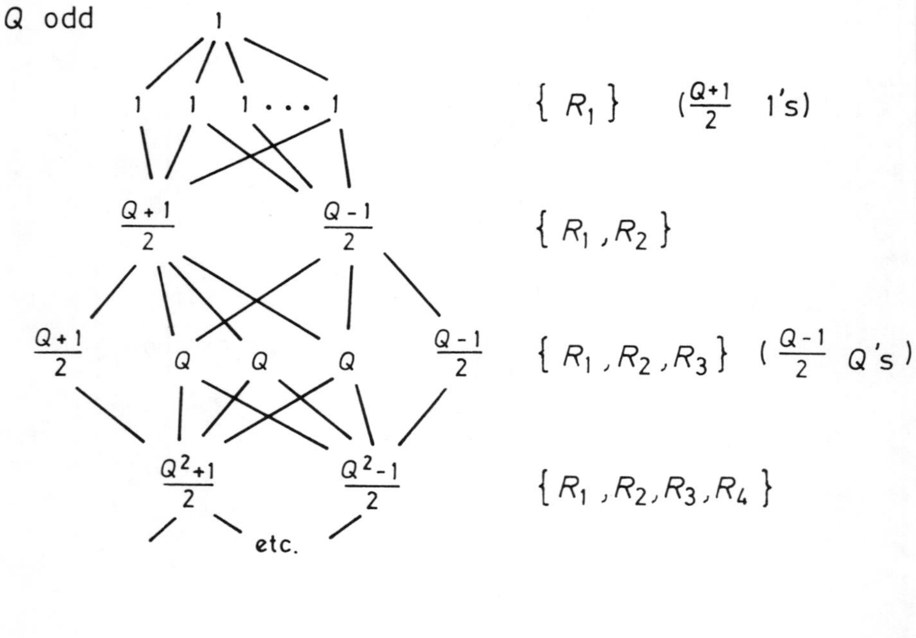

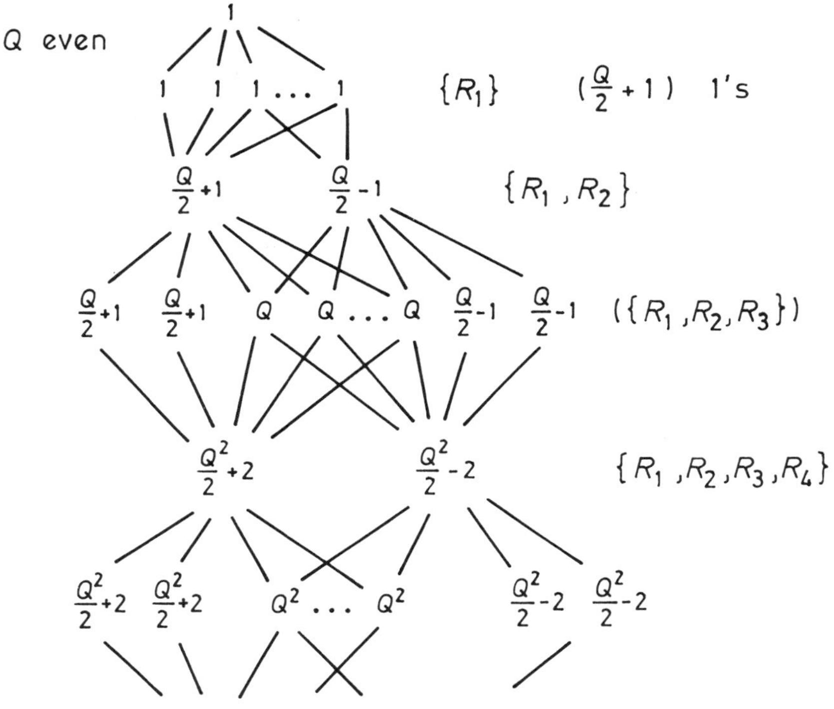

A few comments are perhaps in order since this diagram does not seem to be in the literature (a full proof is available on request).

The key to the analysis is to note that $x_{\rm H}^n(\lambda) = x_{\rm H}^m(\lambda)$ iff $m = -n$. A spectral analysis of R_i then reveals that the algebra generated by R_1 and R_2 is the fixed point algebra for the period 2 automorphism of the $Q \times Q$ matrix algebra defined by $U_1 \longrightarrow U_1^{-1}, U_2 \longrightarrow U_2^{-1}$. Use of the basic construction technique then gives the whole diagram.

Note that if Q is a prime, one may take the limit $\lambda \longrightarrow i\infty$, calling σ_i the limit of R_i. Since $m^2 = n^2 \Longleftrightarrow m = \pm n$ in this case, the algebra generated by the σ_i's is the same as that generated by the R_i's. The σ_i's define the metaplectic representation which may be analyzed by the usual techniques of character theory to give the Bratteli diagram (see [(GJ]). The method fails when Q is not prime for two reasons. The first is that the metaplectic representation is more difficult to analyze if Q is not prime, and the second, more significant, reason is that if Q is not prime, the $R_i(\lambda)$'s do not belong to the algebra generated by the braid group.

One could define a subfactor of the hyperfinite II_1 factor in the following way: the large factor is the von Neumann algebra generated by $R_1(\lambda)$, $R_2(\lambda)$, ..., and the subfactor is the von Neumann algebra generated by $R_2(\lambda)$, $R_3(\lambda)$, ... It is clear from our analysis that the resulting subfactor is of the form $M^{D_n} \subseteq M^{\mathbb{Z}/2\mathbb{Z}}$ where D_n is the dihedral group of order $2n$ acting in some outer fashion on M. This generalizes [J3].

The Bratteli diagram for the Temperley-Lieb algebra in the subfactor context

For each value $4\cos^2 \pi/n$ of the index $[M : N]$ of a subfactor of a II_1 factor, the algebra generated by the e_i's in the tower of II_1 factors, is different. To obtain its structure, it is convenient to begin with the "generic" Bratteli diagram:

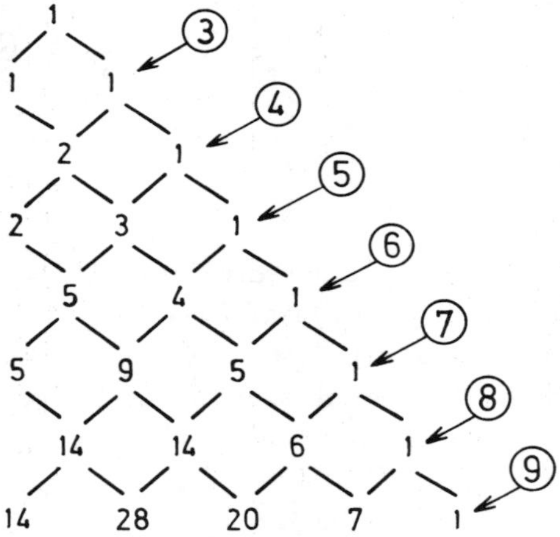

etc.

We have attached integers to the 1's down the diagonal. To obtain the structure of alg$\{1, e_1, e_2, \ldots\}$ when $[M : N] = 4\cos^2 \pi/n$, one eliminates the diagonal "1" with label n and everything below and to the right of it, and adjusts the remaining numbers so that the addition rule holds. Thus for $n = 5$, one obtains the diagram:

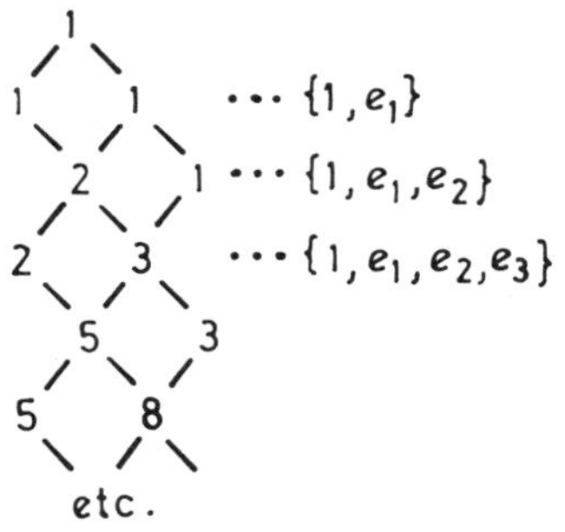

For $n = 4$ and 6, one obtains the same diagram as for the 2 and 3 state Potts models, respectively.

One feature of these Bratteli diagrams is that there is a simple way to represent the corresponding algebras on the vector space whose basis is the descending paths from the initial "1" to a given level. Thus one defines an algebra whose basis elements are pairs (γ_1, γ_2) of such paths, which end at the same point on the diagram. The algebra structure is given by $(\gamma_1, \gamma_2)(\gamma_3, \gamma_4) = \delta_{\gamma_2, \gamma_3}(\gamma_1, \gamma_4)$. It is a trivial exercise to see that this algebra has the right structure. Although this may seem to be a banal observation, it is suggestive in that one might look for simple representations of the appropriate commutation relations (e.g. Temperly-Lieb, Fateev-Zamolodchikov) on the space of paths. The existence of some representation is of course guaranteed by the knowledge that one has the correct Bratteli diagram. It is this line of reasoning that led Pasquier to his *A-D-E* models and Ocneanu to his first combinatorial descriptions of subfactors. Wenzl also used this idea to construct appropriate Hecke algebra representations in [W2].

6. Type III Factors

We return now to the GNS construction of Sec. 2, this time with a view to constructing type III factors, i.e. factors which possess no trace of any sort. So let us begin with an appropriate algebra A which we are trying to "complete" to form a type III factor. It is obviously no use beginning with a trace for the state(positive linear functional) φ, but let us simplify our discussion by assuming that $\varphi(x^*x) \neq 0$ for $x \neq 0$. A fundamental difference arises in the GNS construction if φ is not a trace: it is that, while *left* multiplication by

elements of A is bounded, the same need not be, and often is not, true of right multiplication. Let us immediately give an example.

Example 6.1. The Powers factors

For A we take the infinite tensor product $\bigotimes_{i=1}^{\infty} M_2(\mathbb{C})$ just as in Sec. 2, but we define the (Powers) state φ_λ by

$$\varphi_\lambda(x_1 \otimes x_2 \otimes \ldots \otimes x_n \otimes 1) = \prod_{i=1}^{n} \operatorname{trace}\left(x_i \begin{pmatrix} \frac{1}{1+\lambda} & 0 \\ 0 & \frac{\lambda}{1+\lambda} \end{pmatrix}\right),$$

where "trace" means the sum of the diagonal elements, and $0 < \lambda < 1$. Powers, in [P], showed that, if R_λ is defined to be the von Neumann algebra resulting from the GNS construction, then the R_λ are all of type III and mutually non-isomorphic. Thus type III factors exist.

In this example, only left multiplication operators are automatically bounded and extend to the whole Hilbert space. Treatment of this left-right asymmetry is the Tomita-Takesaki theory which we briefly outline. If $\mathcal{H}_\varphi$ is the GNS Hilbert space, it contains the dense subspace A. We define the (conjugate linear) operator S to be the (unbounded) operator with domain A, $S(x) = x^*$. Under suitable conditions, satisfied in the case of the Powers factor, the closure of the graph of S is again the graph of an operator also denoted S. By the von Neumann theory, S then has a polar decomposition $S = J\Delta$, where J is a conjugate linear isometry and Δ is positive and self-adjoint. If M is the von Neumann algebra generated by left multiplication operators from A on $\mathcal{H}_\varphi$, the main result of Tomita-Takesaki theory (see [Ta]) is that

a) $JMJ = M'$ (the commutant of M)

b) $\Delta^{it} M \Delta^{-it} = M$.

This is a remarkable result since it shows that to any appropriately continuous state φ on a von Neumann algebra, there is a canonical 1-parameter automorphism group σ_t^φ of M defined by applying the above procedure with $A = M$ and letting

$$\sigma_t^\varphi(x) = \Delta^{it} x \Delta^{-it} .$$

Connes ([Co3]) has shown that if the state φ is changed, σ_t^φ changes only by an inner perturbation so that one has the following invariant of a von Neumann algebra:

$$T(M) = \{t \in \mathbb{R} | \sigma_t^\varphi \text{ is an inner automorphism}\} .$$

This invariant is particularly easy to calculate for Powers factors, since the group σ_t^φ is easily seen to be conjugation by

$$\left[\bigotimes_{i=1}^{\infty}\begin{pmatrix}\frac{1}{1+\lambda} & 0\\ 0 & \frac{\lambda}{1+\lambda}\end{pmatrix}\right]^{it}.$$

This is inner iff $\lambda^{it}=1$, i.e. $t\in(2\pi/\log\lambda)\mathbb{Z}$.

In general, the fixed point algebra M^{σ_φ} is called the *centralizer* of φ and it is part of the theory that φ is a trace when restricted to the centralizer (indeed, $\varphi(xy)=\varphi(yx)$ for $x\in M^{\sigma_\varphi}$, $y\in M$). In the case of the Powers factor, this centralizer is the closure of the algebra of Sec. 2, whose Bratteli diagram was Pascal's triangle (page 7).

7. Vertex Models in Statistical Mechanics

The difference between vertex models and the spin models of Sec. 1 is that the states of a vertex model are functions from the edges of the graph to the Q-element set S and the energy contribution is associated with a vertex. Thus if the graph is the two-dimensional lattice, a vertex model will be defined by Boltzmann weights $w(a,b|x,y)$ associated with a vertex in the following state configuration:

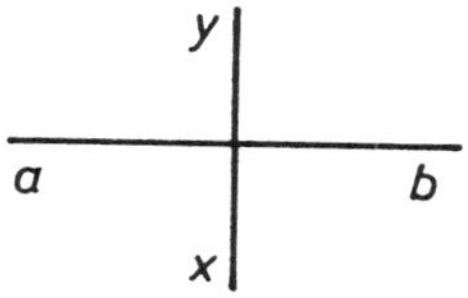

where, as before, a distinction can be drawn between the horizontal and vertical directions.

A simple transfer matrix approach is to use diagonal-to-diagonal transfer matrices. Consider the diagram below:

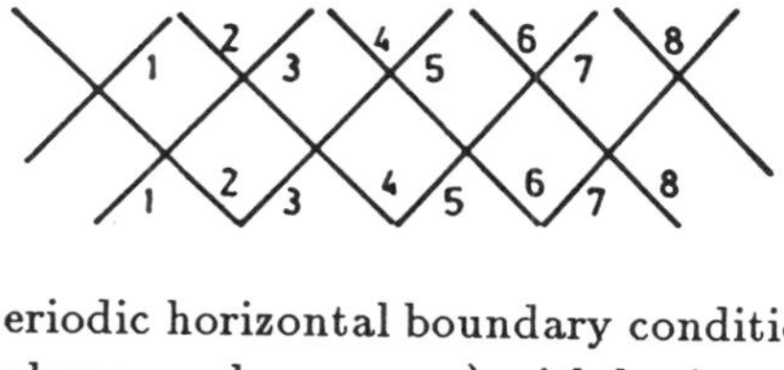

Clearly, if we impose periodic horizontal boundary conditions and define a vector space $\otimes^n V$ ($n=8$ above, n always even) with basis vectors $\otimes_{i=1}^n q_i$, $q_i\in S$,

then the transfer matrix associated with the above diagram is $(R_{23}R_{45}\ldots R_{n,1})$ $(R_{12}R_{34}\ldots R_{n-1,n})$, where

$$R_{i,i+1}\left(\bigotimes_{j=1}^{n} q_i\right) = \sum_{r_i,r_{i+1}=1}^{Q} w(q_i, r_{i+1}|q_{i+1}, r_i)\bigotimes_{j=1}^{n} q'_j$$

with $q'_j = q_j, j \neq i, i+1, q'_j = r_j, q'_{j+1} = r_{j+1}$. The simplest such model that has been extensively studied is the "ice-type" model with $Q = 2$, solved by Lieb (see [L]). In this case, the R matrices can be expressed in the form $R_{i,i+1} = \text{const}(1 + xe_i)$, where e_i (index modulo m) is given by

$$e_i\left(\bigotimes_{i=1}^{n} q_i\right) = \sum_{r_i,r_{i+1}} x(q_i, r_{i+1}|q_{i+1}r_i)\bigotimes_{j=1}^{n} q'_i\ ,$$

with notation as before, and x is the 4×4 matrix below:

$$\frac{1}{1+t}\begin{pmatrix} 0 & 0 & 0 & 0 \\ 0 & 1 & \sqrt{t} & 0 \\ 0 & \sqrt{t} & t & 0 \\ 0 & 0 & 0 & 0 \end{pmatrix}.$$

It follows immediately that for $1 \leq i, j \leq n-1$,

$$e_i^2 = e_i, \quad e_i e_{i\pm1} e_i = \frac{t}{(1+t)^2} e_i, \quad e_i e_j = e_j e_i \quad \text{if } |i-j| \geq 2\ .$$

These are the same relations as we met in the Potts model and this was the key to the equivalence, established first by Temperley and Lieb in [TL], of the Potts and ice-type models. (see also [Ba] Chap. 12).

In fact, we saw in the Potts model that the partition function for the system with periodic boundary conditions in the vertical direction is given by the trace of some power of the row-to-row transfer matrix and that this trace, when normalized so that $\text{tr}(1) = 1$, has the following useful property, called the Markov property,

$$\text{tr}(we_{n+1}) = \frac{1}{Q}\text{tr}(w) \quad \text{if } w \in \text{alg}\{1, e_1, \ldots, e_n\}\ .$$

This property completely defines tr on the algebra generated by the e_i's and it was this fact that was crucial in [J2] for the proof of the restrictions on the index values for subfactors.

It is remarkable that this trace is *not* given in the ice type model by the usual trace but rather by the restriction of the Powers state φ_λ when $2+\lambda+\lambda^{-1} = Q$ (or $\lambda = t$)! The e_i's are easily seen to be in the centralizer of the Powers state. One of the most striking simple facts in the theory of subfactors is that the von Neumann algebra generated by the e_i's is precisely the centralizer of φ_λ. This was proved by Pimsner and Popa in [PP] using entropy considerations. A direct proof is not yet available. It suffices to prove that the element $\begin{pmatrix} \lambda & 0 \\ 0 & 1 \end{pmatrix} \otimes 1 \otimes 1 \otimes$ is in the von Neumann algebra generated by the e_i's.

As far as subfactors are concerned, this example gives an interesting and useful procedure. The obvious subfactor around is the subfactor given by sums of elements of the form $\{1 \otimes x_1 \otimes x_2 \otimes \ldots\}$ in the Powers factor R_λ. This is a trivial subfactor of the form $N \subseteq N \otimes M_2(\mathbb{C})$. But restricting to the centralizer of the Powers state, one gets the interesting subfactor generated by the $\{e_i\text{'s}\}$. This procedure can obviously be generalized to arbitrary finite-dimensional unitary representations of groups. In the case of compact groups, and using the trace for the GNS construction, much analysis has been done by A. Wassermann (see [Wa]). The subfactor has trivial relative commutant precisely when the representation is irreducible.

8. Quantum Groups

Quantum groups in the spirit of Drinfeld ([Dr]) and Jimbo provided vast generalizations of the ice-type model, and in a sense this is what they were invented to do. To be precise, to every simple finite-dimensional Lie algebra, G, and every finite-dimensional representation of G, on V say, one can construct a family $R(\lambda) \in \mathrm{End}(V \otimes V)$ satisfying the Yang-Baxter equation $R_{12}(\lambda) R_{13}(\lambda+\mu) R_{23}(\mu) = R_{23}(\mu) R_{13}(\lambda+\mu) R_{12}(\lambda)$ in $\mathrm{End}(V \otimes V \otimes V)$, with obvious notation.

I do not want to go into the construction of R which is beautifully detailed in [Dr]. The six-vertex model is precisely the case $G = sl_2, V = \mathbb{R}^2$. It is clear that one is also able to use this machine to construct subfactors which generalize the $\{e_i\}$ subfactors. For $G = sl_n, V = \mathbb{C}^n$, these subfactors were considered by Wenzl in [We2]. It is not clear *a priori* what the Powers state should be in this general situation. At this point, we would have to appeal to the knot theory to justify that the Powers state on $\otimes_{r=1}^{\infty} \mathrm{End}(V)$ should be $\varphi_q(\otimes_{i=1}^{\infty} x_i) = \Pi_{i=1}^{\infty}$ trace $(q^{\frac{H}{2}} x_i)$, where H is the element of the Cartan subalgebra of G corresponding via the Killing form to the half sum of the positive roots. This state, which is positive definite for $q \in \mathbb{R}^+$, does have the Markov property with respect to the R_i's as shown by Rosso and Reshetikhin

([Ro], [Re]).

The subfactors created by this process do not seem to have been completely, systematically investigated. They are to be thought of as quantized versions of the subfactors considered by Wassermann.

References

[Ar] W. Arveson, *An Invitation to C^*-algebras, Graduate Texts in Mathematics* **No. 39**, (Springer-Verlag, 1976)

[A-Y BP] H. Au-Yang, R. Baxter and J. Perk, "New solutions of the star-triangle relations for the chiral Potts model", *Phys. Lett.* **A128** (1988) 138-142.

[Ba] R. Baxter, *Exactly solved models in statistical mechanics*, (Academic Press, 1982).

[BW] J. Birman and H. Wenzl, "Braids, link polynomials and a new algebra", to appear in *Trans. A.M.S.*

[Co1] A. Connes, "Classification des facteurs", *Proc. Symp. Pure Math.* **38** (1982), Part 2, 43–110.

[Co2] A. Connes, "Classification of injective factors", *Ann. Math.* **104** (1976) 73-115.

[Co3] A. Connes, "Une classification des facteurs de type III", *Ann. Sci. Ec. Norm-Sup.* **6** (1973) 133-252.

[Dr] V. Drinfeld, "Quantum Groups" in *Proc. ICM 86* Vol. 1, 798-820.

[FZ] V. Fateev and A. Zamolodchikov, "Self-dual solutions of the star-triangle relations in $\mathbb{Z}_N$-models", *Phys. Lett.* **92A** (1982) 37-39.

[GHJ] F. Goodman, P. de la Harpe and V. Jones, "Coxeter graphs and towers of algebras", to appear (Springer-Verlag).

[GJ] D. Goldschmidt and V. Jones, "Metaplectic link invariants", to appear (Geometrica Dedicata).

[J1] V. Jones, "Actions of finite groups on the hyperfinite II_1 factor", *Memoirs A.M.S.*, No. 237 (1980).

[J2] V. Jones, "Index for subfactors", *Invent. Math.* **72** (1983) 1-25.

[J3] V. Jones, "On a certain value of the Kauffman polynomial", to appear in *Comm. Math. Phys.*

[O1] A. Ocneanu, "Subalgebras are canonically fixed point algebras", *A.M.S. Abstracts* **6** (1986) 822-99-165.

[O2] A. Ocneanu, "Quantized groups, string algebras and Galois theory for algebras" — an outline of results, 1987.

[Pa] V. Pasquier, "Etiology of IRF models", Preprint Saclay SPhT/88-20.

[P] R. Powers, "Representations of uniformly hyperfinite algebras and the associated von Neumann algebras", *Ann. of Math.* **86** (1967) 138-171.

[Po] S. Popa, "Classification of subfactors — the finite depth case", UCLA preprint 1989.

[PP] M. Pimsner and S. Popa, "Entropy and index for subfactors" *Ann. Sci. Ec. Norm. Sup.* **19** (1986) 57-106.

[Re] N. Reshetikhin, "Quantized universal enveloping algebra, the Yang-Baxter equation and invariants of links I", Steklov preprint, 1988.

[Ro] M. Rosso, "Quantum groups and V. Jones' vertex models for knots", preprint 1988.

[TL] H. Temperley and E. Lieb, "Relations between the percolation...", *Proc. Roy. Soc. London* (1971) 251-280.

[Ta] M. Takesaki, "Tomita's theory of modular Hilbert algebras and its applications", *Lecture notes in Math*, No. 128 (Springer-Verlag, 1970).

[Wa] A. Wassermann, "Coactions and Yang-Baxter equations for ergodic actions and subfactors", in *Operator algebras and applications* Vol. 2, eds. D. Evans and M. Takesaki, *L.M.S. Lecture Notes* **136** (1988) 202-236.

[We1] H. Wenzl, "On the structure of Brauer's centralizer algebras", *Ann. Math.* **128** (1988) 173-194.

[We2] H. Wenzl, "Hecke algebras of type A_n and subfactors", *Invent. Math.* **92** (1988) 349-383.

[U] H. Umegaki, "Conditional expectation in an operator algebra I", *Tohoku Math. J.* **6** (1954) 358-362.

POLYNOMIAL INVARIANTS IN KNOT THEORY

Louis H. KAUFFMAN

Department of Mathematics, Statistics and Computer Science
University of Illinois at Chicago
Chicago, Illinois 60680, USA

and

Institut des Hautes Etudes Scientifiques
35, route de Chartres
91440 Bures-sur-Yvette, France

ABSTRACT

This article studies polynomial invariants in knot theory, their structure and implications for topology, and their relationship with mathematical physics.

I. INTRODUCTION AND BACKGROUND.

This review article centers on polynomial invariants in knot theory, their structure and implications for topology, and their interrelationships with mathematical physics.

The invariants studied are the original Jones polynomial [56], $V_K(t)$, two two-variable generalizations (the skein polynomials), the Homfly polynomial [38] and the Kauffman polynomial [72,66], and other invariants arising through state models and the Yang-Baxter equation [1,2,3,4,5,58,67,74,78,80,81,129,143]. Along with the newer skein polynomials, the Alexander-Conway polynomial [6,26,46,64,65,66] forms a special case of the Homfly polynomial. The author's bracket polynomial [66,67,69,70,74,75,78] model for the original Jones polynomial is one of the state models.

The relationships with physics begin with a very close connection between state models for knot polynomials and partition functions in statistical mechanics. This relationship has led to the construction of a number of invariants that go beyond the original skein polynomials, and it is not clear at this writing whether these extra invariants may or may not be constructed via geometric operations (e.g. iterated cabling) in conjunction with the skein invariants. This matter of discriminating new invariants is a key problem.

The knot theory may also shed new light on questions about the continuum limit of statistical mechanics models via relations with conformal field theory [1),5),15),39),40),74),91),106),107)].

The state models have led to a deep relation with Lie algebras and their deformations via the Yang Baxter Equations and relations with Quantum Groups [8),12),30),35),52),58),81),89),90),92),104),129),143)]. In the case of the skein polynomials this author [80),81)] has made progress in understanding the structure of these models, and how the basic solutions of Yang-Baxter equations that arise from the A^1 series and B^1, C^1, D^1, A^2 series of the classical simple Lie algebras [52),58),143)] come quite naturally (if intricately) from considerations involving the diagrammatic theory of knots and links. This structure and the questions arising in relation with it form a central part of this research project. In particular, there are candidates for new invariants even in relation to the solution of Yang-Baxter (A^1 series) that yields specializations of the Homfly polynomial for one form of spin spectrum. Changing the spectrum yields other invariants. See section 3 of this paper for more details.

Another part of the connection with statistical mechanics is a relationship [74)] between the bracket model and the classical Potts model - a generalization of the Ising model. This direct relation has led to a new understanding of the Temperley-Lieb Algebra (Jones algebra) in terms of the geometry of diagrams. This gives rise to a number of questions about the structure of the Jones polynomial, the bracket, the Potts model, and representation theory of these algebras. See section 7.

The diagrammatic for the Temperley-Lieb algebra has in turn, led to a generalization of the Artin Braid Group to a Braid Monoid [19),21),69),70),72),74),79),145),152)] and an algebra (The Birman-Wenzl algebra) related with the Kauffman polynomial. Along with understanding the structure of the braid monoid our results [72),79),80)] give interpretations and combinatorial models for the trace on the Birman-Wenzel algebra related to the Kauffman polynomial.

I have constructed *skein models* [80)] for the skein polynomials. These are generalized state models (not of Yang-Baxter form) whose structure is directly related to the original form of skein calculation. The first skein model was given by François Jaeger [51)]. In [80)] I show how to obtain and generalize this model directly as a skein calculation. See section 5.

The skein models generalize to give combinatorial models for the three-variable polynomials for graph embeddings due to Pierre Vogel and the author [79)]. See section 6. It is a good question to inquire the physical significance of the skein models. They share many characteristics of the Yang-Baxter models, but depend more heavily upon global properties of the states.

A recent suggestion due to Witten [150)] has opened another doorway in this theory. Witten proposes to define invariants of links in a three manifold by integrating the trace of the holonomy of a gauge field connection along the link over the space of connections, using a measure weighted by the Chern-Simons Lagrangian. Witten proposes that these invariants include at least specializations of the Jones and Homfly polynomials - he argues for switching identities on geometric grounds. See section 8. This is a very exciting possibility that must be explored, particularly in relation to the states models.

Finally there is a continuing interaction between the subject of knot polynomials and combinatorial considerations such as chromatic polynomials and coloring problems. See particularly the remarks in section 7.

Remark.

Most of the knot theory discussed in this paper is based on the use of diagrams and the fundamental result [128)] of Reidemeister that reduces ambient isotopy to the equivalence relation generated by the three Reidemeister moves (shown below) plus graphical equivalence of the diagrams in the plane. The moves I, II, III show below are representative of these types. They are performed in the context of a larger diagram that is changed only as indicated.

I.

II.

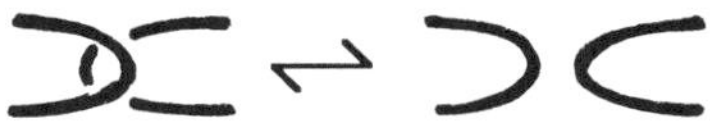

III.

II. KNOT THEORY AND THE BRACKET

The bracket polynomial is defined via the equations

$$\langle O K \rangle = d \langle K \rangle$$

$$\langle O \rangle = 1$$

Here the small diagrams stand for larger diagrams that are otherwise identical. The small circle denotes an isolated component of this form (it may be nested within the diagram). These equations give a well-defined three-variable polynomial on link diagrams.

Well-definedness is easily seen by a state-summation. A "state" S is any choice of splittings of the crossings of K . Each such splitting determines a *vertex weight* according to the rule

governed by the labelling of regions

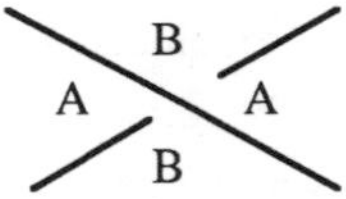

In this convention the regions of A-type are swept out by the overcrossing line as it turns couterclockwise.

Let <K|S> denote the *product* of the vertex weights, and |S| denote the number of circuits in the state S . Then the bracket polynomial is given by the formula

$$\langle K \rangle = \sum_S \langle K \mid S \rangle d^{|S| - 1}$$

In order to obtain invariance under the Reidemeister move of type II it is necessary to take $B = (1/A)$ and $d = -(A^2+A^{-2})$. With this specialization the bracket polynomial becomes an invariant of *regular isotopy* (the equivalence relation generated by the Reidemeister moves of type II and type III).

Normalization in the form

$$f_k = \left(-A^3\right)^{-w(K)} \langle K \rangle$$

where w(K) denotes the writhe or twist number for an oriented link (it is the sum of the crossing signs and is also an invariant of regular isotopy), gives an invariant of all three Reidemeister moves, hence an invariant of ambient isotopy of links in three space.

This model reconstructs the original Jones polynomial. The result is

$$V_K(t) = f_K\left(t^{-1/4}\right)$$

Many results come from the bracket, and good problems remain. Alternating and adequate links [66),67),69),70),74),78),83),118),119),120),121),122),136),137),138),139),144)] have special state properties that allow the calculation of the highest and lowest degree terms. In general more geometric understanding is needed. In joint work with S. Lins (see [78)] for an announcement) I have reformulated the bracket states in terms of embeddings of the link universe (shadow of the diagram) in surfaces of higher genus. For adequate links this genus figures in the formula for the span (difference between highest and lowest degree terms).

It is an open problem whether the bracket detects knottedness. In general we know very little about the phenomena of cancellation of terms in the bracket. This is the *cancellation problem.*

It is not known how the bracket behaves on ribbon knots. (A ribbon knot bounds a smooth disk in upper four-space. Ribbons are conjectured to exhaust the knots with this property). This is the *slice problem.*

Both the cancellation and slice problems can be stated for all the other invariants considered in this article. Both problems are solved only for the classical Alexander polynomial. (There exist non-trivial knots with trivial Alexander polynomial. Slice knots have Alexander polynomial in the form f(t)f(1/t).).

I regard the cancellation and slice problems as the two most significant knot theoretic problems facing these invariants. All the techniques and ideas discussed in the rest of this paper (including the farthest excursions into physics) are associated with these problems.

Remark. It follows from [141)] that it is sufficient to construct invariants of regular isotopy in order to study ambient isotopy. Such invariants need not have any particularly good behaviour under the Reidemeister move of type I. It is fortunate that the skein polynomials behave well in this regard. Other invariants may not do so and still be effective.

Regular isotopy invariants are related to invariants of framed links. M. Hennings [47)] has constructed a polynomial invariant of framed links that appears different from the Homfly polynomial. He is investigating a corresponding structure related to the Kauffman polynomial.

Remark. The Kauffman polynomial in regular isotopy form satisfies the equations

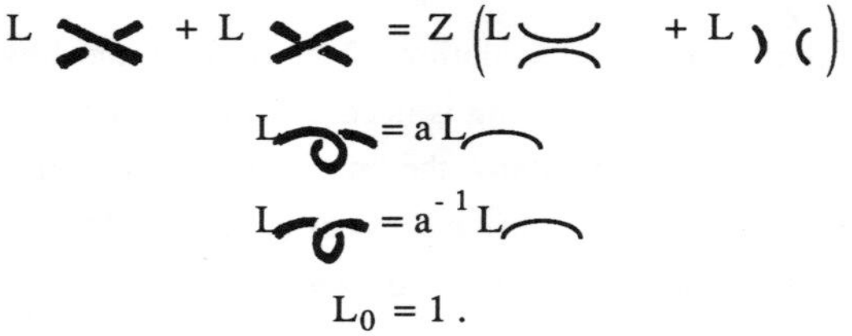

It generalizes the bracket, and is normalized to an ambient isotopy invariant in the same fashion.

Remark. The Homfly polynomial can also be phrased as a regular isotopy invariant. Its equations are

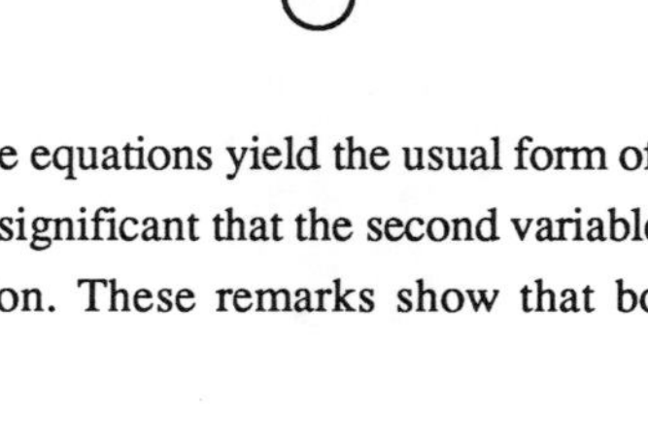

After writhe normalization these equations yield the usual form of the polynomial.

In both these cases it is significant that the second variable (a) can be shifted to the bottom of the skein calculation. These remarks show that both the Homfly and the

Kauffman polynomial can be regarded as invariants of framed links, with the new variable (a) measuring the framing.

Final Remark. The three-variable bracket is not invariant under any of the Reidemeister moves, but it is invariant under *flyping,* an operation that twists a 2-strand tangle as shown below :

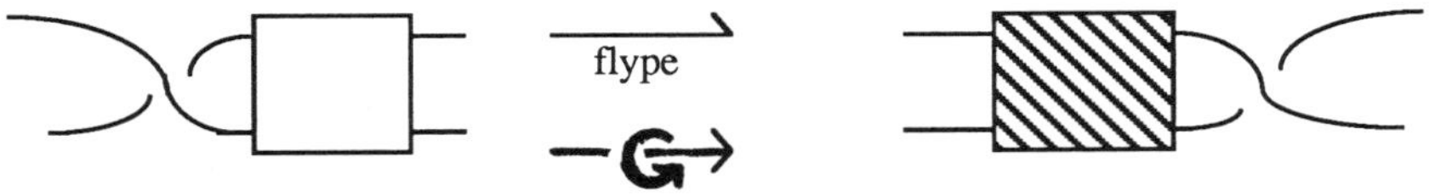

The Tait conjecture (unproved) states that two reduced alternating link diagrams on the 2-sphere that represent ambient isotopic links in three-space are related by a sequence of flypes. A counter-example to the Tait conjecture could be detected by the full bracket.

III. YANG-BAXTER STATE MODELS

The bracket is a special case of what I shall call a *Yang-Baxter state model.* In general what we hope for in a state model for a knot invariant is a combinatorial summation that is well defined on diagrams from the outset, and that contains parameters that can be adjusted to produce an invariant. The author's model for the Conway polynomial [65], and his work on the bracket [67] and Potts models [74] are the first instances of state models in knot theory.

The quest for such models arose originally [65] because the recursive definition of the Conway polynomial via the identity

seemed to involve a very subtle inductive argument to justify its well-definedness. The same situation occurred in even more intensity with the Homfly polynomial and the Kauffman polynomial. Thus the bracket model appeared as a source of great relief - a simple place from which to start.

The idea for a Yang-Baxter state model comes from statistical mechanics. It is a way to parallel the construction of a partition function - with the knot or link diagram replacing the usual lattice whose physical states are catalogued by the partition function.

Vertex weights are assigned to each crossing of the diagram in a given "state" S of that diagram. The state model will then take the form

$$[K] = \sum_{S} [K|S] \lambda^{\|S\|}.$$

where $\|S\|$ is some appropriate global evaluation of the state S, and $[K|S]$ denotes the product of vertex weights corresponding to the state S.

The global evaluation term $\lambda^{\|S\|}$ can be absorbed into the local vertex weights - if we allow more geometry. In particular, with a piecewise linear diagram one can use the local angles between crossing lines (see [74] for the case of the bracket, and [58] for the general case). I will keep the global term in this section and show how models arise as generalizations of the bracket.

I will also speak of these models in the language of the scattering of particles at a vertex (compare [153]). Any two-dimensional statistical mechanics model can be viewed as a 1+1 dimensional model of particle interactions (one dimension of space, one dimension of time) by choosing an "arrow of time" at each vertex. Vertex weight then becomes the analogue of the quantum mechanical scattering amplitude for the process at the vertex. The lines coming into a vertex are labelled with spins a and b , while the lines going out are labelled with spins c and d . Conservation of spin demands that a+b = c+d . Momentum can be indicated via the angle between the lines. The global term is related to total momentum.

To be specific, a state for a model of this type will consist in a labelling of the edges of the universe (the locally 4-valent plane graph underlying the link diagram) by "spins" from an appropriate index set. The elements of the index set are assumed ordered so that we can say of any two spins whether one is less than, equal to, or greater than another. Given such a state, each crossing has two "input" spins and two "output" spins. The vertex weight for a positive crossing will be denoted by R^{ab}_{cd} where a and b are the input spins and c and d are the output spins as indicated below.

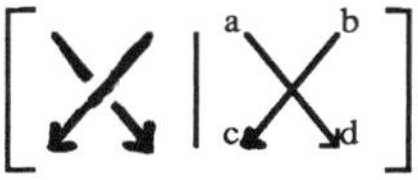

$$= R^{ab}_{cd}\,.$$

It is asssumed the matrix R is invertible with inverse $\bar{R}$. The inverse matrix elements give the vertex weights for negative crossings. This guarantees invariance of the model under the *oriented* type II move :

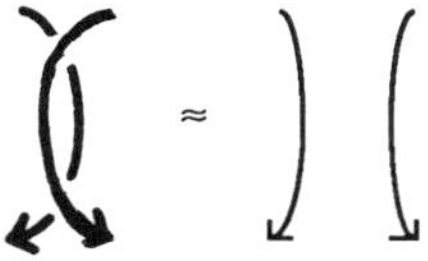

$$\Rightarrow \sum_{i,j} R^{ab}_{ij}\bar{R}^{ij}_{cd} = \delta^{a}_{c}\delta^{b}_{\delta}$$

$$\Rightarrow R\,\bar{R} = I$$

The simplest way that such a model can be invariant under the type III move is if it satisfies

1. The local states S at a triangle all contribute the same factor to the global term ||S|| .

2. The sum of the products of the local vertex weights at a triangle is invariant under the triangle move. That is :

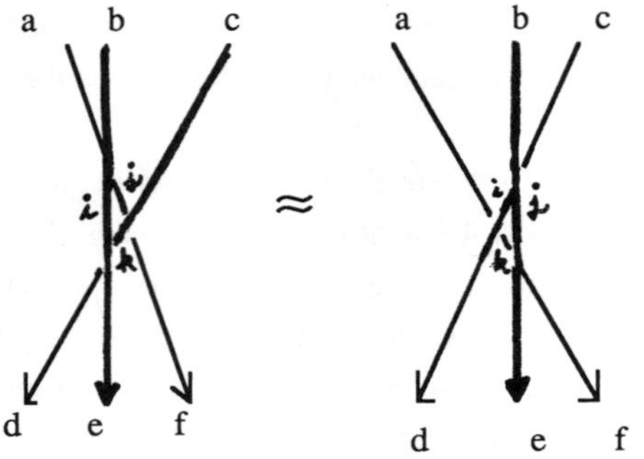

$$\sum_{ijk} R^{ab}_{ij} R^{jc}_{kf} S^{jk}_{de} = \sum_{ijk} R^{bc}_{ij} R^{ai}_{dk} R^{kj}_{ef}$$

The relation indicated in 2. is called the (Quantum) Yang-Baxter Equation. I have deliberately ignored the momentum parameter in this presentation. The momentum can be indicated diagrammatically via an angular parameter θ . See the diagram below :

Figure 1

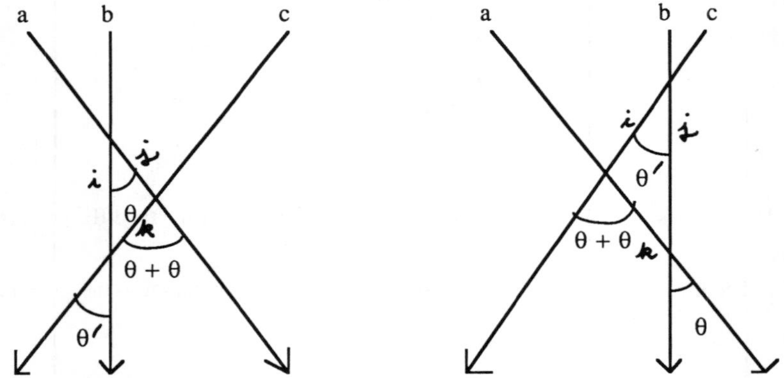

$$\sum_{ijk} S^{ab}_{ij}(\theta)\, S^{jc}_{kj}(\theta+\theta')\, S^{jk}_{de}(\theta') = \sum_{ijk} S^{bc}_{ij}(\theta')\, S^{ai}_{dk}(\theta+\theta')\, S^{kj}_{ef}(\theta).$$

Figure 1

Solutions to the Yang-Baxter equation are a subject of intensive study in mathematical physics. They first arose in statistical mechanics through the work of Baxter, and in the scattering context through the work of Yang. Particular solutions to the Yang-Baxter equations that are useful for constructing models for specializations of the Homfly and Kauffman polynomials first arose in work of Jimbo. See [52),58),129),143)].

In any case, if the vertex weights satisfy the Yang-Baxter equations, then we have a regular isotopy invariant of links *if* the model also satisfies invariance under the type II move with reversed orientation :

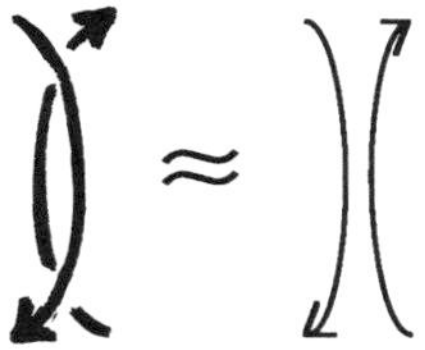

This leads to a condition that can be managed by adjusting the parameter λ in the global term. I will give a specific example shortly, but first a description of the problems and questions that arise from this kind of model : First, the Yang-Baxter model is satisfying in that it isolates the problems of well-definedness. Any vertex weights and global evaluation will give a well-defined function on diagrams. The invariance under the type III move becomes related to a deep problem in mathematical physics (solving the Yang-Baxter equation). This means that the structure of the knot invariants is related to an extraordinary body of mathematics involving everything from bare-hands solutions of these equations to delicate deformations of Lie algebras (quantum groups [52),30)]).

How are these models to be understood ? There are two directions - *combinatorial* and *global*. By global I mean the possibility of a more general, three dimensional definition of the invariants through which the state models will be seen as a mode of expression (just as the determinant or free differential calculus expressions of the Alexander polynomial are expressions of the structure of the first homology of the infinite cyclic covering space of the knot complement). Global definitions for these invariants are just beginning to appear (see section 8). Eventually, the state models will be seen in the perspective of global definitions.

The combinatorial direction is equally important. Here one seeks understanding by reducing the definitions to minimal form with respect to the structure of knot theory as a diagrammatic system. Even in the case of the bracket, much deeper combinatorial insight is needed for further progress. For the Yang-Baxter models I have been working very hard to reformulate them so that the diagrammatic structure is transparent, and the triangle invariance is understandable on its own terms. In the case of the models of Jones and Turaev for specializations of the Homfly and Kauffman polynomials, this has led to a good story for building these invariants and the corresponding Yang-Baxter tensors from link-theoretic grounds [78),80),81)]. We are now in the position of seeing these Yang-Baxter tensors as natural structures in knot theory (analogous to the regular solids as natural structures in geometry) that are contextualized/classified by the theory of quantum groups (as the regular solids are classified via their symmetry groups). In this sense, knot theory is now a key contributor to the theory of deformations of Lie algebras, providing a ground for understanding these ideas.

Here is a quick summary of the models related to specializations of the Homfly and Kauffman polynomials. First the Homfly : A state consists in an oriented splitting 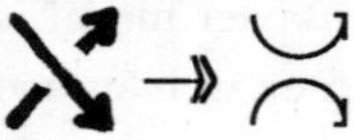 or a projection 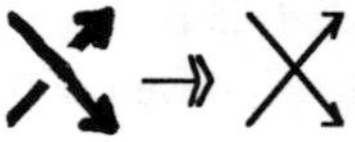 of each crossing so that no circuit in the state has any self-crossings. Each circuit is labelled with a spin from the given index set. I use the following notation for spin relations at a projected crossing or at a split :

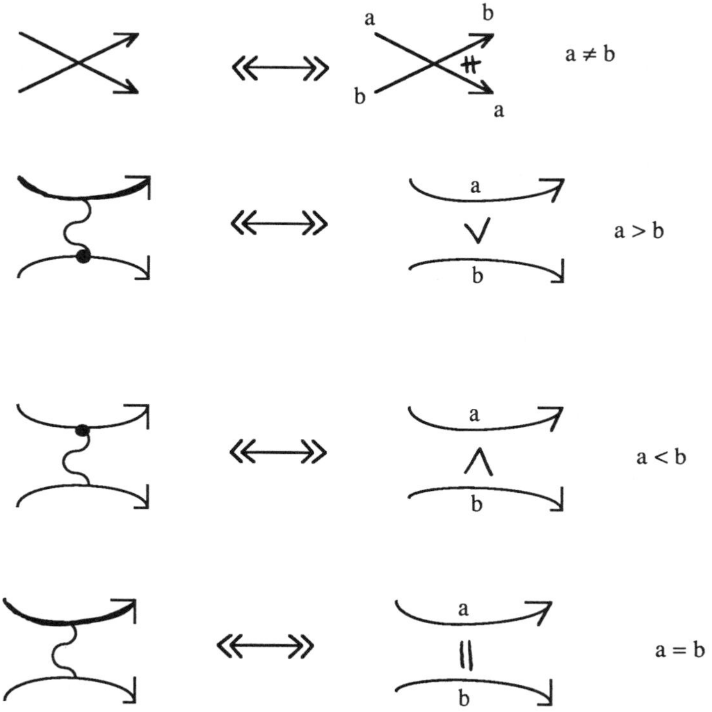

The same forms can be used as generalized "Kronecker deltas". Thus :

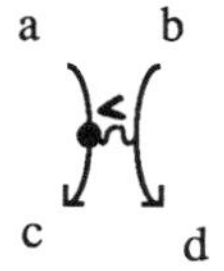

$$= \begin{cases} 1 & \text{if } a = c < b = d \\ 0 & \text{otherwise} \end{cases}$$

The model then has the expansion as a generalized bracket :

$$\left(Z = W - W^{-1}\right)$$

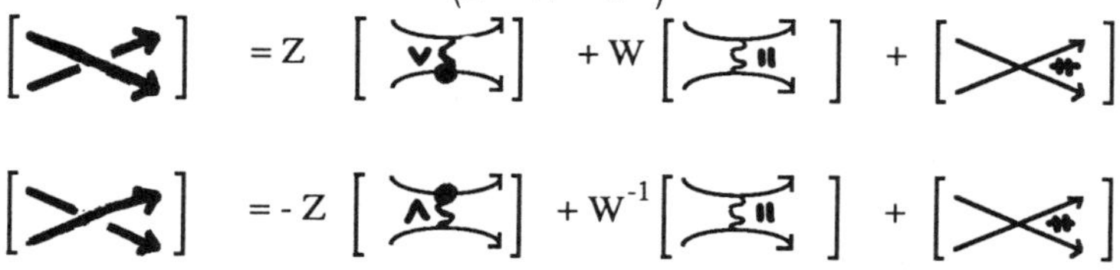

Such an expansion is equivalent to giving the vertex weights. In particular, the Yang-Baxter tensor for this model is given by

$R^{ab}_{cd} = Z$ 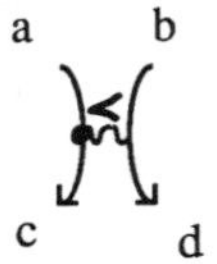 $+ W$ 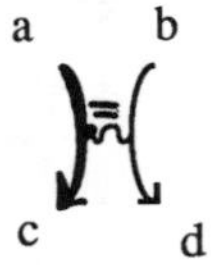 $+$ 

$\overline{R}^{ab}_{cd} = -Z$ 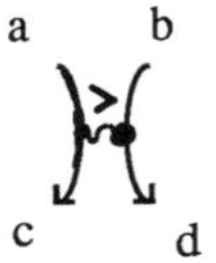 $+ W^{-1}$ 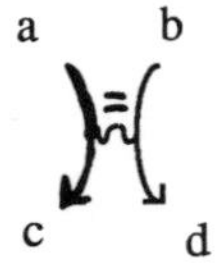 $+$ 

Note that since

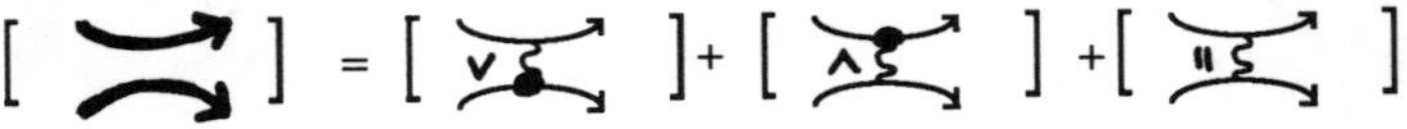

is a tautology, we have the exchange identity

when $Z = W - W^{-1}$.

The global evaluation of a state is given by the formula

$$\|S\| = \sum_{C \text{ a circuit in } S} \text{label}(C) \cdot \text{rot}(C)$$

where

rot() = + 1

rot() = - 1

Thus

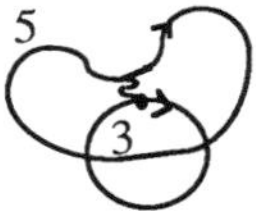

$S \Rightarrow \|S\| = -5 - 3 = -8$.

This global evaluation is a direct generalization of the counting of circuits in the bracket. The rotation numbers rot(C) are special cases of the Whitney degree of a plane curve.

It is an interesting exercise to verify directly that this model satisfies the Yang-Baxter equation (invariance under oriented type III move). One finds that this invariance occurs for *any ordered index set*. For any finite ordered index set regular isotopy invariance then becomes the demand that

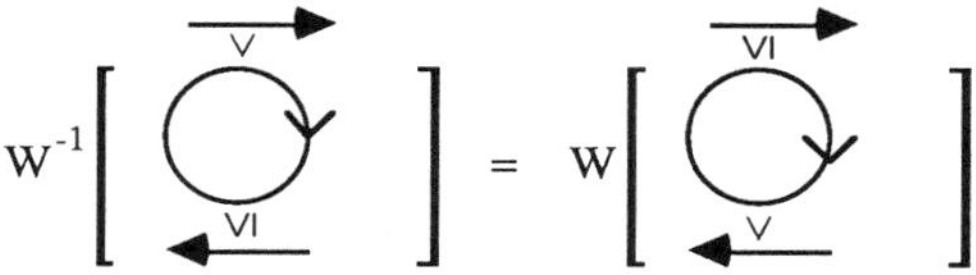

and

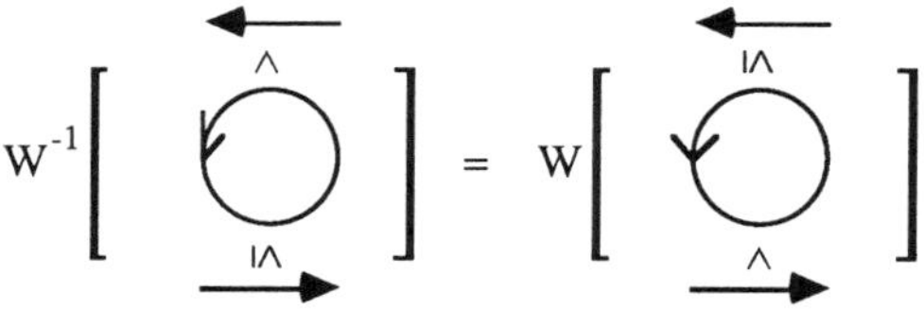

If we require multiplicativity,

Then everything simplifies. The model imposes a very strong requirement on the index set, and I prove [81)] that the index set must be of the form $\{-n,-n+2,\ldots,n-2,n\}$. This is exactly the spectrum of spins that Jones used in his model [58),143)] coming from quantum groups.

In the physical context a restriction to a discrete spectrum of this form occurs in the quantization of angular momentum - and it is mediated by the linear algebra of a sequence of raising and lowering operators. Here the same pattern occurs through topological restrictions. This relationship needs closer investigation.

A similar pattern gives rise to a version of the Turaev model for specializations of the Kauffman polynomial. Here I do not find extra invariants. Indeed the conditions here are quite tight and the natural tensor acquires extra factors to compensate the Whitney degree (the knot theory motivates the extra factors) and it requires a special index set in order to satisfy the Yang-Baxter equation. This situation begs for a closer comparison between the quantum groups and the knot theory.

IV. YANG-BAXTER MODELS - SCATTERING PICTURE.

This section turns to another viewpoint about the Yang-Baxter models. Let R satisfy the Yang-Baxter equation without rapidity parameter θ. Let S be defined by the formula

$$S^{ab}_{cd}(\theta) = R^{ab}_{cd}\lambda^{(d-a)\theta/2\pi} \longleftrightarrow$$

where λ is an extra parameter to be fixed in a given application. Assume that spin is conserved : a+b = c+d whenever R^{ab}_{cd} is non-zero. Then it follows from the geometry of Figure 1 (section 3) that S satisfies the full Yang-Baxter Equation :

$$\sum_{i,j,k} S^{ab}_{ij}(\theta) S^{jc}_{kf}(\theta+\theta') S^{ik}_{de}(\theta') = \sum_{i,j,k} S^{bc}_{ij}(\theta') S^{ai}_{dk}(\theta+\theta') S^{kj}_{ef}(\theta).$$

In a piecewise-linear link diagram we may assume that each crossing is an intersection of two straight line segments, and that there are isolated vertices of the form

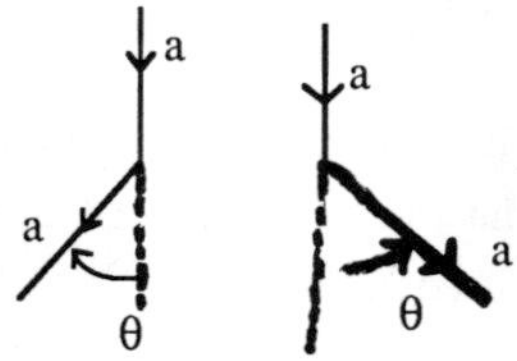

$$\longleftrightarrow \lambda^{a\theta/2\pi}.$$

Here I have indicated the angle conventions for these vertices, and corresponding vertex weights (measuring the angular turn in proportion to the spin). It is then a simple exercise in Euclidean geometry to see that if we use S for positive crossings, $\bar{S}$ for negative crossings :

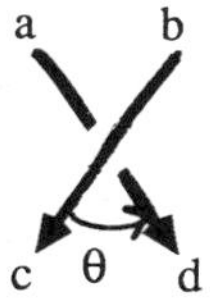

$$\longleftrightarrow \quad \overline{S}^{ab}_{cd}(\theta) = \overline{R}^{ab}_{cd}\, \lambda^{\frac{(d-a)\theta}{2\pi}},$$

and the angle weights for isolated vertices, then the state model [K] obtained by summing over all products of vertex weights for all spin assignments to the edges of the projected diagram, is an invariant of regular isotopy provided that

$$\sum_{ij} R^{ja}_{ib}\lambda^{\left(\frac{b-j}{2}\right)}\overline{R}^{ic}_{jd}\lambda^{\left(\frac{d-i}{2}\right)} = \delta^{a}_{d}\delta^{c}_{b}$$

This condition is equivalent to the statement that $\hat{S}(\pi)$ is inverse to $\tilde{\overline{S}}(\pi)$ where $\hat{S}^{ab}_{cd} = S^{bd}_{ac}$, $\tilde{S}^{ab}_{cd} = S^{ca}_{db}$. This invertibility is compatible with a time-reversal invariance in the scattering picture.

This is the general form of the model as formulated by Jones and Turaev. I sketch it here to raise questions of interpretation and generalization. First, note that the extra isolated vertex terms can be seen as interactions involving the emission or reception of a particle of spin zero. The momentum change is noted by the change in angle. This is a discretization of the "field" that would cause the particle in a smooth knot diagram to curve around in the plane.

By taking a limit to a smooth diagram the model will involve a curvature or Whitney degree term as in [58)]. One would like to know more about the behaviour of this 1+1 dimensional system as the number of crossings in the knot diagram goes to infinity.

A tantalizing question asks how to shift this form of the model into one defined directly for embeddings of links in three-dimensional space. Here the Gauss definition of the linking number of two curves gives a clue :

$$\mathrm{lk}(\alpha, \beta) = \frac{1}{4\pi}\int\!\!\int_{\alpha\times\beta} \frac{\bar{e}\cdot(\overline{T}_{\alpha}\times\overline{T}_{\beta})}{\|\bar{e}\|^{3}}\, dS_{\alpha}\, dS_{\beta}$$

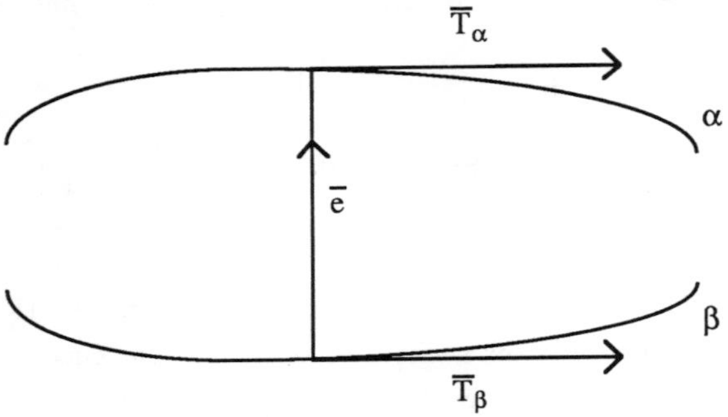

The Gauss kernel

$$\mathcal{G} = \frac{\bar{e} \cdot \left(\bar{T}_\alpha \times \bar{T}_\beta\right)}{\|\bar{e}\|^3}$$

vanishes whenever the tangent vector to the two curves and the direction vector between the two points lie in the same plane. For a nearly planar diagram the Gauss integral becomes a sum of vertex-related contributions. The crossings in such a diagram are special because they are loci of the non-vanishing of the Gauss kernel.

I conjecture that a generalization of the Gauss integral will give rise to a three dimensional definition for the Yang-Baxter models. We can turn the product of vertex weights into a sum of weights by exponentiation. Thus the space-form model will look like

$$[K] = \int_S e^{\int\int_{K \times K} \{K|S\} \mathcal{G} dS_1 dS_2 + \int_K d\theta[S]}$$

where the outer integral runs over spin states, and the inner integrals are the sums of vertex weights modulated by a Gauss kernel, and the total curvature of the space curve. This model is speculative - a way of asking how the vertex models can be moved into three dimensional space.

The model (see section 8) proposed by Witten using gauge fields and the Chern-Simons form has a different character. One wants to know how it is related to the Yang-Baxter models. A movement of the Yang-Baxter models into three-space via the Gauss kernel may be a step in this direction.

V. SKEIN MODELS

So far, the state models discussed for the skein polynomials have produced series of one-variable specializations for these polynomials. It is an open problem whether there are models of Yang-Baxter type that capture the full two-variable polynomials. On the other hand a state model is a combinatorial summation that produces the polynomial. One wants states that correspond to diagrammatic configurations, and vertex weights that are computed from the state.

In fact *there are state models for both of the two-variable polynomials that satisfy this general criterion.* These models are reformulations of the original recursive scheme for calculating the Conway polynomial and its generalizations. I call these "skein models". The first skein model was given by Francois Jaeger in [51], using a matrix inversion technique. In [80], I show how to obtain the model directly as a skein calculation, and I generalize it to the Kauffman polynomial and to the Kauffman-Vogel graph polynomials.

Here is a description of the skein model for the regular isotopy version of the Homfly polynomial: Let K be an oriented link diagram, and U its underlying universe (shadow under planar projection). If U has n edges, label these edges with the integers 1,2,...n in any way. Call this labelling of U a *template* . A state S of K is *any diagram obtained from U by splicing some subset of its crossings.* In order to compute vertex weights, we consider circuits on S with respect to the template. For each circuit C on S (a circuit crosses at each crossing) let i(C) denote the least index on the edges of C . Order the circuits by the order of their indices. Now *travel from the least edge of each circuit in the direction of its orientation, making note of the place of first passsage through each crossing site of K.* Travel the circuits in the order of their least indices. I use the following local notations for these conditions:

 $\leftrightarrow$ first passage on lower leg

 $\leftrightarrow$ first passage on upper leg

[crossing diagram] ↔ first passage on [diagram]

[crossing diagram] ↔ first passage on [diagram]

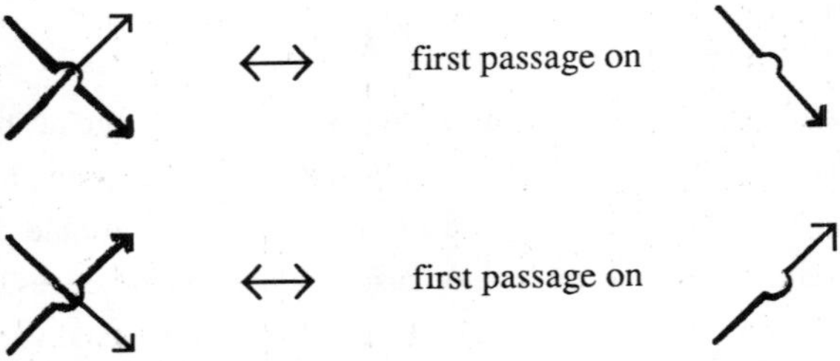

Assign vertex weights to each state as follows:

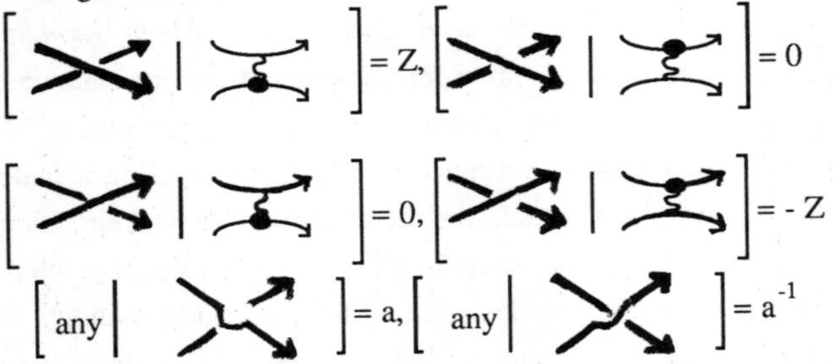

$$\left[\rotatebox{0}{\includegraphics{}} \,\middle|\, \rotatebox{0}{\includegraphics{}}\right] = Z, \quad \left[\rotatebox{0}{\includegraphics{}} \,\middle|\, \rotatebox{0}{\includegraphics{}}\right] = 0$$

$$\left[\rotatebox{0}{\includegraphics{}} \,\middle|\, \rotatebox{0}{\includegraphics{}}\right] = 0, \quad \left[\rotatebox{0}{\includegraphics{}} \,\middle|\, \rotatebox{0}{\includegraphics{}}\right] = -Z$$

$$\left[\text{any} \,\middle|\, \rotatebox{0}{\includegraphics{}}\right] = a, \quad \left[\text{any} \,\middle|\, \rotatebox{0}{\includegraphics{}}\right] = a^{-1}$$

Thus

$$\left[\rotatebox{0}{\includegraphics{}}\right] = Z\left[\rotatebox{0}{\includegraphics{}}\right] + a\left[\rotatebox{0}{\includegraphics{}}\right] + a^{-1}\left[\rotatebox{0}{\includegraphics{}}\right]$$

$$\left[\rotatebox{0}{\includegraphics{}}\right] = -Z\left[\rotatebox{0}{\includegraphics{}}\right] + a\left[\rotatebox{0}{\includegraphics{}}\right] + a^{-1}\left[\rotatebox{0}{\includegraphics{}}\right]$$

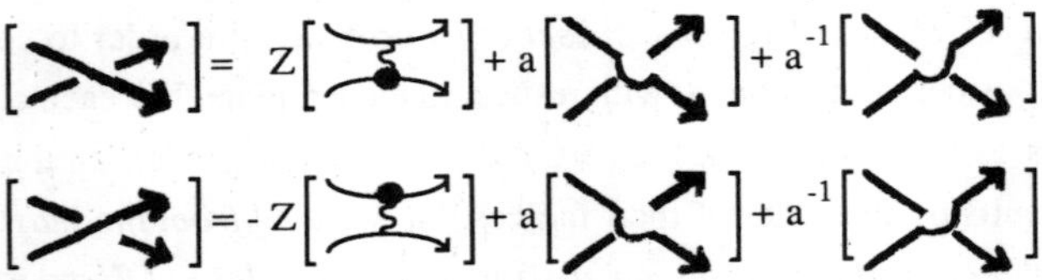

Then the state summation is given by the formula

$$[K] = \sum_S [K|S] \delta^{|S|-1}$$

where $[K|S]$ denotes the product of vertex weights, $|S|$ is the number of circuits in the state, and $\delta = (a - a^{-1})Z^{-1}$.

In particular we have the formula

$$[K] = \sum_S O^{\hat{t}(S)} (-1)^{t_-(S)} a^{w(S)} \delta^{|S|-1}$$

where $w(S)$ is the writhe of the link diagram obtained by repacing each first crossing in the state by an over-crossing. Here $t_-(S)$ denotes the number of split negative crossings in the state, and $\hat{t}(S)$ is the number of splices with a zero vertex weight (inappropriate passage). The contributing states in this model are in one-to-one correspondence with the set of unlinks at the bottom of a skein calculation using standard unknotting procedure relative to the template.

This skein model shares all of the main features of the state models we have considered so far, including a direct expansion formula.

The major difference between skein model and the bracket and Yang-Baxter models is that its well-definedness on diagrams requires proof. The independence of choice of template is equivalent to the usual inductive proof of well definedness for the polynomial.

Nevertheless, skein models give direct formulas for the Homfly and Kauffman polynomials, and show that the structure of state model is implicit in Conway's original scheme.

That the vertex weights are computed from the precedence structure created by the template is a particular feature of this model. It is an issue that is implicit in any global use of the idea of Feynman diagrams. These models should be compared with the structure of normal orderings of operators whose products are interpreted as diagrammatic particle interactions in Feynman's approach.

VI. INVARIANTS OF GRAPH EMBEDDINGS

The two-variable skein polynomials extend to three-variable invariants for rigid isotopy classes of embeddings of (locally 4-valent) graphs in three-space. The extension is obtained by replacing each graph vertex by a crossing or by a splice, and summing over all possibilities. Symbolically this takes the form

and

where A and B are new variables with Z=A-B the usual variable for the regular isotopy Homfly polynomial. For links, [K] denotes the regular isotopy version of the Homfly. A similar formula works for the Kauffman polynomial.

In [79] we turn this procedure upside down by regarding the planar graph evaluations as fundamental. By using the idea of a template (see section 5), one finds that for a plane graph, G, [G] is given by the formula

$$[G] = \sum_{S} (-1)^{t(S)} A^{t_{+}(S)} B^{t_{-}(S)} a^{w(S)} \delta^{|S|-1} .$$

Here t+(S) denotes the number of splits in S with first passage on the top of the two horizontal strands (oriented from left to right) : t-(S) is the number of bottom splits and t(S) is the total number of splits.

There is a skein model state expansion in the form

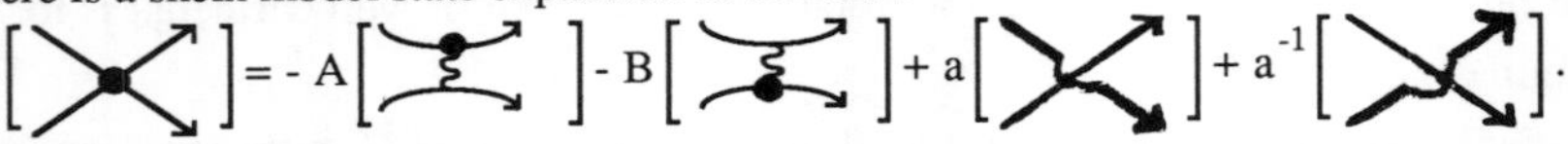

We also develop a calculus for these three variable graph polynomials that shows how to recursively compute the polynomials for plane graphs entirely in that category. The formalism extends traces on the Hecke and Birman-Wenzl algebras to a graph theoretic context.

The graph polynomials are useful for detecting chirality in graph embeddings, and have potential applications in chemistry and molecular biology. It happens naturally that the appropriate equivalence relation for these invariants is *rigid vertex isotopy*, a mixture of ambient isotopy on the strands and mechanical rigidity at the graph vertices. The problem of inventing polynomial invariants for graphs with topological vertices is challenging, due to braiding at the vertex.

VII. THE POTTS MODEL

The q-state Potts model in statistical mechanics [12),74)] has a perspicuous expression in terms of the bracket. Given any plane graph G , there is an associated (medial) alternating link K(G) . The partition function, Z_G , is given by

$$Z_G = W_{K(G)} = q^{N/2}\{K(G)\}$$

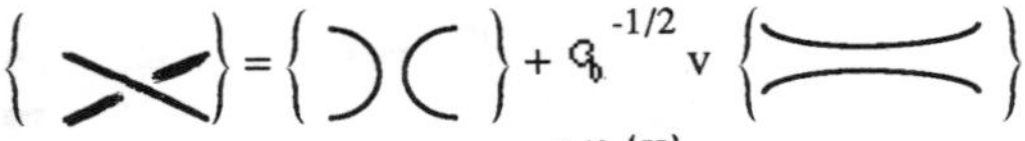

$$\{OK\} = q^{1/2}\{K\}$$

$$\{O\} = q^{1/2}$$

where N denotes the number of vertices of the graph G , T is the temperature variable for the Potts model, $v = \left(e^{-\frac{1}{kT}}\right) - 1$, and k is Boltzmann's constant.

This translation into link diagrams allows a number of features of the physical model to come to view. In particular, for the limit of rectangular lattices it is conjectured that criticality in the anti-ferromagnetic case occurs when

$$\sqrt{q} = v.$$

This corresponds to the case, where the bracket expansion is independent of the choice of over and under-crossing, a particular symmetry in the formulation.

The Potts partition function can be expanded in terms of the Temperley-Lieb algebra. This algebra appears diagrammatically (compare [67),74)]) here via

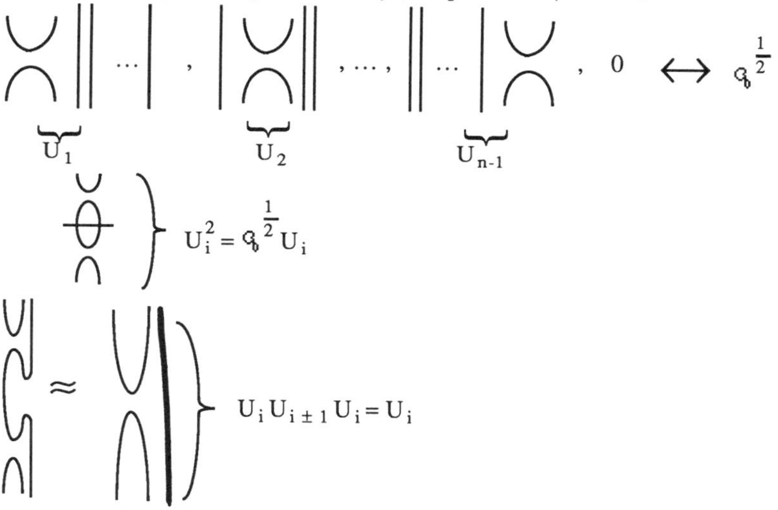

Diagrammatic for Temperley-Lieb Algebra

Figure 2

This diagrammatic can be used to study representations of the Temperley-Lieb algebra (compare [74)]) and it also shows how the Jones algebra [55),56)] (which is formally identical to the Temperley-Lieb algebra) appears via the bracket.

One may hope to use the diagrammatic for the Temperley-Lieb algebra to shed light on the apparent relationship between its representations and representations of the Virasoro algebra. (See [1),15),39),91)]). One expects a picture of this relationship via a dovetailing of the geometry of the Potts models with conformal symmetry at the continuum limit. The relevant geometry of the Potts model is formalized in the diagrammatic relations with the Temperley-Lieb algebra.

This question involving *the relationship of the Potts models with conformal field theory* is significant for mathematical physics, and it affords real possibility of an application from knot theory (diagrammatic) to physics.

A curious feature of the Potts model is that, from the flyping invariance of the bracket (see section 2), it follows that *modulo the Tait conjecture* the Potts partition function is a topological invariant of the associated alternating link K(G) when the graph G is free of isthmuses and loops. This is an unexpected link with the topology that may shed light on the nature of the Tait conjecture.

Potts and the Alexander Polynomial

By generalizing the Potts model to allow different sets of spins at different vertices of the graph (a natural generalization if the physical substrate is a mixture of materials) I have been able to include the classical Alexander polynomial in this context. The result is that the *Conway polynomial is the low temperature limit of a generalized Potts model.* The states model (FKT model) that underlies this statement is constructed in the author's monograph *Formal Knot Theory* [65)]. Here the restriction to alternating knots is lifted. To each oriented link is associated a graph (whose vertices correspond to the regions of the diagram) and a generalized Potts model. This model raises many questions. I would like to know how the topological invariance at low temperature affects the behaviour of the model at higher temperatures and at criticality. Since the FKT model is closely related to the fundamental group, and to techniques of classical knot theory, there is a rich vein of material to investigate here.

Chromatic Polynomial and Coloring

The low temperature limit of the standard q-state Potts model is the chromatic polynomial of the associated graph. Thus the chromatic polynomial for a plane graph can be computed via a bracket evaluation of an associated alternating link. For special values of q (the Beraha numbers $q_n=4\cos^2(\pi/n)$) it has long been conjectured that the zeroes of the chromatic polynomial for large plane graphs cluster at these values. (See [14)].) The same values appear for self-adjoint representations of the Jones algebra, and for unitary representations of the Virasoro algebra. The context under discussion may lead to a clarification of these relationships.

In general there are deep relations here with combinatorics. A guiding problem for this project has been to uncover a more conceptual proof of the Four Colour Theorem. I have found a reformulation of the coloring problem in terms of the vector cross product algebra in three-space [76)]. This reformulation turns the four color problem into a delicate question about a Lie algebra. The idea for the reformulation comes from state models.

The *real* four-color-conjecture would state that *one can extend map colorings over pentagonal regions via two-color switching processes* (see [77]). There is good algorithmic evidence that this is true (and it is everyone's experience that it can always be done in any special case). The direct difficulty in dealing with this problem makes the creation of a context - statistical mechanics, knot theory, spin networks [123], conformal field theory, quantum groups, combinatorial reformulations - of great use in focusing on the central issue of the plane map.

VIII. THREE DIMENSIONS

Knots and links are embedded in three dimensions. It is strange that all the new invariants have emerged through the planar representation of knot theory via diagrams and Reidemeister moves. (The theory of braids is a chapter in this planar theory). The fundamental problem is to give a three-dimensional definition of the invariants.

Witten has recently proposed to define an invariant by the formula

$$W(K) = \int D A\, e^{inLcs}\, \mathrm{Tr}\,(\mathrm{Hol}\,(K))$$

where the integration is performed over all gauge connections (with respect to a given gauge group or representation). Here

$$Lcs = \frac{1}{8\pi^2} \int d^3X\, \mathrm{Tr}\left(A \wedge dA + \frac{2}{3} A \wedge A \wedge A\right)$$

is the Chern-Simons Lagrangian, and Tr Hol (K) denotes the product of the traces of the holonomy of the connection taken along the components of the link K . Witten asserts that this definition yields (at least) specializations of the Jones and Homfly polynomials.

This definition is very close to ideas being pursued by L. Smolin and C. Rovelli [131] in creating states for a theory of quantum gravity. It remains to be seen whether these methods can be made rigourously explicit - integration over all connections is a formal device. Nevertheless, this avenue is very exciting. I believe that it will lead to true three dimensional definitions, and it may help in the creation of combinatorial or algebraic versions more suited to computations.

In particular, there is the possibility of finding our way to invariants of compact three-manifolds. It seems natural to define directly states models on a three manifold (viewed as a cell complex or a decorated graph [82]) . But so far, without a guide, no definition has emerged. The holonomy approach to these invariants may show the way.

References

1) Akutsu, Y. and Wadati, M., "Knot invariants and critical statistical systems", J. Phys. Soc. Japan 56, 839-842 (1987).
2) Akutsu, Y. and Wadati, M., "Exactly solvable models and new link polynomials I.N-state vertex models", J.Phys. Soc . Japan 56, 3039-3051 (1987).
3) Akutsu, Y, Deguchi, T. and Wadati, M., "Exactly solvable models and new link polynomials II. Link polynomials for closed 3-braids", J.Phys.Soc.Japan 56, 3464-3479 (1987).
4) Akutsu, Y., Deguchi, T. and Wadati, M.,"Exactly solvable models and new link polynomials III. Two-variable polynomial invariants" (to appear).
5) Akutsu, Y. and Wadati, M., "Knots, links, braids and exactly solvable models in statistical mechanics", Comm.Math.Phys. 117 , 243-259 (1988).
6) Alexander, J.W., "Topological invariants of knots and links",Trans. Amer. Math.Soc. 20, 275-306 (1928).
7) Anstee, R.P., Przytycki, J.H., and Rolfsen, D., "Knot polynomials and generalized mutation" (to appear).
8) Babelon, O., "Jimbo's q-analogues and current algebras", Letters in Mathematical Physics, 15, 111-117 (1988).
9) Ball, R., Mehta M.L., "Sequence of invariants for knots and links", J.Physique 42, 1193-1199 (1981).
10) Bauer, W.R., Crick, F.H.C., White, J.H., "Supercoiled DNA.", Sci.Amer. Vol. 243, N° 1. (July 1980), 118-133.
11) Baxter, G., "Local weights which determine area, and the Ising Model", J.Math. Physics, Vol.8, No. 3, March 1967.
12) Baxter R.J., "Exactly Solved Models in Statistical Mechanics", Academic Press (1982).
13) Baxter, R.J., "On Zamolodchikov's solution of the tetrahedron equations", Comm. Math.Phys. 88, 185-285 (1983).
14) Baxter, R.J., "Chromatic polynomials of large triangular lattices", J.Phys.A.: Math.Gen. 20, 5241-5261 (1987).
15) Belavin, A.A., Polyakov, A.M., and Zamolodchikov, A.B., "Infinite conformal symmetry in two-dimensional quantum field theory", Nucl.Phys. B241, 333-380 (1984).
16) Birman, J.S., "Braids, links and mapping class groups", Annals of Math. Studies No.82, Princeton University Press. Princeton, N.J. (1976).
17) Birman, J.S., "On the Jones polynomial of closed 3-braids", Invent. Math. 81, 287-294 (1985).
18) Birman, J.S. and Kanenobu, T., "Jones' braid-plat formula, and a new surgery triple", Proc. Amer.Math.Soc. 102 (1988), 687-695
19) Birman, J.S. and Wenzel, H., "Braids, link polynomials and a new algebra", (To appear in Trans. A.M.S).
20) Brandt, R.D., Lickorish, W.B.R., Millett, K.C., "A polynomial invariant for unoriented knots and links", Invent.Math. 84, 563-573 (1986).
21) Brauer, R., "On algebras which are connected with the semisimple continuous groups", Annals of Math. Vol.38, No.4, Oct.1937, 857-872
22) Burde, G. and Zieschang, H., "Knots", de Gruyter (1986).
23) Burgoyne, P.N., "Remarks on the combinatorial approach to the Ising problem", J.Math.Phys.Vol.4, No.10, (Oct. 1963).

24) Casson, A., Gordon, C.,McA., "On slice knots in dimension three". In Geometric Topology, R.J. Milgram ed., Proc. of Symp. Pure Math. XXXII, Amer. Math.Soc., 39-53 (1978).
25) Chew, G.F. and Poenaru, V., "Single-surface basis for topological particle theory", Physical Rev. D.,Vol.32, No.10 (Nov.1985).
26) Conway, J.H., "An enumeration of knots and links and some of their algebraic properties", Computational Problems in Abstract Algebra. Pergamon Press, New York, 329-358 (1970).
27) Cromwell, P.R., "Homogeneous links" (preprint 1988).
28) Crowell, R.H., "Genus of alternating link types", Ann. of Math. 69 , 2558-275 (1959).
29) Crowell, R.H., Fox, R.H., "Introduction to Knot Theory", Blaisdell Pub. Co. (1963).
30) Drinfeld, V.G., "Quantum Groups", Proc.Intl.Congress Math., Berkeley, Calif. USA, 798-820 (1986).
31) Drinfeld, V.G., "Hamiltonian structures on Lie groups, Lie bialgebras and the geometric meaning of the classical Yang-Baxter equations", Soviet Math., Dokl. Vol 27 , No.1 (1983).
32) Drinfeld, V.G., "Hopf algebras and the quantum Yang-Baxter equation", Soviet Math. Dokl. Vol.32, No. 1 (1985).
33) Elliot, G., Choi, M., Yui, N., "Gauss polynomials and the rotation algebra" (preprint 1988).
34) Ernst C. and Sunners, D.W., "The growth of the number of prime knots", Math. Proc. Camb.Phil.Soc.102, 303-315 (1987).
35) Faddeev, L.D., Reshetikhin, N.Yu., Takhtajan, L.A., "Quantization of Lie groups and Lie algebras", LOMI Preprint E-14-87, Steklov Mathematical Institute, Leningrad,USSR.
36) Fox, R.H., Milnor, J.W., "Singularities of 2-spheres in 4-space and cobordism of knots", Osaka J. Math. 3, 257-267 (1966).
37) Franks, J. and Williams, R., "Braids and the Jones polynomial", Trans. Amer. Math. Soc. 303, 97-108 (1987).
38) Freyd, P., Yetter, D., Hoste, J., Lickorish, W.B.R., Millet, K.C., Ocneanu, A., [HOMFLY], "A new polynomial invariant of knots and links". Bull.Amer.Math.Soc.12, 239-246 (1985) .
39) Friedan, D., Qiu, Z., Shenker, S., "Conformal invariance, unitarity and two-dimensional critical exponents", *Vertex operators in Mathematics and Physics* - Prooceedings of a conference November 10-17, 1983. Publ. MSRI #3, Springer Verlag (1984).
40) Fröhlich, J., "Statistics of fields, the Yang-Baxter equation and the theory of knots and links" (preprint 1987).
41) Fröhlich, J. and Marchetti, P., "Quantum field theories of vertices and anyons" (preprint 1988).
42) Fuller, F.B., "Decomposition of the linking number of a closed ribbon: a problem from molecular biology", Proc.Natl.Acad.Sci.USA, 75, 3557 (1978), 3557-3561.
43) Giller, C.A., "A family of links and the Conway calculus", Trans.Amer.Math.Soc. 270, 75-109 (1982).
44) Glasser, M.L., "The free energy of the three-dimensional Zamolodchikov model", J.Math.Phys. 27 (11), Nov. 1986.
45) de la Harpe, P., Kervaire, M., and Weber C., "On the Jones polynomial", L'Enseign.Math. 32 , 271-335 (1986).

46) Hartley, R., "Conway potential functions for links", Comment.Math.Helv. 58, 365-378 (1983).
47) Hennings, M.A., "A polynomial invariant for banded links" (preprint 1988).
48) Hitt, L.R., and Silver, D.S., "Stalling's twists and the Jones polynomial" (to appear).
49) Ho, C.F., "A new polynomial invariant for knots and links" - preliminary report, AMS Abstracts, Vol. 6, No. 4, Issue 39, p. 300 (1985).
50) Hoste, J., "A polynomial invariant of knots and links", Pacific J. Math. 124, 295-320 (1986).
51) Jaeger, F., "A combinatorial model for the Homfly polynomial" (preprint 1988).
52) Jimbo, M., "A q-difference analogue of U(q) and the Yang-Baxter equation", Lect. in Math. Physics 10, 63-69 (1985).
53) Jimbo, M., "Quantum R-matrix for the generalized Toda systems", Comm. Math. Phys. 102, 537-547 (1986).
54) Jonish, D., Millett, K.C., "Isotopy invariants of graphs" (preliminary report 1986) .
55) V.F.R. Jones, "A new knot polynomial and von Neumann algebras", Notices of AMS 33, 219-225 (1986).
56) Jones, V.F.R., "A polynomial invariant for links via von Neumann algebras", Bull. Amer. Math. Soc. 12 , 103-112 (1985).
57) Jones, V.F.R., "Hecke algebra representations of braid groups and link polynomials", Ann. of Math. 126, 335-388 (1987).
58) Jones, V.F.R., "On knot invariants related to some statistical mechanics models" (preprint 1988).
59) Kac, M. and Ward, J.C., "A combinatorial solution of the two-dimensional Ising model", Physical Review, 88, No. 6, Dec. 15, 1952.
60) Kanenobu, T., "Infinitely many knots with the same polynomial", Proc. Amer. Math. Soc. 97, 158-161 (1986).
61) Kanenobu, T. "Examples on polynomial invariants of knots and links", Math. Ann., 275, 555-572 (1986) .
62) Kanenobu, T., Sakuma, M. "A note on Kauffman polynomial" (preprint 1986).
63) Kaufman, Bruria, "Crystal statistics II. partition function evaluated by spinor analysis", Physical Reviw, Vol.76, No. 8, Oct. 15, 1949.
64) Kauffman, L.H., "The Conway polynomial", Topology 20 (1981), 101-108.
65) Kauffman, L.H., "*Formal Knot Theory*", Princeton University Press Mathematical Notes #30 (1983).
66) Kauffman, L.H., "*On Knots*", Annals of Mathematics Study, 115, Princeton University Press (1987).
67) Kauffman, L.H., "State models and the Jones polynomial", Topology 26, 395-407 (1987).
68) Kauffman, L.H., "Invariants of Graphs in Three-Space" (To appear in Trans AMS).
69) Kauffman, L.H., "New invariants in the theory of knots" (lectures given in Rome, June 1986) (to appear in Asterisque).
70) Kauffman, L.H., "New invariants in the theory of knots", Amer. Math. Monthly. Vol.95, No 3, 195-242 (March 1988).
71) Kauffman, L.H., "An invariant of regular isotopy" (announcement 1985).
72) Kauffman, L.H., "An invariant of regular isotopy" (to appear in Trans. Amer. Math. Soc.)
73) Kauffman, L.H. "*Knots and Physics*" (book in preparation - based on lectures given at Universita di Bologna and Politechnico di Torino - 1985 and subsequent developments).

74) Kauffman, L.H., "Statistical mechanics and the Jones polynomial" AMS Contemp. Math. 78 (1988), 263-297.

75) Kauffman, L.H.," A Tutte polynomial for signed graphs" (to appear in *Combinatorics and Complexity* - a special issue of Discrete Applied Mathematics).

76) Kauffman, L.H., "Map coloring and the vector cross product" (to appear in Journal of Combinatorial Theory).

77) Kauffman, L.H., "*Map Reformulation*", Princelet Edition (1987).

78) Kauffman, L.H., "State models for knot polynomials - an introduction" (to appear in the proceedings of the meeting of the Brasilian Mathematical Society - July 1987).

79) Kauffman, L.H. and Vogel, P., "Link polynomials and a graphical calculus" (to appear).

80) Kauffman, L.H., "State models for link polynomials" (to appear in l'Enseign. Math.)

81) Kauffman, L.H., "Knots, abstract tensors and the Yang-Baxter equations" (to appear).

82) Kauffman, L.H. and S. Lins, "Decomposition of the vertex group for 3-manifolds (to appear).

83) Kauffman, L.H., "Statistical mechanics and the Alexander polynomials" (to appear)

84) Kidwell, M. "On the degree of the Brandt-Lickorish-Millet polynomial of a link", Proc. AMS 100 (1987), 755-762.

85) Kidwell, M., Richter, R.B., "Trees and Euler tours in a planar graph and its relatives", Amer. Math. Montly 94 (1987), 618-630.

86) Kirkman, T.P., "The enumeration, description and construction of knots with fewer than 10 crossings" , Trans. Royal Soc. Edin. 32, 281-309 (1865).

87) Kobayashi, K., "On the genus of a link and the degree of the new polynomial" (to appear).

88) Kobayashi, K., "Coded graph of oriented links and Homfly polynomials", *Topology and Computer Science* . North Holland Math. Series, 277-294 (1987).

89) Kohno, T., "Monodromy representations of braid groups and Yang-Baxter equations". Ann. Inst. Fourier, Grenoble, 37, 4 (1987), 139-160.

90) Kulish, P.P., Reshetikhin, N.Yu. and Sklyanin, E.K., "Yang-Baxter equation and representation theory: I. Letters in Math. Physics, 5, 393-403 (1981).

91) Kuniba, A., Y. Akutsu and Wadati, M., "Virasoro algebra, von Neumann algebra and critical eight-vertex SOS models", J. Phys. Soc. Japan 55, No. 10, 3285-3288 (1986).

92) Lawrence, R., "A universal link invariant using quantum groups" (preprint 1988).

93) Levine, J., "Estimating bridge number" (preprint 1986).

94) Lickorish, W.B.R., "A relationship between link polynomials", Math. Proc.Camb. Phil.Soc. 100, 109-112 (1987).

95) Lickorish, W.B.R., "Linear skein theory and link polynomials", Topology and its Applications, 27, 265-274 (1987).

96) Lickorish, W.B.R., and Lipson, A.S., "Polynomials of 2-cable-like links", Proc. Camb. Phil. Soc. 100 , 355-361 (1987).

97) Lickorish, W.B.R., Millett, K.C., "A polynomial invariant for oriented links", Topology 26, 107-141 (1987).

98) Lickorish, W.B.R., and Millett, K.C., "The reversing result for the Jones polynomial", Pacific J. Math. 124, 173-176 (1986).

99) Lickorish, W.B.R., and Millett, K.C., "Some evaluations of link polynomials", Comment. Math. Helv. 61, 349-359 (1986).

99a) Lickorish, W.B.R., "Polynomials for links", Bull. London Math. Soc. 30 (1988), 558-588.

100) Lipson A.S., "An evaluation of a link polynomial", Math. Proc. Camb. Phil. Soc. 100, 361-364 (1986).
101) Lipson, A.S., "Some more states models for link invariants", Pacific J. Math. (to appear).
102) Little, C.N., "Non-alternate +- knots", Trans. Royal Soc. Edin. 35, 663-664 (1889).
103) Majid, S., "Non-commutative A/G : a new approach to quantization of photons" (preprint 1988).
104) Manin, Yu.I., "Quantum groups and non-commutative geometry" (Lecture Notes, Montreal, Canada, July 1988).
105) Manin, Yu.I., "Some remarks on Koszul algebras and Quantum groups", Ann. Inst. Fourier, Grenoble 37, 4, 191-205 (1987).
106) Martin, P.P., "On representations of an operator algebra and the transfer matrix spectrum of the q-state Potts model" (preprint 1986).
107) Martin, P.P., and Westbury, B.W., "Representations of the Temperley-Lieb Algebras (preprint 1988).
108) Morton, H.R., "Seifert circles and knot polynomials", Math. Proc. Camb. Phil. Soc., 99, 107-109 (1986).
109) Morton, H.R., "Threading knot diagrams". Math. Proc. Camb. Phil. Soc., 99, 247 (1986).
110) Morton, H.R., "The Jones polynomial for unoriented links" Quart. J. Math. Oxford (2), 37, 55-60 (1986).
111) Morton, H.R., Short, H.B., Calculating the 2-variable polynomial for knots presented as closed braids" (preprint 1986).
112) Morton H.R., Short, H.B., "The 2-variable polynomial of cable knots", Math. Proc. Camb. Phil. Soc., 101 , 267-278 (1987).
113) Morton, H.R. and Traczyk, "The Jones polynomials of satellite links around mutants (to appear).
114) Murakami, H., "A fomula for the two-variable Jones polynomial", Topology 26, 409-412 (1987).
115) Murakami, H., "On the derivatives of the Jones polynomial", Kobe J. Math. 3, 61-64 (1986).
116) Murakami, H., "A recursive calculation of the Arf invariant of a link", J. Math. Soc. Japan, 38, 335-338 (1986).
117) Murakami, J., "The Kauffman polynomial of links and representation theory" (preprint 1986).
118) K. Murasugi, "On the genus of the alternating knot I, II", J. Math. Soc. Japan 10 (1958), 94-105, 235-248.
119) K. Murasugi, "Jones polynomials of alternating links", Trans.Amer.Math.Soc. 295 147-174 (1986).
120) Murasugi, K., "Jones polynomials and classical conjectures in knot theory", Topology 26, 187-194 (1987).
121) Murasugi, K., "Jones polynomials of periodic links", Pacific J. Math., 131, 319-329 (1988)
122) Murasugi, K., "On invariants of graphs with applications to knot theory" (to appear in Trans. AMS).
123) Penrose, R., "Applications of negative dimensional tensors" *Combinatorial Mathematics and its Applications* edited by D.J.A. Welsh. Academic Press (1971).
124) Prztycki, J., "Conway formulas for Jones-Conway and Kauffman polynomials" (preprint 1986).
125) Przytycki, J. "Equivalences of cables of mutants of knots" (to appear).

126) Przytycki and Traczyk, P., "Invariants of links of Conway type", Kobe J. Math. 4, 115-139 (1987).
128) Reidemeister, K., "*Knotentheorie*" , Chelsea Publishing Co., New York (1948),Copyright 1932, Julius Springer, Berlin.
129) Reshetikhin, N.Y., "Quantized universal enveloping algebras, the Yang-Baxter equation and invariants of links, I and II. LOMI reprints E-4-87 and E-17-87, Steklov Institute, Leningrad, USSR.
130) Rolfsen, D., "Knots and Links", Publish or Perish Press (1976).
131) Rovelli, C., and Smolin, L., "Knot theory and quantum gravity (preprint 1988).
132) Schultz, C., "Solvable q-state models in lattice statistics and quantum field theory", Phys. Rev. Letters, 46, No. 10, March 1981.
133) Simon, J., "Topological chiralty of certain molecules", Topology,Vol. 25, No. 2, 229-235 (1986).
134) Tait, P.G., "On Knots I, II, III". Scientific Papers Vol. I, Cambridge University Press, London, 1898, 273-347.
135) Thistlethwaite, M.B., "Knot tabulations and related topics", *Aspects of Topology*, Ed. I.M. James and E.H. Kronheimer, Cambridge University Press (1985) 1-76.
136) Thistlethwaite, M.B., "A spanning tree expansion of the Jones polynomial", Topology 26, 297-309 (1987).
137) M. Thistlethwaite, "Kauffman's polynomial and alternating links", Topology 27 (1988), 311-318.
138) Thistlethwaite, M., "An upper bound for the breadth of the Jones polynomial", Math. Proc. Camb. Phil. Soc. 103 (1988), 451-456.
139) Thistlethwaite, M. "On the Kauffman polynomial for an adequate link", Inventiones Math. Vol. 93 Fasc. 2(1988), 285-296.
140) Traczyk, P. "10_{101} has no period 7 : a criterion for periodicity of links" (to appear).
141) Trace, B., "On the Reidemeister moves of a classical knot", Proc. Amer. Math. Soc., Vol. 89, No. 4 (1983), 722-724.
142) Truong, T.T., "Some nouvel aspects of Bethe-Ansatz methods for vertex systems", Physica 124A, 603-612 (1984).
143) Turaev, V.G., "The Yang-Baxter equations and invariants of links", LOMI preprint E-3-87, Steklov Institute, Leningrad, USSR. Inventiones Math, 92 (1988), 527-553.
144) Turaev, V.G., "A simple proof of Murasugi and Kauffman theorems on alternating links", L'Enseign. Math. 33, 203-225 (1987).
145) Turaev, V.G., "The Conway and Kauffman modules of the solid torus with an appendix on the operator invariants of tangles", LOMI Preprint E-6-88, Steklov Mathematical Institute, Leningrad, USSR.
146) Tutte, W.T., "A contribution to the theory of chromatic polynomials", Canadian J. Math. 6, 80-91, (1953).
147) Tze, C.H., "Manifold-splitting regularization, self-linking, twisting, writhing numbers of spacetime ribbons and Polyakov's Fermi-Bose transmutations" (preprint 1988).
148) White, J.H., "Self-linking and the Gauss integral in higher dimensions", Amer. J. Math., July, Vol XCI, (1969), 693-728.
149) Whitney, H., "On regular closed curves in the plane", Comp. Math. 4 (1937), 276-284.
150) Witten, E., "Quantum field theory and the Jones Polynomial", Commun. Math. Phys. 121, 351-399 (1989).
151) Wolcott, K., "The knotting of theta curves and other graphs in S^3" (to appear in proceedings of 1985 Georgia Topology Conference).

152) Yetter, D. "Markov algebras", AMS Contemp. Math. 78 (1988), 705-730.
153) Zamolodchikov, A.B., Factorized S matrices and lattice statistical systems", Soviet Sci. Reviews, Part A (1979-1980).
154) Zamolodchikov, A.B., "Tetrahedron equations and the relativistic S-matrix of straight strings in 2+1 dimensions", Comm. Math. Phys. 79, 489-505 (1981).

ALGEBRAS OF LOOPS ON SURFACES, ALGEBRAS OF KNOTS, AND QUANTIZATION

V.G.Turaev

ABSTRACT

We study a Lie bialgebra generated by homotopy classes of loops on an oriented surface F. A biquantization of this Lie bialgebra is constructed in terms of skein equivalence classes of links in $F \times \mathbb{R}$. A short overview of related notions of Lie-Poisson Algebra is included.

INTRODUCTION

1. The aim of the present paper is to quantize certain Poisson algebras of loops on an oriented surface F. These algebras are intimately related to Poisson algebras of smooth functions on the spaces of conjugacy classes of linear representations of $\pi_1(F)$. Therefore, the quantization problem for the algebra of loops is a natural version of the quantization problem for algebras of functions on Poisson manifolds. It is especially interesting that in our setting the role of quantum observables is played by knots and links in $F \times \mathbb{R}$.

The constructions of the present paper have been inspired by the theory of polynomial invariants of links in S^3 which treats the Alexander polynomial and the newly developed Jones-Conway and Kauffman polynomials. (Formally, these polynomials are not used in the paper.) Note that one approach to building these polynomials is based on the theory of quantum R-matrices (see [3], [4], [15], [17]). Some algebraic notions introduced in the frameworks of this theory play important role in the present paper. On the other hand an adequate treatment of the geometric situation demanded to introduce some new algebraic notions, interesting in themselves.

Of course, the quantum character of knots and links can not some as a surprise after all the work done on relationships between the knot theory, statistical mechanics, R-matrices and so on. Probably, the main novelty here is the systematic study of knots in the cylinder over a surface, rather than in S^3 .

It should be emphasized that quantization is considered in this paper from a purely algebraic point of view, the analytical aspects of the notion being ignored.

2. In this paper F denotes an oriented surface and K denotes a commutative associative ring with unit.

W.Goldman [6] introduced a Lie bracket in the free K-module Z , generated by free homotopy types of noncontractible loops in F . The resulting Lie algebra acts in a canonical way on certain Poisson manifolds of linear representations of $\pi_1(F)$. (A related Lie bracket in the module, generated by free homotopy types of non-oriented loops in F , was implicit in the earlier paper of S.Wolpert [18]). Definitions of both brackets are reproduced in §1).

Let $S=S(F)$ be the symmetric algebra of Z . The additive free generators of S correspond to finite non-ordered collections of non-contractible loops in F , considered up to free homotopy of loops. The product of generators corresponds to the union of collections; the empty collection represents $1 \in S$. The Goldman bracket in Z extends by the Leibniz rule

$$[ab, c] = a[b,c] + [a,c]b \qquad (1)$$

to a Lie bracket in S . This makes S a Poisson algebra. (A Poisson algebra is a commutative associative algebra equipped with a Lie bracket so that (1) holds).

3. The following well known construction enables one under certain conditions to build Poisson algebras from non-commutative algebras over the polynomial ring $K[h]$. Let A be an associative algebra over this ring, which is free as a $K[h]$-module. Assume that the algebra A/hA is commutative so that $ab-ba \in hA$ for any $a, b \in A$. The formula

$$[a \bmod hA,\ b \bmod hA] = h^{-1}(ab-ba) \bmod hA \qquad (2)$$

equips A/hA with a Lie bracket which satisfies (1). The inverse to this construction is called a quantization of Poisson algebras (see, for instance, [3]), [17])). Speaking non-formally, the quantization is a non-commutative extension (or h-deformation) of a commutative Poisson algebra so that the first approximation to non-commutativity is determined by the Lie bracket.

In this paper I present a $K[h]$-algebra $A(F)$ which is additively generated by isotopy classes of oriented links in $F \times \mathbb{R}$ modulo some equivalence relation. It turns out that $A(F)$ quantizes $S(F)$ (though in a weaker sense then it was described above: I do not know whether $A(F)$ is free as the $K[h]$-module). The equivalence relation for links essentially imitates the skein equivalence relation for oriented links in S^3 , introduced by J.Conway [1]) in the course of his study of the Alexander polynomial. The relation says that if 3 oriented links L_+, L_-, L_0 coincide outside a ball and look as in Fig.1 inside the ball then one must put (formally) $L_+ - L_- = hL_0$. If

$L_+, L_-, L_0 \subset F \times \mathbb{R}$ and $h \to 0$ then this relation degenerates to the equality $L_+ = L_-$ which mirrors the fact that the projections of L_+ and L_- in F are homotopic. This obvious argument enables one to define the quantization homomorphism $A(F) \to S$.

Fig. 1

A quantum character of the Conway relation was foreseen by L.Kauffman [9]) who wrote: "I always thought that the Conway identity bore a striking resemblance to the exchange identities of quantum physics such as the Heisenberg form of the uncertainty principle $PQ-QP=\hbar i$". It seems that the present paper provides a justification for this hint.

The following relation is a counterpart of the Conway relation in the theory of non-oriented links:

$$L_+ - L_- = h(L_0 - L_\infty). \tag{3}$$

Here L_+, L_-, L_0, L_∞ are 4 links which coincide outside a ball and look as in Fig.2 inside the ball.

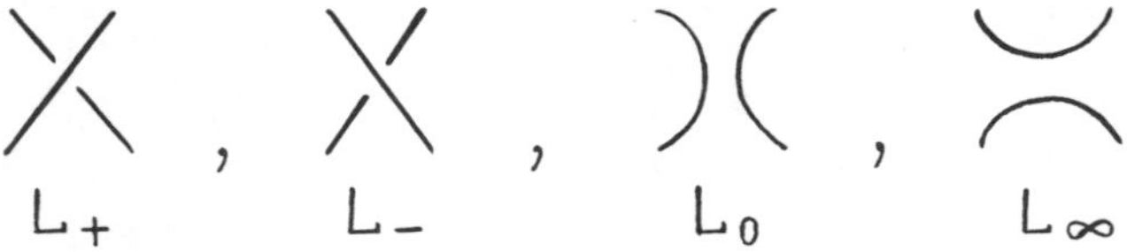

Fig. 2

For links in S^3 the relation (3) and the close relation, obtained by trading both minuses in (3) for pluses, have been studied by several outhors, see [9]), [10]). In this paper (3) is used to build an algebra of non-oriented knots in $F \times \mathbb{R}$ which quantizes the Poisson algebra of non-oriented loops on F.

4. Besides the Goldman bracket, the module Z may be equipped with a Lie cobracket $Z \to Z \otimes Z$ so that Z becomes a Lie bialgebra in the sense of V.Drinfel'd [2]). This cobracket induces in S a structure of "bipoisson bialgebra" (see § 3). For bipoisson bialgebras it is natural to consider a "biquantization" which produces a bialgebra (Hopf algebra) over the polynomial ring on two variables $K[h, \hbar]$. Here h and $\hbar$ are two independent "Planck constants" which are used to deform respectively the multiplication and the comultiplication of the bipoisson bialgebra. Note that the notion of bipoisson bialgebra is self-dual. The duality transforms the multiplication and the Lie bracket into the comultiplication and the Lie cobracket of the dual object (and vice versa) and exchanges the roles of $h, \hbar$.

In this paper $A(F)$ is provided with a structure of bialgebra which turns out to be a biquantization of S.

The Lie bialgebra Z is an infinitesimal analogue of an infinite dimensional Poisson Lie group $\operatorname{Exp} Z^*$. The algebra $A(F)$ also quantizes certain Poisson algebra of functions on $\operatorname{Exp} Z^*$.

5. The paper is self-contained, though most of the proofs are omitted. I also didn't include the relative version of the constructions, which deals with proper arcs in F, tangles in $F \times \mathbb{R}$ and in the algebraic part with modules over bipoisson bialgebras. A detailed exposition will be presented elsewhere.

6. Development of this work was stimulated by two lectures on quantization given by L.D.Faddeev in 1987/88 in the Steklov Mathematical Institute. I am also indebted to S.P.Novikov for critical remarks and to M.A.Semenov-Tian-Shansky for numerous talks on Poisson groups.

7. <u>Plan of the paper</u>. In §§ 1, 2 I recall the definition of the Goldman-Lie bracket and introduce the Lie cobracket in Z. In §§ 3, 4 the Poisson and bipoisson struc-

tures, associated with Lie brackets and cobrackets are briefly discussed. § 5 is concerned with quantization and biquantization. In § 6 the knot algebras are introduced and the main theorems are stated. § 7 deals with a central extension of Z by the Lie bialgebra of regular homotopy classes of loops on F and the corresponding extensions of knot algebras. In the Appendix an example of biquantization in an algebraic setting is presented.

7. Conventions and notation. If the contrary is not stated explicitly the word "module" means a K-module and the sign $\otimes$ denotes the tensor product over K.

1. ALGEBRAS OF LOOPS ON F

1.1. The Goldman bracket. Denote by $\hat{\pi}$ the set of free homotopy classes of non-contractible loops $S^1 \to F$. Denote by Z the free K-module with the basis $\hat{\pi}$. For a non-contractible loop $\alpha : S^1 \to F$ denote the class of α in $\hat{\pi}$ by $[\alpha]$ so that $[\alpha] \in \hat{\pi} \subset Z$. For a contractible loop α put $[\alpha] = 0 \in Z$. Fix an orientation in S^1.

Let α and β be non-contractible loops on F lying in general position. The set $\alpha(S^1) \cap \beta(S^1)$ is finite; denote it by $\alpha \# \beta$. For $x \in \alpha \# \beta$ denote by $\varepsilon_x = \pm 1$ the intersection index of α and β in x. Denote by $(\alpha\beta)_x$ the loop obtained by the orientation preserving smoothing of $\alpha(S^1) \cup \beta(S^1)$ in the point x (see Fig.3).

Fig. 3

Set

$$[[\alpha], [\beta]] = \sum_{x \in \alpha \# \beta} \varepsilon_x [(\alpha\beta)_x] \in Z. \quad (4)$$

According to [6], Theorems 5.2 and 5.9 the bilinear pairing $[\, ,\,]: Z \otimes Z \to Z$, given by (4) on the generators, is correctly defined and provides Z with the Lie algebra structure. It should be noticed that Goldman considers the slightly larger Lie algebra $Z \oplus K \cdot \{1\}$, where $\{1\}$ is the homotopy classes of contractible loops on F and $[Z, \{1\}] = 0$.

Let $i: Z \to Z$ be the involution transforming $[\alpha]$ into $[\alpha^{-1}]$ where α^{-1} denotes the loop α traversed in the opposite direction. According to [6] i is an automorphism of the Lie algebra Z . Put $Z_{\square} = \{a \in Z \mid i(a) = a\}$. Clearly, $Z_{\square}$ is a subalgebra of Z freely generated as a module by the set $\{a + i(a) \mid a \in \hat{\pi}\}$. These generators bijectively correspond to free homotopy types of non-oriented non-contractible loops on F . The Lie algebra $Z_{\square}$ is implicit in [18].

1.2. Poisson algebras of loops. It is easy to verify that the Lie bracket in a Lie algebra $\mathfrak{g}$ uniquely extends to a Lie bracket in the symmetric algebra $S(\mathfrak{g})$ of the module $\mathfrak{g}$, so that the Leibniz rule (1) is satisfies. Note that $S(\mathfrak{g}) = \bigoplus_{n \geq 0} S^n(\mathfrak{g})$ where $S^0(\mathfrak{g}) = K$, $S^1(\mathfrak{g}) = \mathfrak{g}$ and $S^n(\mathfrak{g})$ is the n-th symmetric tensor power of $\mathfrak{g}$. The resulting Poisson algebra is denoted by $PS(\mathfrak{g})$. It is called the symmetric Poisson algebra of the Lie algebra $\mathfrak{g}$.

In particular, the Lie algebras Z and $Z_{\square}$ give rise to symmetric Poisson algebras $PS(Z)$ and $PS(Z_{\square})$. The inclusion $Z_{\square} \hookrightarrow Z$ induces an (injective) Poisson algebra homomorphism $PS(Z_{\square}) \to PS(Z)$ (i.e. an algebra homomorphism which preserves the Lie bracket).

The Lie bracket $[\, ,\,]$ in $PS(Z)$ admits a deformation depending on one parameter $k \in K$. (This deformation is implicit in [6], Theorem 3.15). For $a, b \in \hat{\pi}$ put $[a, b]_k = [a, b] - k\,(a \cdot b)\, ab$ where $a \cdot b$ is the integer intersection number of a and b, and ab is the product of a and b in $S(Z)$. The bracket $[\,,\,]_k$ extends to $S(Z)$ by the Leibniz rule which makes $S(Z)$ a Poisson algebra. It is denoted by $P_k S(Z)$. Clearly, $P_0 S(Z) = PS(Z)$.

1.3. Poisson actions of algebras of loops. For the sake of completeness I briefly formulate some results of Goldman on the Poisson actions of Z and $Z_\square$. These results are not used in this paper.

Assume that F is connected and denote its fundamental group by π. As usual, the set $\hat{\pi}$ is identified with the set of conjugacy classes of non-trivial elements of π.

A Poisson manifold is a smooth finite-dimensional manifold N equipped with a Lie bracket in the algebra $C^\infty(N)$ of smooth $\mathbb{R}$-valued functions on N, so that the Leibniz rule (1) is satisfied. A Poisson action of the Poisson algebra S on the Poisson manifold N is a Poisson algebra homomorphism $S \to C^\infty(N)$.

In [5]) for a Lie group G satisfying fairly general conditions, Goldman constructs a symplectic structure on the smooth part N_G of the space $\mathrm{Hom}(\pi, G)/G$ of conjugacy classes of representations $\pi \to G$. This symplectic structure generalizes the Weil-Petersson Kähler form on the Teichmüller space (taking $G = SL(2;\mathbb{R})$). In general a symplectic structure on N_G induces a Poisson structure on N_G. For some classical Lie groups Goldman [6]), § 3 constructed Poisson actions of Poisson algebras of loops considered above. Take $K=\mathbb{R}$. In the case $G = GL(n;*)$, $* = \mathbb{R}, \mathbb{C}, \mathbb{H}$ the algebra $PS(Z)$ acts on N_G. This action associates with a generator $a \in \hat{\pi}$ the function $f \mapsto 2\,\mathrm{Re}(\mathrm{tr} f(a))$ on N_G. If $G = SL(n;\mathbb{R})$ it is $P_{1/n} S(Z)$ which acts on N_G. For $G = O(p,q), O(n;\mathbb{C}), O(n;\mathbb{H}), U(p,q), Sp(p,q), Sp(n;\mathbb{R})$ the algebra $PS(Z_\square)$ acts on N_G.

With the Lie groups $SL(n;\mathbb{C}), SL(n;\mathbb{H}), SU(p,q)$ one also may associate some Poisson algebras of loops. Here the underlying commutative algebra is $S \otimes S$ and the definition of Lie brackets mimics the formulas given in [6], Theorems 3.16 and 3.17. Note that the Poisson algebra of loops corresponding to $SL(n;\mathbb{H}), SL(n;\mathbb{C})$ is a real form of the complex Poisson algebra $P_{i/2n} S(Z) \otimes_K P_{-i/2n} S(Z)$, $i=\sqrt{-1}$, $K=\mathbb{C}$; cf. Remark 1.4. The action of the latter complex algebra on N_G maps $a \otimes 1$ into $(f_a + i\tilde{f}_a)/2$, and $1 \otimes a$ into $(f_a - i\tilde{f}_a)/2$, where $a \in \hat{\pi}$, f_a and $\tilde{f}_a$ are functions on N_G described in [6], Sec.3.16.

1.4. Remark. The tensor product of Poisson algebras A and B is the algebra $A \otimes B$ equipped with the Lie bracket given by the formula

$$[a \otimes b, a' \otimes b'] = aa' \otimes [b, b'] + [a, a'] \otimes bb' .$$

2. LIE BIALGEBRA OF LOOPS ON F

2.1. Lie coalgebras and Lie bialgebras. A Lie cobracket in a module $\mathfrak{g}$ is a linear homomorphism $\nu : \mathfrak{g} \to \mathfrak{g} \otimes \mathfrak{g}$ such that $Perm_{\mathfrak{g}} \circ \nu = -\nu$ and $(\tau^2 + \tau + 1) \circ (\nu \otimes id_{\mathfrak{g}}) \circ \nu = 0$ where $Perm_{\mathfrak{g}}$ is the permutation $a \otimes b \mapsto b \otimes a$ in $\mathfrak{g} \otimes \mathfrak{g}$ and τ is the permutation $a \otimes b \otimes c \mapsto b \otimes c \otimes a$ in $\mathfrak{g} \otimes \mathfrak{g} \otimes \mathfrak{g}$. Of course, the dual homomorphism $\nu^* : \mathfrak{g}^* \otimes \mathfrak{g}^* \to \mathfrak{g}^*$ will be a Lie bracket in $\mathfrak{g}^*$.

A Lie bialgebra is a module $\mathfrak{g}$ with a Lie bracket $[\ ,\ ]$ and a Lie cobracket $\nu : \mathfrak{g} \to \mathfrak{g} \otimes \mathfrak{g}$, so that ν is a 1-cocycle. (This means that $\nu([a,b]) = a\nu(b) - b\nu(a)$ for all $a, b \in \mathfrak{g}$; here $\mathfrak{g}$ acts in $\mathfrak{g} \otimes \mathfrak{g}$ by the rule $a(b \otimes c) = [a, b] \otimes c + b \otimes [a, c]$.

In the class of free modules of finite rank the notion of Lie bialgebra is self-dual: any Lie bialgebra structure in $\mathfrak{g}$ induces a Lie bialgebra structure in $\mathfrak{g}^*$ and vice versa.

2.2. The Lie cobracket in $\mathcal{Z}$. Let $\mathcal{Z}$ be the module introduced in Sec.1.1. Let α be a non-contractible generic loop in F. Denote by $\#\alpha$ its (finite) set of double points $\{x \in \alpha(S^1) \mid \operatorname{card} \alpha^{-1}(x) > 1\}$. Each point $x \in \#\alpha$ is traversed by α twice, the tangent vectors of α in x being linearly independent. Assume that these vectors are numerated u_1, u_2 so that the pair u_1, u_2 is positively oriented. For $r = 1, 2$ denote by α_x^r the loop with starts from x in direction u_r and goes along α till the first return to x. Clearly, up to a choice of parametrization $\alpha = \alpha_x^1 \alpha_x^2$. Set

$$\nu([\alpha]) = \sum_{x \in \#\alpha} \left([\alpha_x^1] \otimes [\alpha_x^2] - [\alpha_x^2] \otimes [\alpha_x^1]\right). \qquad (5)$$

2.2.1. Theorem. The linear homomorphism $\nu : \mathcal{Z} \to \mathcal{Z} \otimes \mathcal{Z}$ given on generators by (5) is a correctly defined Lie cobracket. The Goldman-Lie algebra $\mathcal{Z}$ with the cobracket ν is a Lie bialgebra.

2.3. Remarks. 1. If α is a simple loop then $\nu([\alpha]) = 0$ and, moreover, $\nu([\alpha^n]) = 0$ for any integer n. Conjecture: if $\nu([\alpha]) = 0$ then α is homotopic to a power of a simple loop. A similar assertion for intersection of 2 loops is proven in [6]): If α is a simple loop and β is a loop on F then α, β are homotopic to non-intersected loops iff $[[\alpha], [\beta]] = 0$.

2. The Goldman bracket and the cobracket ν may be computed purely algebraically from the operations λ, μ which were introduced in [13]), Supplement 2. Put $\pi = \pi_1(F)$ and denote by $\mathcal{Y}$ the group ring $K\pi$. With each pair $a, b \in \pi$ the "intersection" λ associates $\lambda(a, b) \in \mathcal{Y}/((a-1)\mathcal{Y} + \mathcal{Y}(b-1))$. With each $a \in \pi$ the "self-intersection" μ associates an element $\mu(a)$ of the quotient module of $\mathcal{Y}$ by the submodule generated (as a module) by $\{1\}$ and the set $\{c + c^{-1}a^{-1} \mid c \in \pi\}$. Let $q : \pi \setminus \{1\} \to \hat{\pi}$ be the natural projection. Put $q(\{1\}) = 0$. It is easy

to compute that if $\lambda(a,b)=\sum_i k_i c_i \mod (a-1)\mathcal{Y}+\mathcal{Y}(b-1)$ where $k_i \in K$ and $c_i \in \pi$ then $[q(a),q(b)] = \sum_i k_i q(ac_i bc_i^{-1})$. If $\mu(a)$ is represented by $\sum_i k_i c_i \in \mathcal{Y}$ then $\nu(q(a)) = \sum_i k_i (q(c_i^{-1})\otimes q(c_i a) - q(c_i a)\otimes q(c_i^{-1}))$.

3. POISSON AND COPOISSON BIALGEBRAS

3.1. Coalgebras and bialgebras. A coalgebra is a module A equipped with a linear homomorphism (comultiplication) $\Delta : A \to A\otimes A$ such that $(id_A \otimes \Delta)\Delta = (\Delta\otimes id_A)\Delta : A \to A^{\otimes 3}$. The coalgebra A is called cocommutative if $Perm_A \circ \Delta = \Delta$.

A bialgebra (Hopf algebra) is a module equipped with a structure of associative algebra and of coalgebra so that the comultiplication is an algebra homomorphism. A bialgebra is called bicommutative if it is commutative as the algebra and cocommutative as the coalgebra.

For an arbitrary module g the symmetric algebra $S(g)$ may be canonically provided with a structure of bialgebra using the comultiplication which transforms $a\in g$ into $a\otimes 1 + 1\otimes a$. This bialgebra is denoted by $S_b(g)$. It is clearly bicommutative.

3.2. Poisson bialgebras. By a Poisson bialgebra we mean a module A equipped with the structure of commutative bialgebra and of Poisson algebra with one and the same multiplication so that the bialgebra comultiplication $A\to A\otimes A$ is a Poisson algebra homomorphism (cf. Remark 1.4).

If g is a Lie algebra then the bialgebra $S_b(g)$ and the symmetric Poisson algebra $PS(g)$ coincide as algebras with $S(g)$ and make $S(g)$ a Poisson bialgebra. It is denoted by $PS_b(g)$.

3.3. Copoisson bialgebras. The theory of copoisson coalgebras is dual to the theory of Poisson algebras, cf. [3]). A copoisson coalgebra is a cocommutative coalgebra A equipped with a Lie cobracket $\nu : A \to A \otimes A$ which is related to the comultiplication $\Delta : A \to A \otimes A$ by formula

$$(id_A \otimes \Delta)\circ\nu = [\nu \otimes id_A + (Perm_A \otimes id_A)(id_A \otimes \nu)]\circ\Delta . \quad (6)$$

A copoisson bialgebra is a module A provided with the structure of cocommutative bialgebra and copoisson coalgebra with one and the same comultiplication Δ, so that for all $a, b \in A$

$$\nu(ab) = \nu(a)\,\Delta(b) + \Delta(a)\,\nu(b) \quad (7)$$

where ν is the Lie cobracket $A \to A \otimes A$. Conditions (6) and (7) are dual respectively to (1) and the condition of preserving the Lie bracket by the comultiplication.

3.4. Theorem. Let $\mathfrak{g}$ be a Lie coalgebra and $S = S_\ell(\mathfrak{g})$. The Lie cobracket in $\mathfrak{g}$ uniquely extends by (7) (where Δ is the comultiplication in S) to a linear homomorphism $\nu : S \to S \otimes S$. The pair S, ν is a copoisson bialgebra.

The copoisson bialgebra S, ν is denoted by $CS_\ell(\mathfrak{g})$.

3.5. Bipoisson bialgebras. A bipoisson bialgebra is a module A equipped with the structure of Poisson bialgebra and of copoisson bialgebra with the same underlying bicommutative bialgebra, so that the Lie cobracket $\nu : A \to A \otimes A$, the comultiplication $\Delta : A \to A \otimes A$ and the Lie bracket $[\,,\,]$ in A (and in $A \otimes A$) are related by the formula

$$\nu([a,b]) = [\nu(a), \Delta(b)] + [\Delta(a), \nu(b)]$$

for all $a, b \in A$. This formula and the notion of bipoisson bialgebra are self-dual.

If $\mathfrak{g}$ is a Lie bialgebra then the Poisson bialgebra $PS_\ell(\mathfrak{g})$ and the copoisson bialgebra $CS_\ell(\mathfrak{g})$ make a bipoisson bialgebra. It is denoted by $BS_\beta(g)$.

4. SPIRAL LIE COALGEBRAS AND POISSON GROUPS

4.1. Spiral Lie coalgebras. Let g be a Lie coalgebra with the Lie bracket $\nu: g \to g \otimes g$. For $n \geq 2$ put

$$\nu_n = (id_g^{\otimes(n-2)} \otimes \nu)(id_g^{\otimes(n-3)} \otimes \nu) \dots (id_g \otimes \nu)\nu : g \to g^{\otimes n}$$

g is called spiral if $g = \bigcup_{n \geq 2} \operatorname{Ker} \nu_n$. For spiral g the dual Lie algebra g^* has the following completeness property. Let $g^* = g_1^* \supset g_2^* \supset \dots$ be the lower central series, where $g_{n+1}^* = [g^*, g_n^*]$ for $n \geq 1$. Let $z_1, z_2, \dots \in g^*$ be a sequence such that for any n all terms of the sequence starting from a certain place belong to g_n^*. Clearly, if $a \in \operatorname{Ker} \nu_n$, then $g_n^*(a) = 0$. Since $g = \bigcup \operatorname{Ker} \nu_n$ the formula $a \mapsto z_1(a) + z_2(a) + \dots$ determines a linear homomorphism $g \to K$, i.e. an element of g^* which is "an infinite sum $z_1 + z_2 + \dots$ ". (A similar argument shows that $\bigcap_n g_n^* = 0$).

If g has an increasing filtration $h_0 \subset h_1 \subset \dots$ such that

$$g = \bigcup_{n \geq 0} h_n \quad \text{and} \quad \nu(h_n) \subset \sum_{i=0}^{n-1} h_i \otimes h_{n-i-1}$$

for all n then g is spiral. This shows that the Lie coalgebra Z introduced in § 2 is spiral. One should take h_n to be the submodule of Z generated by the homotopy classes of loops with $\leq n$ self-intersections.

4.2. The group $\operatorname{Exp} g^*$. The bialgebra $S_e(g)$. Let $K \supset \mathbb{Q}$. Let g be a spiral Lie coalgebra over K. In view of the results of Sec.4.1 for any elements x, y of the Lie algebra g^* one may consider the infinite sum

$m(x,y) = x+y+(1/2)[x,y]+\ldots$ where the right hand side is the classical Campbell-Hausdorff series. As usual, the mapping $x, y \mapsto m(x,y)$ is a group multiplication in g^*. Here $x^{-1} = -x$ and 0 is the group unit. The resulting group is denoted by $\mathrm{Exp}\, g^*$.

The group multiplication m induces in the symmetric algebra $S = S(g)$ a bialgebra structure as follows. Assume for simpicity that g is a free K-module. The natural embedding $g \to (g^*)^*$ extends to an imbedding of S into the algebra of polynomial K-valued functions on g^*. Indentify S with the corresponding algebra of functions on g^*, and identify $S \otimes S$ with the corresponding algebra of functions on $g^* \times g^*$. It is easy to show that $a \circ m \in S \otimes S$ for any $a \in S$. The algebra S with the comultiplication $a \mapsto a \circ m$ is a commutative bialgebra. It is denoted by $S_e(g)$. If the Lie cobracket in g is equal to zero, then $S_e(g) = S_b(g)$.

4.3. <u>Poisson bialgebra</u> $PS_e(g)$. A Lie bialgebra is said to be spiral if it is free as module and spiral as coalgebra.

4.3.1. <u>Theorem</u>. Let $K \supset \mathbb{Q}$. Let g be a spiral Lie bialgebra over K with the Lie bracket $[\, ,\,]$. Then there exists a unique Lie bracket $\langle\, ,\, \rangle$ in $S(g)$ such that: (i) the bialgebra $S_e(g)$ with the bracket $\langle\, ,\, \rangle$ is a Poisson bialgebra; (ii) for any $a, b \in g$

$$[a, b] = \langle a, b\rangle \mod \bigoplus_{n \geq 2} S^n(g) .$$

The Poisson bialgebra $S_e(g), \langle\, ,\, \rangle$ is denoted by $PS_e(g)$. If the Lie cobracket in g is equal to zero then $PS_e(g) = PS_b(g)$.

4.4. <u>Remarks</u>. 1. It is natural to consider Theorem 4.3.1 in the contex of the theory of Poisson Lie groups, see [2]), [3]), [12]), [17]). Recall, that a Poisson Lie group is a Lie group H provided with a multiplicative Poisson structure; the multiplicativity means that the group multiplication $H \times H \to H$ induces a Poisson algebra homomorphism $C^\infty(H) \to C^\infty(H \times H)$. The infinitesimal analogue of a Poisson Lie group is a Lie bialgebra over $\mathbb{R}$.

The group $\mathrm{Exp}\, g^*$ is "the simply connected Lie group with the Lie algebra g^*". The Lie bracket $\langle\, , \rangle$ is "the Poisson bracket in the algebra $S(g)$ of functions on $\mathrm{Exp}\, g^*$". Condition (i) of Theorem 4.3.1 means that this Poisson bracket is multiplicative. Condition (ii) is a variant of the formula $d\langle a, b\rangle = [da, db]$ which relates the Poisson bracket $\langle\, , \rangle$ on a Lie group H and the corresponding Lie bracket $[\, ,]$ in h^*, where: h is the Lie algebra of H; $a, b \in C^\infty(H)$ and d is the differential in the unit of H.

If g is a spiral Lie bialgebra over $\mathbb{R}$ of finite dimension then the Lie bracket in $PS_e(g)$ is a restriction of the multiplicative Poisson bracket in $C^\infty(\mathrm{Exp}\, g^*)$ associated by the Drinfel'd theory with the Lie bracket in g.

2. If the genus of the surface F is non-zero then the Goldman bracket in Z can not be induced by any homomorphism $Z^* \to Z^* \otimes Z^*$. On the other hand the Lie cobracket in Z induces a Lie bracket in Z^*. The Lie algebra Z^* can be shown to be a projective limit of nilpotent Lie algebras, free and of finite rank as K-modules. Therefore, if $K = \mathbb{R}$ then the group $\mathrm{Exp}\, Z^*$ is a projective limit of nilpotent Lie groups.

5. QUANTIZATION OF ALGEBRAS AND COALGEBRAS. BIQUANTIZATION

Fix a commutative associative K-algebra Q with unit and with augmention φ (i.e. φ is a homomorphism of K-algebras $Q \to K$ with $\varphi(1) = 1$). An additive homomorphism f of a Q-module A into a K-module is said to be φ-linear (or linear over φ) if $f(qa) = \varphi(q) f(a)$

for arbitrary $q \in Q$, $a \in A$.

5.1. Quantization of Poisson algebras. Let $h \in \operatorname{Ker} \varphi$. A quantization over (Q, φ, h) of a Poisson K-algebra A_0 is a pair (a Q-algebra A, a surjective φ-linear ring homomorphism $p: A \to A_0$) such that for any $a, b \in A$

$$ab - ba = h p^{-1}([p(a), p(b)]) \mod h \operatorname{Ker} p, \tag{8}$$

where $[\ ,\ ]$ is the Lie bracket in A_0. For the sake of brevity the quantization over (Q, φ, h) is also called a quantization over Q. The homomorphism $p: A \to A_0$ is called a quantization homomorphism. Clearly, $\operatorname{Ker} \varphi \supset hQ$ and $\operatorname{Ker} p \supset hA$. If $\operatorname{Ker} \varphi = hQ$ and $\operatorname{Ker} p = hA$ then the quantization is said to be reduced. If A, A_0 and p belong to a certain category $\mathcal{C}$ then we say that (A, p) is a quantization of A_0 in the category $\mathcal{C}$.

5.2. Remarks. 1. Our definition of quantization is very broad. For example, if I is an ideal of the algebra A lying in the kernel of the quantization homomorphism $A \to A_0$ then the pair (Q-algebra A/I, induced homomorphism $A/I \to A_0$) is also a quantization of A_0. If one restricts himself to the class of reduced quantizations (A, p) where A is a free Q-module then such a factorization becomes impossible.

2. Quantizations over the polynomial ring $K[h]$ (with the usual augmentation) are universal: each such quantization $p: A \to A_0$ induces a quantization $\varphi \otimes p: Q \otimes_{K[h]} A \to A_0$ over (Q, φ, h).

5.3. Quantization of copoisson coalgebras. The following definition is dual the one given in Sec.5.1. Let $\hbar \in \operatorname{Ker} \varphi$. A quantization over $(Q, \varphi, \hbar)$ of a copoisson K-coalgebra A_0 is a pair (a Q-coalgebra A, a φ-linear epimorphism of coalgebras $p: A \to A_0$) such that for any $a \in A$

$$\Delta(a) - \mathrm{Perm}_A(\Delta(a)) = \hbar\,(p\otimes p)^{-1}(\nu(p(a))) \mod \hbar\,\mathrm{Ker}(p\otimes p)$$

where Δ is the comultiplication in A and ν is the Liecabracket in A_0. The terminology which followed the definition of quantization in Sec.5.1 will be also applied to quantizations of copoisson coalgebras.

5.4. Two dual examples of quantization. 1. Let $\mathfrak{g}$ be a Lie algebra over K. The Poisson bialgebra $PS_b(\mathfrak{g})$ admits the following quantization over $K[h]$. Consider the tensor algebra $T = \oplus_{n\geq 0}\, \mathfrak{g}^{\otimes n}$. Let $V = V_h(\mathfrak{g})$ be the algebra over $K[h]$ which is the quotient of $K[h]\otimes_K T$ by the twosided ideal generated by elements $ab - ba - h[a,b]$ with $a, b \in \mathfrak{g}$. The comultiplication $V \to V\otimes_{K[h]} V$ which sends $a\in\mathfrak{g}$ into $a\otimes 1 + 1\otimes a$ makes V a bialgebra. The multiplicative homomorphism $v_h : V \to S(\mathfrak{g})$ which transforms $k(h)\otimes a$ with $k(h)\in K[h]$, $a\in\mathfrak{g}$ into $k(0)a$ is a reduced quantization of the Poisson algebra $PS_b(\mathfrak{g})$ is the category of bialgebras.

If $\mathfrak{g}$ is a Lie bialgebra then the Lie cobracket of $\mathfrak{g}$ extends uniquely to a $K[h]$-linear homomorphism $V \to V\otimes_{K[h]} V$, which amplifies V to a copoisson bialgebra. The quantization homomorphism $v_h : V \to S(\mathfrak{g}) = CS_b(\mathfrak{g})$ is a homomorphism of copoisson bialgebras.

Note that $V/(h-1)V$ is the universal envelopping algebra $U(\mathfrak{g})$ of the Lie algebra $\mathfrak{g}$. If $\mathfrak{g}$ is a free K-module then V may be completely described in terms of $U(\mathfrak{g})$ and its standard filtration $K = U^0 \subset U^1 \subset \ldots$ where $U^n = (K+\mathfrak{g})^n$. Namely, the formula $a \mapsto h\otimes a$ where $a\in\mathfrak{g}$ defines a bialgebra homomorphism $\psi : V \to K[h]\otimes U(\mathfrak{g})$. Its image is equal to $\oplus_{n\geq 0}(h^n \otimes U^n)$. If $\mathfrak{g}$ is a free module then ψ may be shown to be an injection.

2. Let $K \supset \mathbb{Q}$ and let $\mathfrak{g}$ be a spiral Lie coalgebra

over K. The copoisson bialgebra $CS_b(g)$ admits the following quantization over the polynomial ring $K[\hbar]$. It follows from the definition of comultiplication Δ in the bialgebra $S_e(g)$ (see Sec.4.2) that

$$\Delta(S^n) \subset \bigoplus_{\substack{i,j \geqslant 0 \\ i+j \geqslant n}} (S^i \otimes S^j)$$

where $S^n = S^n(g)$. This implies that the $K[\hbar]$-submodule

$$\bigoplus_{n \geqslant 0,\, m \geqslant -n} (\hbar^m \otimes S^n)$$

of the bialgebra $K[\hbar, \hbar^{-1}] \otimes S_e(g)$ is a subbialgebra. Denote this subbialgebra by $E_\hbar(g)$. Define the homomorphism $e_\hbar : E_\hbar(g) \to S(g)$: if $a \in \hbar^m \otimes S^n$ with $m > -n$ then $e_\hbar(a) = 0$; if $a = \hbar^{-n} \otimes b$ with $b \in S^n$ then $e_\hbar(a) = b$. It turns out that $(E_\hbar(g), e_\hbar)$ is a reduced quantization of the copoisson coalgebra $CS_b(g)$ in the category of bialgebras.

If g is a spiral Lie bialgebra then the Lie bracket $\langle\,,\rangle$ in $S_e(g)$ induces a Lie bracket in $E_\hbar(g)$ by the formula

$$\{\hbar^m \otimes a, \hbar^{m'} \otimes a'\} = \hbar^{m+m'+1} \otimes \langle a, a' \rangle .$$

This makes $E_\hbar(g)$ a Poisson bialgebra. Theorem 4.3.1 implies that $e_\hbar : E_\hbar(g) \to S(g) = PS_b(g)$ is a Poisson bialgebra homomorphism. Notice that in the category of Poisson bialgebras $PS_e(g) = E_\hbar(g)/(\hbar - 1)E_\hbar(g)$.

5.5. Biquantization of bipoisson bialgebras. Let $h, \hbar \in Ker\,\varphi$. It is convenient to describe first the construction whose inversion suggests the notion of biquantization. Let A be a bialgebra over Q, free as a Q-module. Assume that for any $a, b \in A$

$$ab-ba \in hA \quad \text{and} \quad \Delta(a)-\mathrm{Perm}_A(\Delta(a)) \in \hbar A \otimes_Q A \qquad (9)$$

where Δ is the comultiplication in A. Put $A_0 = K \otimes_Q A = A/\mathrm{Ker}\,\varphi \cdot A$. The bialgebra structure of A factorizes to a bialgebra structure in A_0. Denote the projection $A \to A_0$ by p. Introduce a Lie bracket in A_0 by the formula

$$[p(a), p(b)] = p(h^{-1}(ab-ba))$$

where $a, b \in A$. Introduce a Lie cobracket ν in A_0 by the formula

$$\nu(p(a)) = (p \otimes p)(\hbar^{-1}(\Delta(a)-\mathrm{Perm}_A(\Delta(a)))).$$

It is easy to check up that A_0 is a bipoisson bialgebra over K. Similarly, one provides $A_\hbar = A/hA$ with a Poisson bialgebra structure over $Q_\hbar = Q/hQ$ and provides $A_h = A/\hbar A$ with a copoisson bialgebra structure over $Q_h = Q/\hbar Q$. The homomorphisms φ and p are included in the commutative diagrams respectively in the categories of K-algebras and bialgebras

$$\begin{array}{ccc} Q & \xrightarrow{\varphi_\hbar} & Q_h \\ {\scriptstyle \varphi_h}\downarrow & \searrow^{\varphi} & \downarrow{\scriptstyle \psi_h} \\ Q_\hbar & \xrightarrow{\psi_\hbar} & K \end{array} \quad ; \quad \begin{array}{ccc} A & \xrightarrow{p_\hbar} & A_h \\ {\scriptstyle p_h}\downarrow & \searrow^{p} & \downarrow{\scriptstyle q_h} \\ A_\hbar & \xrightarrow{q_\hbar} & A_0 \end{array} \qquad (10)$$

where $\varphi_h, \varphi_\hbar, p_h, p_\hbar$ are projections. The homomorphisms of these diagrams have the following properties: (i) all of them are surjective and $h \in \mathrm{Ker}\,\varphi_h$, $\hbar \in \mathrm{Ker}\,\varphi_\hbar$;

(ii) the homomorphisms of the second diagram are linear over the corresponding homomorphisms of the first diagram; (iii) p_h and q_h are quantization homomorphisms (respectively over (Q, φ_h, h) and $(Q_h, \psi_h, \varphi_{\hbar}(h))$) of Poisson algebras $A_{\hbar}$ and A_0 in the categories of bialgebras and copoisson bialgebras respectively; (iv) $p_{\hbar}$ and $q_{\hbar}$ are quantization homomorphisms (respectively over $(Q, \varphi_{\hbar}, \hbar)$ and $(Q_{\hbar}, \psi_{\hbar}, \varphi_h(\hbar))$) of copoisson coalgebras A_h and A_0 in the categories of bialgebras and Poisson bialgebras respectively.

Let now A_0 be an arbitrary bipoisson bialgebra over K. A biquantization over $(Q, \varphi, h, \hbar)$ of A_0 is a pair (10) of commutative diagrams respectively in the categories of K-algebras and bialgebras such that: A is a Q-bialgebra, satisfying (9) for all $a, b \in A$; A_h is a copoisson bialgebra over Q_h; $A_{\hbar}$ is a Poisson bialgebra over $Q_{\hbar}$; the homomorphisms of the diagrams satisfy the conditions (i-iv).

The biquantization (10) is called reduced if the kernels of $\varphi, \varphi_h, \varphi_{\hbar}, p, p_h, p_{\hbar}$ are equal respectively to $hQ+\hbar Q$, hQ, $\hbar Q$, $hA+\hbar A$, hA, $\hbar A$. A reduced biquantization (10) is determined by φ and p except the Lie bracket in $A_{\hbar}$ and the Lie cobracket in A_h. If A is a free Q-module (or at least none element of A is annulated by h or $\hbar$) then these two brackets can be also reconstructed from φ, p.

If $Q = K[h, \hbar]$ then every biquantization (10) gives rise to a two-parameter family of K-bialgebras

$$A(k, k') = A / (h-k)A + (\hbar - k')A$$

where $k, k' \in K$. This plane in the class of bialgebras traverses $A(0,0) = A_0$, $A(1,0) = A_h/(h-1)A_h$ and $A(0,1) = A_{\hbar}/(\hbar-1)A_{\hbar}$. Note that the homomorphisms p_h and $p_{\hbar}$ induce quantization homomorphisms

$$A/(\hbar-1)A \to A(0,1) \quad \text{and} \quad A/(h-1)A \to A(1,0)$$

where $A(0,1)$ inherits a Poisson bialgebra structure from $A_\hbar$ and $A(1,0)$ inherits a copoisson bialgebra structure from A_h.

5.6. Biquantization of Lie bialgebras. A biquantization of a Lie bialgebra g is a reduced biquantization (10) of the bipoisson bialgebra $BS_\ell(g)$ such that A_h is the copoisson bialgebra $Q_h \otimes_{K[h]} V_h(g)$ and $q_h = \psi_h \otimes v_h$. Notice that if $Q = K[h,\hbar]$ then $Q_h = K[h]$, $A_h = V_h(g)$, $q_h = v_h$ and $A(1,0) = U(g)$.

Let $K \supset \mathbb{Q}$. A biquantization (10) of a spiral Lie bialgebra g is said to be normalized, if $A_\hbar$ is the Poisson algebra $Q_\hbar \otimes_{K[\hbar]} E_\hbar(g)$ and $q_\hbar = \psi_\hbar \otimes e_\hbar$ (see Sec. 5.4.2). If $Q = K[h,\hbar]$, then $Q_\hbar = K[\hbar]$, $A_\hbar = E_\hbar(g)$, $q_\hbar = e_\hbar$ and $A(0,1) = PS_e(g)$.

To define normalized biquantization for non-spiral Lie bialgebras one has to introduce for them an analogue of $E_\hbar$. There are two natural approaches to this. First, one may use instead of $S(g)$ its completion with respect to the powers of the ideal $\bigoplus_{n>0} S^n(g)$. Second, one may use a Poisson algebra of functions on a Lie group associated with g^* (cf. the Appendix).

6. ALGEBRAS OF KNOTS. TOPOLOGICAL QUANTIZATION

6.1. The module $A(M;Q)$. Let M be an oriented 3-manifold. By a link in M we mean a finite family of non-intersected smooth imbedded circles in $\mathrm{Int}\, M$. Two links in M are isotopic if they may be smoothly deformed in each other in the class of links. The empty set $\emptyset$ is considered as a unique up to isotopy link in M with 0 components. The number of components of a link L is denoted by $|L|$.

A non-empty link $L \subset M$ is said to be semi-local if

there exists an imbedded 3-ball $B \subset M$ such that $L \cap \partial B = \emptyset$ and $L \cap \mathrm{Int}\, B \neq \emptyset$. A triple of oriented links L_+, L_-, $L_0 \subset M$ is called a Conway triple if L_+, L_-, L_0 are identical outside some ball $B \subset M$ and look as in Fig.1 inside B. Clearly, $|L_+| = |L_-| = |L_0| \pm 1$. We define the type of Conway triple L_+, L_-, L_0 to be 1 if $|L_+| = |L_0| + 1$ and to be 2 in the opposite case.

Fix a commutative associative K-algebra Q with unit, an augmentation $\varphi : Q \to K$ and a pair $h, \hbar \in \mathrm{Ker}\, \varphi$.

Let $\mathcal{X}$ be the set of isotopy classes of oriented links in M. Denote by $A(M;Q)$ the quotient of the free Q-module $Q\mathcal{X}$ with the basis $\mathcal{X}$ by the submodule generated by elements of 3 kinds; (i) the elements $L \in \mathcal{X}$ corresponding to arbitrary semi-local links $L \subset M$; (ii) the elements $L_+ - L_- - hL_0$ corresponding to arbitrary Conway triples of type 1; (iii) the elements $L_+ - L_- - \hbar L_0$ corresponding to arbitrary Conway triples of type 2. An element of this quotient module represented by an oriented link L is denoted by $[L]$. In particular, if L is semi-local then $[L] = 0$.

It is evident that $A(M;Q)$ is a direct sum of the free Q-module $Q \cdot [\emptyset]$ of rank 1 with $\sum_{L \neq \emptyset} Q \cdot [L]$. For example, $A(S^3;Q) = Q \cdot [\emptyset]$ since all non-empty links in S^3 are semi-local.

Denote by $\hat{\pi}$ the set of non-trivial free homotopy classes of loops in M. Let S be the symmetric algebra of the free K-module with basis $\hat{\pi}$. Each oriented link L with components $L_1, \ldots, L_l$ gives rise to an element

$$p_M(L) = \prod_{i=1}^{l} p_M(L_i) \in S$$

where $p_M(L_i)$ is the class of L_i in $\hat{\pi}$ if L_i is not contractible and $p_M(L_i) = 0$ if L_i is contractible. The inclusion $h, \hbar \in \mathrm{Ker}\, \varphi$ implies that the formula $q \cdot [L] \longmapsto$

$\varphi(q)\, p_M(L)$ (with $q \in Q$) determines a φ-linear epimorphism

$$p_M : A(M;Q) \longrightarrow S .$$

6.2. Remarks. 1. Clearly,

$$A(M;Q) = Q \otimes_{K[h,\hbar]} A(M; K[h,\hbar]).$$

2. If there exists an invertible element q of Q such that $\hbar = q^2 h$ then the distinction between the types of Conway triples is useless. Indeed, in generators $\widetilde{L} = q^{|L|} L$ both elements (ii), (iii) described above look like $q^{-|L_+|}(\widetilde{L}_+ - \widetilde{L}_- - qh\widetilde{L}_0)$. This is the reason why the types of Conway triples play no role in the present day theory of polynomial invariants of links in S^3. In our setting, distinction between these types is most essential.

3. Factorizing out the semi-local links we decrease the Q-torsion. Indeed, an application of the Conway relation to the triple exhibited on Fig.4 shows that $\hbar L_0 = 0$ modulo the Conway relation. This implies that modulo the Conway relations $\hbar$ annulates all semi-local links except the trivial knot.

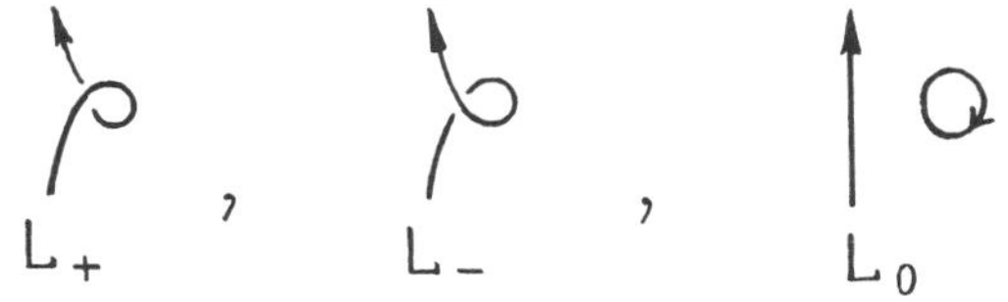

Fig. 4

6.3. The knot algebra A . Consider the 3-manifold $F \times I$ where F is an oriented surface and $I = [0,1]$, and provide $F \times I$ with the product orientation of the given orientation in F and the standard orientation in I . For links L, L' in $F \times I$ put

$$LL' = \{(x,t)\in F\times I \mid t \geq \frac{1}{2}, (x,2t-1)\in L \text{ or } t\leq \frac{1}{2}, (x,2t)\in L'\}.$$

Clearly LL' is a link in $F\times I$. The formula $L, L' \mapsto LL'$ define a structure of associative Q-algebra in $A(F\times I; Q)$ with unit $[\emptyset]$. Denote this algebra by A. It is easy to see that classes of knots generate A.

Since all loops in $F\times I$ are contractible into $F\times 0 = F$ the homomorphism $p_{F\times I}$ constructed in Sec.6.1 maps A onto $S(Z)$ where Z is the Goldman-Lie algebra of F. In terms of the geometric interpretation of the generators of $S(Z)$ discussed in Introduction, $p_{F\times I}([L])$ is represented by the projection of $L\subset$ $\subset F\times I$ into F.

6.3.1. Theorem. The pair $(A, p_{F\times I}: A \to S(Z))$ is a quantization of $PS(Z)$ over (Q,φ,h). If $\operatorname{Ker}\varphi = hQ$ then the quantization is reduced.

This Theorem is actually very easy, and I present a (sketch of a) proof. Let L, L' be oriented knots in $F\times I$. Let α and β be loops in F parametrizing the projections of L, L' into F. It is obvious that $L'L$ may be obtained from LL' by moving L down through L'. During this process L intersects L' several times. For each such intersection we have the Conway relation $[L_+]-$ $-[L_-]=h[L_0]$. Combining them together we get

$$[L]\cdot[L'] - [L']\cdot[L] = h\sum_{x\in\alpha\#\beta} \varepsilon_x [L_x]$$

where L_x is an oriented knot in $F\times I$, whose projection into F is parametrized by $(\alpha\beta)_x$ (see Sec.1.1). This implies (8).

6.4. Remarks. 1. Theorem 6.3.1 provides us with a large family of quantization of $PS(Z)$. For example, if $Q = K[h]$ then one may take as $\hbar$ an arbitrary element of hQ. All these quantizations are induced by the universal quantization corresponding to $Q=K[h,\hbar]$. Moreover, an analogue of Theorem 6.3.1 holds for the extension of $\mathcal{A}$ defined like $\mathcal{A}$ but without factorizing out the elements of type (iii). The resulting algebra seems to be very hard to observe even in the simplest case when F is the disk. In contrast to $\mathcal{A}$ it does not bear a natural bialgebra structure.

2. It seems reasonable to conjecture that the Q-module $\mathcal{A}(F\times I;Q)$ is free. If F is S^2 or the 2-disc, this is evident. If $F=S^1\times I$ this follows from the results of [7], [16]. For a further discussion of related modules see [11].

6.5. <u>A deformation of</u> $\mathcal{A}$. Assume that $K\supset \mathbb{Q}$ and that Q is an extension of the ring of formal power series $K[[h]]$. (For example, $Q=K[\hbar][[h]]$ or $Q=K[[h]]$ with $\hbar\in hQ$). For $k\in K$ define a new multiplication $*_k$ in $\mathcal{A}(F\times I;Q)$ by the rule

$$[L]*_k[L'] = \exp((L\cdot L')kh)\,[LL']$$

where $L\cdot L'$ is the integer intersection number of the elements of $H_1(F)$ represented by L and L'. (If $L=\emptyset$ or $L'=\emptyset$ then $L\cdot L'=0$).

6.5.1. <u>Theorem</u>. The multiplication $*_k$ makes $\mathcal{A}(F\times I;Q)$ an associative Q-algebra. This algebra together with $p_{F\times I}$ is a quantization of $P_{2k}S(Z)$ over (Q,φ,h). If $\operatorname{Ker}\varphi=hQ$ then this quantization is reduced.

6.6. <u>A comultiplication in</u> $\mathcal{A}$. We shall provide the algebra $\mathcal{A}$ introduced in Sec.6.3 with a comultiplication Δ. This comultiplication is closely related to the composition product of polynomial invariants of links in S^3 introduced by F.Jaeger [8]. We shall use the technique of link diagrams.

Oriented links in $F\times I$ may be presented by oriented

link diagrams on F exactly in the same fashion in which links in $\mathbb{R}^3$ are presented by plane diagrams. A diagram on F is a finite general position collection of oriented loops on F provided with additional structure: at each self-intersection point of the diagram one branch is cut and considered as the lower one (the undercrossing), the second branch being considered as the upper one (the overcrossing). The oriented link presented by an oriented link diagram $\mathcal{D}$ is denoted by $L(\mathcal{D})$.

The union of loops of a link diagram $\mathcal{D}$ is a 4-valent graph on F. Its vertices and edges are called vertices and edges of $\mathcal{D}$. The set of edges is denoted by $\mathrm{Edg}\,\mathcal{D}$. Each vertex has type 1 or -1: see Fig.1 where the depicted vertex of L_+ (resp. L_-) has type 1 (resp. -1).

Let $m \geqslant 1$. By a m-labelling of the (oriented) link diagram $\mathcal{D}$ we mean a function $f: \mathrm{Edg}\,\mathcal{D} \to \{1, 2, \ldots, m\}$ such that the following condition holds: For each vertex v of $\mathcal{D}$ if a, b are resp. upper and lower edges which look in v and c, d are resp.upper and lower edges which look out of v then either $f(a) = f(c)$ and $f(b) = f(d)$ or $f(a) = f(d) > f(b) = f(c)$. The number of vertices of type 1 (resp.-1) in which the last possibility occurs is denoted by $|f|_+$ (resp. $|f|_-$). Set $|f| = |f|_+ + |f|_-$.

The set of m-labellings of $\mathcal{D}$ is denoted by $\mathrm{Lbl}_m(\mathcal{D})$. If $f \in \mathrm{Lbl}_m(\mathcal{D})$ then for each $i = 1, \ldots, m$ the edges of $\mathcal{D}$ lying in $f^{-1}(i)$ make an oriented link diagram, denoted by $\mathcal{D}_{f,i}$. (It is understood that in the self-crossing points of $\mathcal{D}_{f,i}$ the choice of the lower/upper branches is the same as in $\mathcal{D}$). Put

$$\|f\| = |L(\mathcal{D})| - \sum_{i=1}^{m} |L(\mathcal{D}_{f,i})| \in \mathbb{Z}$$

Clearly, $|f| \geqslant \|f\| \geqslant -|f|$ and $|f| \equiv \|f\| \bmod 2$.

6.6.1. <u>Theorem</u>. There exists a unique Q-linear homomorphism $\Delta: A \to A\otimes A$ such that for each oriented link diagram $\mathcal{D}$ on F

$$\Delta([L(\mathcal{D})]) = \sum_{f\in Lbl_2(\mathcal{D})} (-1)^{|f|_-} h^{\frac{|f|+\|f\|}{2}} \hbar^{\frac{|f|-\|f\|}{2}} [L(\mathcal{D}_{f,1})]\otimes[L(\mathcal{D}_{f,2})].$$

The pair A, Δ is a bialgebra. The homomorphism $p_{F\times I}: A\to \to S_\ell(\mathbb{Z})$ is a bialgebra homomorphism.

6.7. Remark. The projection $A\to Q\cdot[\emptyset]$ along $\sum_{L\neq\emptyset} Q\cdot[L]$ is the counit of the bialgebra A. I believe that the Q-linear homomorphism $A\to A$, which transforms $[L]$ into $(-1)^{|L|}[L']$ where $L'=\{(x,t)\in F\times I \mid (x,1-t)\in L\}$ is an antipode (a conjugation) of A.

6.8. Biquantization of $\mathbb{Z}$. Let A be the $K[h,\hbar]$-module $A(F\times I; K[h,\hbar])$ with the bialgebra structure introduced in Sec.6.3, 6.6. It is easy to show that lifting a loop in F to a knot in $F\times I$ which projects onto this loop determines a $K[h]$-algebra isomorphism $V_h(\mathbb{Z})\to A/\hbar A$. Denote by $p_\hbar$ the composition of the projection $A\to A/\hbar A$ and the inverse isomorphism $A/\hbar A\to V_h(\mathbb{Z})$. Denote by p_h the homomorphism $A\to E_\hbar(\mathbb{Z})$ which sends the class $[L]\in A$ of the link L presented by a diagram $\mathcal{D}$ into

$$\hbar^{-|L|}\otimes \sum_{\substack{m\geq 1\\ f\in Lbl_m(\mathcal{D}),\ |f|=\|f\|}} \frac{(-1)^{|f|_-}}{m!} \prod_{i=1}^{m} p_{F\times I}([L(\mathcal{D}_{f,i})]).$$

6.8.1. Theorem. The diagrams

$$\begin{array}{ccc} K[h,\hbar] & \xrightarrow{\hbar\mapsto 0} & K[h] \\ \downarrow{\scriptstyle h\mapsto 0} & \searrow & \downarrow{\scriptstyle h\mapsto 0} \\ K[\hbar] & \xrightarrow{\hbar\mapsto 0} & K \end{array} \quad ; \quad \begin{array}{ccc} A & \xrightarrow{p_\hbar} & V_h(Z) \\ {\scriptstyle p_h}\downarrow & \searrow{\scriptstyle p_{F\times I}} & \downarrow{\scriptstyle v_h} \\ E_\hbar(Z) & \xrightarrow{e_\hbar} & S_b(Z) \end{array}$$

are commutative and make a normalized biquantization of the Lie bialgebra Z.

6.9. <u>A module</u> $A_\square(M;Q)$ <u>and a quantization of</u> $PS(Z_\square)$. Let M, S, Q, φ, h be the same objects as in Sec.6.1. Denote by $X_\square$ the set of isotopy classes of links in M. Denote by $A_\square(M;Q)$ the quotient of the free Q-module $QX_\square$ with basis $X_\square$ by the submodule generated by elements $L_+ - L_- - hL_0 + hL_\infty$ corresponding to arbitrary sequences of four links which are identical outside a ball in M, look as in Fig.2 inside the ball, and such that $|L_+| = |L_0| + 1$. One associates with a link L of l components $L_1, \dots, L_l$ the element

$$p_M^\square(L) = \prod_{i=1}^{l} p_M^\square(L_i) \in S$$

where for a non-contractible L_i $p_M^\square(L_i) = \alpha_i + \alpha_i^{-1}$, α_i being the free homotopy class of a loop parametrizing L_i, and for contractible L_i $p_M^\square(L_i) = 0$. This gives rise to a φ-linear epimorphism $p_M^\square : A_\square(M;Q) \to S$.

The formula $L, L' \mapsto LL'$ defines an associative multiplication in $A_\square(F\times I;Q)$. The image of $p_{F\times I}^\square$ is equal to $S(Z_\square) \subset S(Z)$.

6.9.1. <u>Theorem</u>. The pair $(A_\square(F\times I;Q), p_{F\times I}^\square)$ is a quantization of $PS(Z_\square)$.

Of course, it is natural to factorize $A_\square$ further by elements associated with sequences L_+, L_-, L_0, L_∞ as above but with $|L_+|=|L_0|-1$. This factorization and operations in the resulting algebra similar to Δ will be considered elsewhere.

7. AN EXTENSION OF $Z=Z(F)$ AND OF $A=A(F)$

The algebras of loops and knots constructed in §1, §6 have canonical extensions, which will be briefly discussed in this section. I concentrate here on the case of oriented loops and links, the non-oriented case will be considered in a detailed exposition.

7.1. The set $\hat{\Pi}$. By an immersed loop in F we mean an immersion $S^1 \to F$, where the circle is oriented. Two immersed loops are regularly homotopic if they may be smoothly deformed to each other in the class of immersed loops. Let $\hat{\Pi}$ be the set of regular homotopy classes of non-contractible immersed loops. The infinite cyclic group $T=\{t^n | n\in \mathbb{Z}\}$ acts in $\hat{\Pi}$ as follows: the action of t (resp. t^{-1}) adds one small curl to the right (resp. to the left) of the loop. Clearly, $\hat{\Pi}/T$ is the set $\hat{\pi}$ considered in §1.

7.1.1. Theorem. Let $E\to F$ be the bundle of unitary tangent vectors of F (where F is equipped with a Riemannian metric). Each immersed loop $\alpha: S^1 \to F$ lifts to a loop $\tilde{\alpha}: x \mapsto (\alpha(x), \alpha'(x)/|\alpha'(x)|)$ in E. The formula $\alpha \mapsto \tilde{\alpha}$ defines a bijection of $\hat{\Pi}$ onto the set of non-trivial conjugacy classes in $\pi_1(E)$.

7.1.2. Corollary. The action of T in $\hat{\Pi}$ is free.

7.2. The Lie bialgebra $\tilde{Z}$. Let $\tilde{Z}$ be the free K-module with the basis $\hat{\Pi}$. The action of T in $\hat{\Pi}$ induces a structure of a free $K[T]$-module in $\tilde{Z}$. The same constructions as in §1 make $\tilde{Z}$ a Lie bialgebra over $K[T]$. Clearly, $Z=\tilde{Z}/(t-1)$. The Lie bracket in $PS(\tilde{Z})$ may be deformed as in Sec.1.2 which gives a Poisson $K[T]$-algebra $P_kS(\tilde{Z})$ for every $k\in K[T]$. The Lie bialgebra $\tilde{Z}$ is spiral so that the constructions of §4 apply. The projection $\tilde{Z}\to Z$ induces a group epimorphism $\mathrm{Exp}\,\tilde{Z}^* \to \mathrm{Exp}\,Z^*$

and a Poisson bialgebra epimorphism $PS_e(\tilde{Z}) \to PS_e(Z)$ linear over $aug: K[T] \to K$ $(aug(T) = 1)$.

If F is parallelizable then each parallelization of F induces an isomorphism $K[T] \otimes_K Z \to \tilde{Z}$ of Lie bi algebras over $K[T]$. This isomorphism transforms $t^n \otimes [\alpha]$ where α is a loop on F, into the class of an immersed loop which is homotopic to α and has the total rotation angle of the tangent vector $2\pi n$, $\pi = 3,14\ldots$. Existence of such an isomorphism, of course, reduces the quantization problem for $\tilde{Z}$ to the same problem for Z. However, in the case of closed F of genus >1 there seem to be no such reductions.

7.3. <u>The extended link modules</u>. Let M be an oriented 3-manifold equipped with a non-singular vector field ω. A link in M is said to be ω-regular if it is never tangent to ω. Let X_ω be the set of oriented ω-regular links in M considered up to smooth isotopy in the class of ω-regular links. The free abelian group $\mathbb{Z}^2$ with generators t_1, t_2 acts in X_ω: see Fig.5, where it is understood that ω is orthogonal to the plane of the picture and is directed upwards (towards the reader). One defines Conway triples of ω-regular links as in Sec.6.1 with the same convention on ω as above.

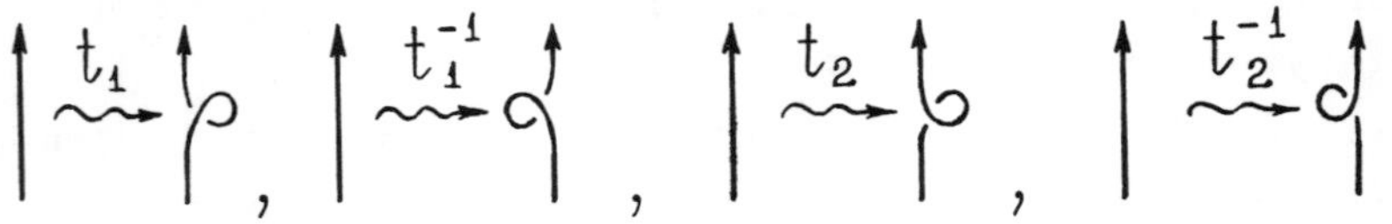

Fig. 5

Let $Q, \varphi, h, \hbar$ be the same objects as in Sec.6.1. Let C be the submodule of the free Q-module QX_ω with basis X_ω, generated by elements qL, where $L \in X_\omega$ and $q \in (\mathrm{Ker}\,\varphi)^{\|L\|}$, $\|L\|$ being the number of homotopy trivial components of L. (This C should be considered as a "positive part" of QX_{ω}). Let $A(M,\omega;Q)$ be the quotient of C by the submodule, generated by $(t_1-t_1^{-1})\phi-$ $-\hbar O$, $(t_2-t_2^{-1})\phi-\hbar O$ where O is a trivial knot lying in a small disc transversal to ω, and by elements qL_+- $-qL_--hqL_0$ (resp. $qL_+-qL_--\hbar qL_0$) with $q\in(\operatorname{Ker}\varphi)^{\|L_+\|}$ corresponding to arbitrary Conway triples of type 1 (resp. 2). The action of t_1, t_2 makes $A(M,\omega;Q)$ a module over $\widetilde{Q}=Q[t_1,t_1^{-1},t_2,t_2^{-1}]/(t_1-t_1^{-1}-t_2+t_2^{-1})$. Forgetting ω we obtain a homomorphism $A(M,\omega;Q)\to A(M;Q)$ linear over the augmentation $t_1\mapsto 1, t_2\mapsto 1:\widetilde{Q}\to Q$.

7.4. <u>The extended knot algebras</u>. Let ω be the constant non-singular vector field on $F\times I$ whose trajectories are segments $x\times I$. Put $\widetilde{A}=A(F\times I,\omega;Q)$. Note that X_{ω} is exactly the set of oriented immersed link diagrams on F considered up to the second and third Reidemeister moves. The multiplication $L,L'\mapsto LL'$ makes $\widetilde{A}$ an associative algebra over $\widetilde{Q}$ with $Q\otimes_{\widetilde{Q}}\widetilde{A}=A$. Theorems 6.3.1 and 6.5.1 may be transferred to the present setting with the following changes. Fix a φ-linear homomorphism $\psi:\operatorname{Ker}\varphi\to K[T]$ such that $\psi(\hbar)=t-t^{-1}$ and $\operatorname{Im}\psi\subset$ $\subset(t-t^{-1})K[T]$. The quantization homomorphism $P_{\psi}:\widetilde{A}\to$ $\to S(\widetilde{Z})$ is defined to satisfy the following conditions: P_{ψ} is linear over the homomorphism $t_1\mapsto t$, $t_2\mapsto t:\widetilde{Q}\to$ $K[T]$; if an oriented knot L is presented by an ascending (monotonely increasing) diagram $\mathcal{D}$ then $P_{\psi}([L])$ is the regular homotopy class of an immersed loop parametrizing $\mathcal{D}$ if this loop is not contractible, otherwise $P_{\psi}(q[L])=\psi(q)$ for any $q\in\operatorname{Ker}\varphi$. The diagram of quantization and forgetting homomorphisms

$$\begin{array}{ccc} \widetilde{A} & \longrightarrow & A \\ \downarrow & & \downarrow \\ S(\widetilde{Z}) & \longrightarrow & S(Z) \end{array}$$

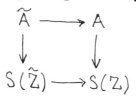

is commutative.

The formula given in the statement of theorem 6.6.1 defines a ρ-twisted comultiplication in $\widetilde{A}$ and makes $\widetilde{A}$ a bialgebra over $(\widetilde{Q},\rho)$ where ρ is the involution in $\widetilde{Q}$ exhanging t_1 and t_2. This means in particular that one should consider the tensor product $\widetilde{A}\otimes_\rho\widetilde{A}$ so that $qa\otimes b = a\otimes\rho(q)b$. Theorem 6.8.1 with the necessary changes transfers to this setting.

7.5. <u>Remarks</u>. 1. It is well known that each oriented compact 3-manifold M admits a non-singular vector field ω. If B is a ball in $\operatorname{Int} M$ then the inclusion homomorphism $A(M\setminus B,\omega;Q)\to A(M,\omega;Q)$ is clearly an isomorphism. Local changes of ω in balls generate an equivalence relation in the class on non-singular vector fields on M. The group $H_1(M)$ acts freely in the set of equivalence classes (see [14])). If we fix $\omega|_{\partial M}$ then the action becomes transitive. In particular, if M is a homology 3-sphere then $A(M,\omega;Q)$ does not depend on the choice of ω.

2. Let a be an invertible element of Q. Denote by A^*_a the Q-module of additive homomorphisms $\widetilde{A}\to Q$ linear over the ring homomorphism $t_1\mapsto a, t_2\mapsto a^{-1}:\widetilde{Q}\to Q$. Each homomorphism $f\in A^*_a$ gives rise to an isotopy invariant $\widetilde{f}$ of oriented links in $F\times I$ defined by the rule

$$\widetilde{f}(L(\mathcal{D})) = a^{-w(\mathcal{D})} f(h^{\|L(\mathcal{D})\|}[L(\mathcal{D})]),$$

where $w(\mathcal{D})$ is the writhe number of $\mathcal{D}$. If $h=\hbar$ and h is invertible in Q we may consider a simpler invariant $\hat{f}(L)=h^{-\|L\|}\widetilde{f}(L)$ which satisfies the formula

$$a\hat{f}(L_+) - a^{-1}\hat{f}(L_-) = h\hat{f}(L_0)$$

for an arbitrary Conway triple. For example, let $F=S^2$, $Q=\mathbb{C}[h,h^{-1},a,a^{-1}]$, $\hbar=h$. Clearly the action of t_1t_2

in $\widetilde{A}=\widetilde{A}(S^2)$ is trivial so that $\widetilde{A}$ is actually a module over $Q[t,t^{-1}]$. It follows from known results that this module is freely generated by $[\emptyset]$. If f is the generator of A_a^* which sends $[\emptyset]$ into $1 \in Q$ then $\hat{f}$ is the Jones-Conway polynomial, mutliplied by $h^{-1}(a-a^{-1})$.

3. If F is parallelized then the formula $\nabla([L(\mathcal{D})])=$

$$=\sum_{f \in Lbl_2(\mathcal{D})} (-1)^{|f|_-} h^{\frac{|f|+\|f\|}{2}} \hbar^{\frac{|f|-\|f\|}{2}} \left(t_2^{-\mathrm{rot}\,\mathcal{D}_{f,2}}[L(\mathcal{D}_{f,1})]\right) \otimes \left(t_1^{-\mathrm{rot}\,\mathcal{D}_{f,1}}[L(\mathcal{D}_{f,2})]\right)$$

defines a K-linear comultiplication $\nabla : \widehat{A} \to \widehat{A} \otimes \widehat{A}$ which makes $\widehat{A}$ a bialgebra over K. This comultiplication seems to be not of a quantum type since the image of $\mathrm{Perm} \circ \nabla - \nabla$ does not lie in $\hbar \widehat{A} \otimes \widehat{A}$. However, ∇ has another interesting feature. Namely, the identity $\nabla(t_i x) = (t_i \otimes t_i)\nabla(x)$ with $i=1,2$ shows that for arbitrary invertible $a, b \in Q$ the homomorphism ∇ induces a pairing $A_a^* \times A_b^* \to A_{ab}^*$. In the case when $F=S^2$, $Q=\mathbb{C}[h,h^{-1},a,a^{-1},b,b^{-1}]$, $\hbar = h$ this pairing was introduced in [8]).

APPENDIX. AN ALGEBRAIC EXAMPLE OF BIQUANTIZATION

Let $g=gl_n(K)$ be the Lie algebra of $n \times n$-matrices over K. Let r be an element of $g \otimes g$ such that $\mathrm{Perm}_g(r) = -r$ and the homomorphism $\partial r : g \to g \otimes g$ sending $a \in g$ into $[r, a \otimes 1 + 1 \otimes a]$ is a Lie cobracket and makes g a Lie bialgebra. (Here 1 is the unit $n \times n$-matrix). In the terminology of Drinfel'd [2]) $(g, \partial r)$ is a coboundary Lie bialgebra. Here I present a biquantization of the dual bialgebra g^*. The main idea is to modify the equation $RT_1T_2 = T_2T_1R$ which was used in [4]) to quantize certain Lie groups and Lie algebras. A similar, though more complicated technique enables to construct a biquantization of the Lie bialgebra $(g, \partial r)$ itself (cf. [4])).

Let $e_{i,j}$ be the $n \times n$-matrix whose only non-zero term equals 1 and stayes in the i -th row and the j -th column. Let $\{u_{i,j}\}$ be the basis in g^* dual to the basis $\{e_{i,j}\}$ in g .

Denote by E the polynomial ring with coefficients in $K[\hbar]$ of n^2 commuting variables $x_{i,j}$, where $i,j=1,\ldots,n$. Equip E with the comultiplication, sending $x_{i,j}$ into $x_{i,j}\otimes 1+1\otimes x_{i,j}+\hbar\sum_{l=1}^{n}x_{i,l}\otimes x_{l,j}$. Equip E with the Lie bracket such that the Leibniz rule is satisfied and

$$[x_{i,k}, x_{j,l}] =$$

$$= \sum_{s=1}^{n}(r_{i,j}^{s,l}x_{s,k}+r_{i,j}^{k,s}x_{s,l}-r_{i,s}^{k,l}x_{j,s}-r_{s,j}^{k,l}x_{i,s})+$$

$$+\hbar\sum_{s,t=1}^{n}(r_{i,j}^{s,t}x_{t,l}x_{s,k}-r_{t,s}^{k,l}x_{i,t}x_{j,s})$$

where $r_{i,j}^{k,l}=(u_{i,k}\otimes u_{j,l})(r)\in K$. This makes E a Poisson bialgebra. Denote by e the algebra homomorphism $E\to S(g^*)$ sending $f(\hbar)x_{i,j}$ into $f(0)u_{i,j}$. Note that if $K=\mathbb{R}$ then the formula $x_{i,j}\mapsto u_{i,j}-\delta_{i,j}$ (where $\delta_{i,j}$ is the Kronecker symbol) determines an imbedding of the Poisson bialgebra $\mathbb{R}\otimes_{\mathbb{R}[\hbar]}E$ into the Poisson bialgebra of polynomial functions on the group $GL_n(\mathbb{R})$ with the Poisson bracket, corresponding to the Lie cobracket ∂r in g .

Let

$$R=\sum_{i,j,k,l=1}^{n}R_{i,j}^{k,l}\,e_{i,k}\otimes e_{j,l}\in gl_n(K[h\hbar])\otimes gl_n(K[h\hbar])$$

where $R_{i,j}^{k,l}$ is a polynomial of $h\cdot\hbar$ with the free term $r_{i,j}^{k,l}$. Let $A=A(R)$ be the algebra over $K[h,\hbar]$, generated by n^2 non-commuting variables $X_{i,j}$ subject to quadratic relations which are united into one matrix relation

$$[X_1, X_2] + h[R, X_1 + X_2] = h\hbar(X_2 X_1 R - R X_1 X_2). \quad (11)$$

Here $X_1 = [X_{i,j}]_{i,j} \otimes 1$; $X_2 = 1 \otimes [X_{i,j}]_{i,j}$; $[X,Y] = XY - YX$. Provide A with the comultiplication sending $X_{i,j}$ into $X_{i,j} \otimes 1 + 1 \otimes X_{i,j} + \hbar \sum_{l=1}^{n} X_{i,l} \otimes X_{l,j}$. Note that the substitution $Y = \hbar X + 1$ and $R' = h\hbar R + 1$ transforms (11) into $R' Y_1 Y_2 = Y_2 Y_1 R'$.

Consider the diagram of ring homomorphisms and the diagram of algebra homomorphisms, linear over the ring diagram:

$$\begin{array}{ccc} K[h,\hbar] & \xrightarrow{\hbar \mapsto 0} & K[h] \\ {\scriptstyle h\mapsto 0}\downarrow & \searrow & \downarrow{\scriptstyle h \mapsto 0} \\ K[\hbar] & \xrightarrow{\hbar \mapsto 0} & K \end{array} \quad ; \quad \begin{array}{ccc} A & \xrightarrow{p_\hbar} & V_h(g^*) \\ {\scriptstyle p_h}\downarrow & \overset{p}{\searrow} & \downarrow{\scriptstyle v_h} \\ E & \xrightarrow{e} & BS_\beta(g^*) \end{array} \qquad (12)$$

Here $p_h(X_{i,j}) = x_{i,j}$ and $p_\hbar(X_{i,j}) = u_{i,j}$, $p = v_h \circ p_\hbar = e \circ p_h$.

<u>Theorem</u>. The diagrams (12) present a biquantization of the Lie bialgebra g^*.

Proof: direct calculations.

Note that the family of bialgebras $\{A(k,k') \mid k, k' \in K\}$ associated with this biquantization is "a cone over the projective line": for each invertible $q \in K$ the K-bialgebras $A(qk, q^{-1}k')$ and $A(k,k')$ are canonically isomorphic. In particular, if K is a field then this family of bialgebras is essentially equivalent to the line $A(1,k)$, $k \in K$ completed by two points $A(0,1)$ and $A(0,0)$. The biquantization of $\mathbb{Z}$ constructed in § 6 has similar properties.

References

1. Conway, J.H., "An enumeration of knots and links and some of their algebraic properties", Computational problems in abstract algebra. New York, Pergamon Press. 329-358 (1970).
2. Drinfel'd, V.G., "Hamiltonian structures on Lie groups, Lie bialgebras and the geometric meaning of the classical Yang-Baxter equation", Doklady AN SSSR 268, 285-287 (1982). (In Russian).
3. Drinfel'd, V.G., "Quantum groups", Zap.nauchn.sem. LOMI 155, 18-49 (1986). (In Russian; See also Proc. ICM, Berkeley, 1986).
4. Faddeev, L.D., Reshetikhin, N.Yu., Takhtajan, L.A., "Quantization of Lie groups and Lie algebras", Preprint LOMI E-14-87, Leningrad (1987).
5. Goldman, W., "The symplectic nature of fundamental groups of surfaces", Adv.Math. 54, 200-225 (1984).
6. Goldman, W., "Invariant functions on Lie groups and Hamiltonian flows of surface group representations", Inv.Math. 85, 263-302 (1986).
7. Hoste, J., Kidwell, M., "Dichromatic Link Invariants", Preprint (1988).
8. Jaeger, F., "Composition products and models for the homfly polynomial", Preprint. Grenoble (1988).
9. Kauffman, J.H., "New Invariants in the Theory of Knots", Preprint (1986).
10. Lickorish, W.B.R., Millett, K.C., "The New Polynomial Invariants of Knots and Links", Math.Magazine 61, 3-23 (1988).
11. Przytycki, J.H., "Skein modules of 3-manifolds", Preprint (1987).
12. Semenov-Tian-Shansky, M.A., "Dressing Transformations and Poisson Group Actions", Publ.Res.Inst.Math.Sci. Kyoto Univ. 21, 1237-1260 (1985).

13. Turaev, V.G., "Intersections of loops in two-dimensional manifolds", Matem.Sbornik 106, 566-588 (1978). (English translation: Math.USSR Sbornik 35, 229-250 (1979)).
14. Turaev, V.G., "Euler structures, non-singular vector fields and torsion invariants", Preprint LOMI R-9-86 (1986). (In Russian).
15. Turaev, V.G., "The Yang-Baxter equation and invariants of links", Invent.Math. 92, 527-553 (1988).
16. Turaev, V.G., "The Conway and Kauffman modules of the solid torus", Zap.nauchn.sem.LOMI 167, 79-89 (1988). (English translation: Preprint LOMI E-6-88, Leningrad (1988)).
17. Verdier, J.-L., "Groupes quantiques (d'aprés V.G. Drinfel'd)", Asterisque 152-153, 305-319 (1987).
18. Wolpert, S., "On the symplectic geometry of deformations of a hyperbolic surface", Ann.Math. 117, 207-234 (1983).

LOMI, Fontanka 27
Leningrad 191011, USSR

QUANTUM GROUPS

L.FADDEEV, N.RESHETIKHIN, L.TAKHTAJAN

Leningrad Branch of V.A.Steklov
Mathematical Institute of the Academy of
Sciences of the USSR

ABSTRACT

Elementary exposition of the concept of quantum double is presented and analog of Gauss decomposition for quantum groups is derived.

Quantum groups or q-analogues (deformations) of Lie groups and Lie algebras emerged as an abstraction of some formulas and objects from the quantum inverse scattering method [F1] , [S-T-F] , [T-F] . In particular papers [K-R], [S1-S3], were mostly revelant.

The term "quantum group" was introduced by Drinfeld in [D1-D2] , where one can find many new algebraic ideas, not evident from the point of view of mathematical physics. Independently the q-analogues of the classical Lie algebras were obtained by Jimbo [J1-J2] . Both these authors present their views in this volume, so we shall not comment on this more.

Being nearer to the origin of the notion of quantum group we decided to present our own definitions in [F-R-T]. The detailed elaboration of them together with discussion is published in a paper [R-T-F] in new journal "Journal of Algebra and Analysis" launched in Leningrad in 1989.

During the last year we advertized our point of view in many talks and research papers, see i.e. [F2] , [T] , [R] .

For this volume we decided to repeat our definitions and supplement them with general algebraic background based on the notion of "quantum double" introduced by Drinfeld in [D2] , but barely used there. Some new properties of the quantum groups will emerge as a result of this discussion.

We hope that this article will add to the tribute of the Yang-Baxter relations to which this volume is dedicated.

The Yang-Baxter relation is formulated for matrix R acting in tensor product of two linear spaces V

$$R \in \mathrm{Mat}\,(V \otimes V)$$

and can be written in the form

$$R_{12}\, R_{13}\, R_{23} = R_{23}\, R_{13}\, R_{12}\ .$$

Here matrices R_{12}, R_{13}, R_{23} act in the space $V \otimes V \otimes V$ their action coinciding with R in two spaces with corresponding numbers and being trivial in the third.

Given solution R of the Yang-Baxter relation we define two associative algebras with comultiplication dual to each other in a formal sence (we do not discuss corresponding questions of topology).

Algebra $A(R)$ is defined in terms of unity 1 and generators t_{ij} constituting a matrix T formally associated with the space V and satisfying the commutation relations written most naturally in the matrix form

$$R\, T_1\, T_2 = T_2\, T_1\, R\ ,$$

where T_1 and T_2 as R are associated with $V \otimes V$ and

$$T_1 = T \otimes I\ , \quad T_2 = I \otimes T\ .$$

In terms of matrix elements t_{ij} the main relations look as follows

$$R_{ik|pq}\, t_{pm}\, t_{qn} = t_{kq}\, t_{ip}\, R_{pq|mn}$$

where $R_{ik|pq}$ are matrix elements of R in a natural basis of $V \otimes V$.

The Yang-Baxter relation is a guarantee that the higher order relations on t_{ij} which follows from rearranging of the product $T_1 \ldots T_n$ using the main quadratic relation are in fact the consequence of it. Only cubic case is to be checked as follows from the general cathegory - theoretical considerations.

The comultiplication Δ (i.e. homomorphism $A \to A \otimes A$) in $A(R)$ is introduced on generators 1, t_{ij} as follows

$$\Delta(1) = 1 \otimes 1 ,$$

$$\Delta(t_{ij}) = \sum_k t_{ik} \otimes t_{kj} ,$$

or in matrix notations

$$\Delta(T) = T \dot{\otimes} T ,$$

dot being the matrix multiplication. It is easy to check from the main relation that the matrix $\Delta(T)$ satisfies the same relations as T.

The algebra $B(R)$ is generated by 1 and a set of generators $\ell_{ij}^{(\pm)}$ comprising two matrices $L^{\pm}$ and satisfying the relations

$$R_{21} L_1^+ L_2^+ = L_2^+ L_1^+ R_{21} ,$$

$$R_{21} L_1^+ L_1^- = L_2^- L_1^+ R_{21} ,$$

$$R_{21} L_1^- L_2^- = L_2^- L_1^- R_{21} .$$

The comultiplication in $B(R)$ is defined by

$$\Delta(L^{\pm}) = L^{\pm} \dot{\otimes} L^{\pm} .$$

Moreover a further restriction is introduced by the duality condition

$$< L_1^{\pm} \; ; \; T_2 > = R_{12}^{\pm}$$

where $\langle , \rangle$ means pairing between $A(R)$ and $B(R)$ and $\dot{\otimes}$ symbolizes the matrix multiplication in V . Here

$$R_{12}^{+} \equiv R_{21} = P R_{12} P \quad , \quad R_{12}^{-} = R_{12}^{-1}$$

where P is the permutation matrix in $V \otimes V$ and matrix R is supposed to be nonsingular.

The Yang-Baxter relation shows that the pairing is consistent with the main commutation relations.

It is expected that the condition of pairing effectively reduces the number of nonzero generators $\ell_{ij}^{(\pm)}$, so that the total number of generators of $A(R)$ and $B(R)$ is equal.

It is indeed the case in particular examples of R-matrices associated with the classical Lie groups. As was shown by Jimbo [J1-J2] and Bazhanov [B] there exists a solution R of Yang-Baxter relation corresponding to any classical Lie group of $A, B, C, \mathcal{D}$ type, the space V being that of the vector representation. The solution depends on a complex parameter q normalized in such a way that

$$R_q \big|_{q=1} = I .$$

Corresponding algebras $A(R)$ and $B(R)$ with additional constraints can be considered as q-deformations of algebras of functions on a classical Lie group and on a universal enveloping algebra of its Lie algebra. The explicit formulas for R-matrices, determinants, involutions, etc. can be found in [R-T-F] .

Having introduced the main formulas of our approach we shall describe the universal objects which satisfy them in a rather trivial fachion. The concrete algebras $A(R)$ and $B(R)$ will appear as a representation of these universal objects.

We begin with the introduction of an abstract Hopf algebra. Having in mind the mathematical physicists as readers we shall do it in the retro style using generators, structure constants, etc.

The Hopf algebra A is a linear space over a given field k (being a field $\mathbb{C}$ of complex numbers in what follows) with the operations of multiplication $m: A \otimes A \to A$, comultiplication $\Delta: A \longrightarrow A \otimes A$, antipode $\gamma: A \to A$ and counit $\varepsilon: A \to k$. In a chosen linear basis e_s, $s \in I$ (I being a set of indicies, e.g. $I = \mathbb{Z}_+$, $\mathbb{Z}^K$, etc.) these structures are defined by means of the corresponding structure constants

$$m(e_s \otimes e_t) = e_s e_t = m^p_{st} e_p ,$$

$$\Delta(e_s) = \mu_s^{pq} e_p \otimes e_q ,$$

$$\gamma(e_s) = \gamma_s^t e_t ,$$

$$\varepsilon(e_s) = \varepsilon_s ,$$

where m^p_{st}, μ_s^{pq}, γ_s^t, ε_s are numbers in k and summation over repeating indices is understood. We suppose that A contains unit element 1 and put

$$1 = \varepsilon^s e_s$$

with coefficients ε^s in k .

Thus introduced structures in a Hopf algebra satisfy many properties; let us list then out and express them in terms of the corresponding structure constants:

a) unit1:

$$m(a \otimes 1) = m(1 \otimes a) = a,$$

$$m^s_{pt}\, \varepsilon^p = m^s_{tp}\, \varepsilon^p = \delta^s_t\ ;$$

b) associativity:

$$m(m \otimes id) = m(id \otimes m),$$

$$m^s_{pt}\, m^t_{qr} = m^t_{pq}\, m^p_{tr}\ ;$$

c) coassociativity:

$$(\Delta \otimes id)\,\Delta = (id \otimes \Delta)\,\Delta,$$

$$\mu^{tr}_s\, \mu^{pq}_t = \mu^{pt}_s\, \mu^{qr}_t\ ;$$

d) counit:

$$(\varepsilon \otimes id)\Delta = (id \otimes \varepsilon)\Delta = id,$$

$$\varepsilon(ab) = \varepsilon(a)\,\varepsilon(b),$$

$$\mu^{tr}_s\, \varepsilon_r = \mu^{rt}_s\, \varepsilon_r = \delta^t_s,$$

$$m^s_{rt}\, \varepsilon_s = \varepsilon_r\, \varepsilon_t\ ;$$

e) Δ -homomorphism of multiplication:

$$\Delta(a)\,\Delta(b) = \Delta(ab);\quad \Delta(1) = 1 \otimes 1,$$

$$\mu^{ij}_p\, \mu^{uv}_q\, m^r_{iu}\, m^t_{jv} = m^s_{pq}\, \mu^{rt}_s,$$

$$\varepsilon^s \mu^{rt}_s = \varepsilon^r \varepsilon^t;$$

f) antipode:

$$\gamma(ab) = \gamma(b)\,\gamma(a), \quad \Delta \circ \gamma = (\gamma \otimes \gamma) \circ \sigma \circ \Delta,$$

(where σ is a permutation in $A \otimes A$: $\sigma(a \otimes b) = b \otimes a$)
and

$$m(\gamma \otimes id)\Delta(a) = m(id \otimes \gamma)\Delta(a) = \varepsilon(a)\cdot 1,$$

$$m^s_{pq}\, \gamma^t_s = \gamma^u_q\, \gamma^v_p\, m^t_{uv}\,,$$

$$\mu^{pq}_s\, \gamma^s_t = \gamma^p_v\, \gamma^q_u\, \mu^{uv}_t\,,$$

$$m^p_{us}\, \gamma^s_v\, \mu^{uv}_t = m^p_{sv}\, \gamma^s_u\, \mu^{uv}_t = \varepsilon_t\, \varepsilon^p$$

This essentially finishes the definition of a Hopf algebra.

It is clear from the relations written above that the set of structure constants

$$\tilde{m}^{pq}_s = \mu^{pq}_s\,,$$

$$\tilde{\mu}^p_{st} = m^p_{ts}\,,$$

$$\tilde{\gamma}^s_t = (\gamma^{-1})^s_t\,,$$

$$\tilde{\varepsilon}^s = \varepsilon^s\,,$$

$$\tilde{\varepsilon}_s = \varepsilon_s\,,$$

also defines a Hopf algebra with a linear basis e^s, $s \in I$, and a pairing $\langle e^s, e_t \rangle = \delta^s_t$. It can be considered as a dual algebra to A with an additional permutation in comultiplication. We shall denote this algebra as A^o. Also we have, of course, the usual dual A^* and $(A^o)^*$ which we shall denote as A^t .

Now we are ready to introduce two more Hopf algebras: double $\mathcal{D}(A)$ of Hopf algebra A and its dual $\mathcal{D}(A)^*$. We shall begin with the definition of $\mathcal{D}(A)$. As a linear space and as a coalgebra it is a tensor product of

A and A°. This means that $E_s^t = e_s \hat{\otimes} e^t$ constitute a linear basis in $\mathcal{D}(A)$ and

$$\Delta_{\mathcal{D}}(E_s^t) = \mu_s^{pq} m_{ur}^t E_p^r \otimes E_q^u .$$

The multiplication in $\mathcal{D}(A)$ is more complicated. It is defined by the "commutation relation"

$$\mu_s^{rp} m_{pt}^q (e_r \hat{\otimes} 1)(1 \hat{\otimes} e^t) =$$

$$= m_{tp}^q \mu_s^{pr} (1 \hat{\otimes} e^t)(e_r \hat{\otimes} 1) .$$

We can rewrite it also in a more traditional way

$$E_s^i E_j^q = M_{s\,j\,e}^{i\,q\,\kappa} E_\kappa^e ,$$

$$M_{s\,j\,\ell}^{i\,q\,\kappa} = \mu_j^{sn} (\gamma^{-1})_n^v m_{vr}^i \mu_s^{pq} m_{qt}^r \mu_e^{tq} m_{sp}^\kappa .$$

To derive this one is to multiply the last relation by $(e_i \hat{\otimes} 1)$ from the left and by $(1 \hat{\otimes} e^j)$ from the right and use the properties of the structure constants to invert the matrix

$$Q_{ij}^{\alpha\kappa} = \mu_j^{\kappa\ell} m_{\ell i}^\alpha$$

in the form

$$(Q^{-1})_{\alpha s}^{tj} = m_{p\alpha}^t (\gamma^{-1})_n^p \mu_s^{jn} .$$

The counit and antipode in $\mathcal{D}(A)$ are defined as follows

$$\varepsilon_{\mathcal{D}}(E_s^t) = \varepsilon^t \varepsilon_s ,$$

$$\gamma_{\mathcal{D}}(e_s \hat{\otimes} 1) = \gamma_s^t (e_t \hat{\otimes} 1),$$

$$\gamma_{\mathcal{D}}(1 \hat{\otimes} e^t) = (\gamma^{-1})_s^t (1 \hat{\otimes} e^s),$$

and their properties are easily checked. To define $\mathcal{D}(A)^*$ and distinguish it from $\mathcal{D}(A)$ we shall use the notation f_s and f^t for the basis in A^t and A^*, so that the basis in $\mathcal{D}(A)^* = A^* \otimes A^t$ can be written as

$$F_q^p = f^p \otimes f_q .$$

The multiplication in $\mathcal{D}(A)^*$ is given by

$$F_q^p F_t^s = m_{tq}^r \mu_k^{ps} F_r^k$$

and comultiplication is defined by means of the formula

$$\Delta_{\mathcal{D}^*}(F_k^\ell) = M_{j\,q\,k}^{i\,p\,\ell} F_j^i \otimes F_q^p$$

(cf. multiplication in $\mathcal{D}(A)$).

Now we shall introduce several universal elements in these algebras

$$R = \sum_s (e_s \hat{\otimes} 1) \otimes (1 \hat{\otimes} e^s) \in \mathcal{D} \otimes \mathcal{D},$$

$$T = \sum_{s,t} (e_s \hat{\otimes} e^t) \otimes (f^s \otimes f_t) \in \mathcal{D} \otimes \mathcal{D}^*,$$

$$Z^+ = \sum_s (1 \hat{\otimes} e^s) \otimes (e_s \hat{\otimes} 1) \in \mathcal{D} \otimes \mathcal{D},$$

$$Z^- = \sum_s (\gamma(e_s) \hat{\otimes} 1) \otimes (1 \hat{\otimes} e^s) \in \mathcal{D} \otimes \mathcal{D}.$$

It follows from the formulas above that the introduced elements satisfy the relations (with evident notations)

1. $R_{12} R_{13} R_{23} = R_{23} R_{13} R_{12}$

in $\mathcal{D} \otimes \mathcal{D} \otimes \mathcal{D}$;

2. $R_{12} \mathcal{T}_1 \cdot \mathcal{T}_2 = \mathcal{T}_2 \mathcal{T}_1 R_{12}$

in $\mathcal{D} \otimes \mathcal{D} \otimes \mathcal{D}^*$, where $\mathcal{T}_1 = \mathcal{T} \otimes I$, etc.

3. $R_{21} \mathcal{Z}_1^+ \mathcal{Z}_2^+ = \mathcal{Z}_2^+ \mathcal{Z}_1^+ R_{21}$,

$R_{21} \mathcal{Z}_1^+ \mathcal{Z}_2^- = \mathcal{Z}_2^- \mathcal{Z}_1^+ R_{21}$,

$R_{21} \mathcal{Z}_1^- \mathcal{Z}_2^- = \mathcal{Z}_2^- \mathcal{Z}_1^- R_{21}$,

$(id \otimes \Delta_{\mathcal{D}})(\mathcal{Z}^{\pm}) = \mathcal{Z}_{12}^{(\pm)} \mathcal{Z}_{13}^{(\pm)}$,

$(id \otimes \gamma)(\mathcal{Z}^{(\pm)}) = (\gamma^{-1} \otimes id)(\mathcal{Z}^{(\pm)}) = \left(\mathcal{Z}^{(\pm)}\right)^{-1}$

also in $\mathcal{D} \otimes \mathcal{D} \otimes \mathcal{D}$;

4.

$$< \mathcal{T}_1 ; \mathcal{Z}_2^{\pm} > = \mathcal{R}_{12}^{\pm} , \ (id \otimes \Delta_{\mathcal{D}^*}) \mathcal{T} = \mathcal{T}_{12} \mathcal{T}_{13}$$

where

$$\mathcal{R}_{12}^{+} = \mathcal{R}_{21}$$

and

$$\mathcal{R}_{12}^{-} = \mathcal{R}_{12}^{-1} .$$

The comparision of these formulas with those from definitions of quantum groups and Lie algebras, given above, is quite stricking. One can say that the latter are particular representations of the former.

Indeed, let ρ be a representation of algebra $\mathcal{D}(A)$ in a vector space V. Then

$$T = (\rho \otimes id)\,\mathcal{T},$$

$$L^{+} = (\rho \otimes id)\,\mathcal{L}^{+},$$

$$L^{-} = (\rho \otimes id)\,\mathcal{L}^{-},$$

realize the relations of the quantum group.

In the meaningfull example when A is a quantum universal enveloping algebra of a maximal Borel subalgebra of a simple Lie algebra $\mathfrak{g}$ its double $\mathcal{D}(A)$ coincides with a quantum universal enveloping algebra of a Lie algebra $\mathfrak{g}$ itself (after the natural elimination of a second Cartan subalgebra). When ρ is a vector representation of $\mathcal{D}(A)$ the corresponding matrix T gives a set of generators of a quantum group attached to a simple Lie group G (see [R-T-F] . Of course the main nontrivial problem here is in constructing of this representation ρ of $\mathcal{D}(A)$.

It is clear from the possibilities in our disposal that the universal objects

$$\mu^{+} = \sum_{s} (e_s \hat{\otimes} 1) \otimes (f^{s} \otimes 1) \in \mathcal{D} \otimes \mathcal{D}^{*},$$

$$\mu^{-} = \sum_{s} (1 \hat{\otimes} e^{s}) \otimes (1 \otimes f_s) \in \mathcal{D} \otimes \mathcal{D}^{*}$$

in $\mathcal{D} \otimes \mathcal{D}^{*}$ are as natural as $\mathcal{L}^{\pm}$. Let us list their properties: commutation relations

$$R_{12}\mathcal{M}_1^{\pm}\mathcal{M}_2^{\pm} = \mathcal{M}_2^{\pm}\mathcal{M}_1^{\pm}R_{12},$$
$$\mathcal{M}_1^{+}\mathcal{M}_2^{-} = \mathcal{M}_2^{-}\mathcal{M}_1^{+}$$

and pairing between the second $\mathcal{D}$ and $\mathcal{D}^*$ in $\mathcal{M}$ and $\mathcal{L}$

$$\langle \mathcal{L}_1^{+} ; \mathcal{M}_2^{+} \rangle = I ,$$
$$\langle \mathcal{L}_1^{+} ; \mathcal{M}_2^{-} \rangle = R_{21} ,$$
$$\langle \mathcal{L}_1^{-} ; \mathcal{M}_2^{+} \rangle = R_{12}^{-1} ,$$
$$\langle \mathcal{L}_1^{-} ; \mathcal{M}_2^{-} \rangle = I .$$

Moreover we have the simple connection between $\mathcal{M}^{\pm}$ and $\mathcal{T}$:

$$\mathcal{T} = \mathcal{M}^{+}\mathcal{M}^{-}.$$

This means that the matrix T of the generators of a quantum group can be substituted by two matrices

$$M^{\pm} = (\rho \otimes id)\,\mathcal{M}^{\pm} .$$

In examples of simple Lie groups matrices $M^{\pm}$ (as well as $L^{\pm}$) are triangular (or rather of Borel type) and the decomposition

$$T = M^{+}M^{-}$$

can be considered as a quantization of the Gauss decomposition.

Hovewer the comultiplication in terms of $M^{\pm}$ has a more complicated form than written through T .

Both matrix elements t_{ij} or $m_{ij}^{\pm}$ can be used as the generators of a quantum group. The comultiplication is written more simply in terms of t_{ij} , but the commutation relations between t_{ij} are more complicated than those for $m^{\pm}_{ij}$. In particular case of $SL(2)$ the matrices $M^{\pm}$ can be written as

$$M^{+} = \begin{pmatrix} a & X \\ 0 & a^{-1} \end{pmatrix}, \qquad M^{-} = \begin{pmatrix} a^{-1} & 0 \\ Y & a \end{pmatrix}$$

and the only relations are of the Weyl type:

$$XY = YX ,$$

$$Xa = qaX ,$$

$$Ya = q^{-1}aY .$$

The usefulness of the Gauss generators in general case is still to be investigated.

Thus the quantum double has led us to the definition of a Gauss decomposition in quantum groups which is a main new result in this paper. The related results concerning factorization are presented also in [R-S].

References

[B] Bazhanov V., Comm.Math.Phys., 1987, v.113, p.471-503.

[D1] Drinfeld V. Doklady AN SSSR, 1985, v.283, p.1060-1064 (in Russian).

[D2] Drinfeld V. Proceedings of the International Congress of Mathematicians, Berkeley, California: Acad. Press, 1986, v.1, p.798-820.

[F1] Faddeev L. Le Houches Lectures 1982. Amsterdam: Elsevier, 1984.

[F2] Faddeev L. Lecture at Landau memorial conference. Tel-Aviv, 1988. To be published by Pergamon press.

[F-R-F] Faddeev L., Reshetikhin N., Takhtajan L. LOMI preprint E-14-87, 1987, Leningrad.

[J1] Jimbo M. Lett.Math.Phys., 1986, v.11, p.247-252.

[J2] Jimbo M. Comm.Math.Phys., 1986, v.102,p.537-548.

[K-R] Kulish P., Reshetikhin N. Zap.nauchn.seminarov LOMI, 1981, v.101, p.101-110 (in Russian).

[R] Reshetikhin N. LOMI preprints E-4-87, E-17-87, 1988, Leningrad.

[R-S] Reshetikhin N.Yu., Semenov-Tian-Shansky M.A.: Journal of Geometry and Physics, v.V (1988).

[R-T-F] Reshetikhin N., Takhtajan L., Faddeev L. Algebra i analiz, 1989, v.1, N 1 (in Russian).

[S-T-F] Sklyanin E., Takhtajan L., Faddeev L. Teor.Math. Phys., 1979, v.40, p.194-220 (in Russian).

[S1] Sklyanin E., Funkt.analiz i ego pril., 1982, v.16, N 4, p.27-34 (in Russian).

[S2] Sklyanin E. Funkt. analiz i ego pril., 1983, v.17, N 4, p.34-48 (in Russian).

[S3] Sklyanin E. Uspechi Mat.Nauk., 1985, v.40, N 2, p.214 (in Russian).

[T-F] Takhtajan L., Faddeev L. Uspechi Mat.Nauk, 1979, v.34, N 5, p.13-63 (in Russian).

[T] Takhtajan L. Lecture at Taniguchi Symposium, Kyoto, 1988. To be published by Kinokuniya Co.

INTRODUCTION TO THE YANG-BAXTER EQUATION

Michio Jimbo
Department of Mathematics
Faculty of Science
Kyoto University
Kyoto 606, Japan

0. Introduction

For over two decades the Yang-Baxter equation (YBE) has been studied as the master equation in integrable models in statistical mechanics and quantum field theory. Recent progress in other fields — C^*-algebras, link invariants, quantum groups, conformal field theory, etc. — shed new light to the significance of YBE, and has aroused interest among many people.

In the literature YBE first manifested itself in the work of McGuire[1)] in 1964 and Yang[2)] in 1967. They considered a quantum mechanical many body problem on a line having $c\sum_{i<j}\delta(x_i - x_j)$ as the potential. Using a technique — known as Bethe's Ansatz — of building exact wavefunctions, they found that the scattering matrix factorized to that of the two body problem, and determined it exactly. Here YBE arises as the consistency condition for the factorization.

In statistical mechanics, the source of YBE probably goes back to Onsager's star-triangle relation, briefly mentioned in the introduction to his solution of the Ising model[3)] in 1944. Hunt for solvable lattice models has been actively pursued since then[4),5)], culminating in Baxter's solution of the eight vertex model[6)] in 1972. Another line of development was the theory of factorized S-matrix in two dimensional quantum field theory[7)]. Zamolodchikov pointed out[8)] that the algebraic mechanism working here is the same as that in Baxter's and others' works.

In 1978-79 Faddeev, Sklyanin and Takhtajan proposed the quantum inverse method[9),10)] as a unification of the classical integrable models (=soliton theory) and the quantum ones mentioned above. In their theory the basic commutation relaiton of operators is described by a solution of YBE (this terminology itself is due to them). In the beginning of 1980s, the study of YBE has been actively performed in Leningrad, Moscow and other places[11),12)]. These works led to the idea of introducing certain deformations of groups or Lie algebras[13)−16)], as called quantum groups by Drinfeld[17)]. At about the same time there appeared the discovery of new invariants of links[18)], and subsequently the aspect of YBE as the braid-type relation has been brought to attention. Closely related structures have also been revealed in conformal field theory[19)−21)].

The present article is aimed to be an introduction to YBE for non-specialists. About the state of the matter up to 1982 good review papers are available[11),12)]. I have tried here to include some of the more recent developments (to within my limited knowledge), with the emphasis on the role of quantum groups.

The text is organized as follows. Section 1 is devoted to basic definitions, properties and elementary examples of solutions of YBE. In Section 2 the classical YBE is defined, and the structure of its solutions is depicted following Belavin-Drinfeld[22)]. In Section 3 the quantized universal enveloping algebra $U_q\mathfrak{g}$ is introduced. Known facts about its representations and Drinfeld's universal R matrix are briefly summarized. As an application, in Section 4 the trigonometric solutions of YBE related to the vector representation of classical Lie algebras are described. Solutions corresponding to 'higher' representatons can be obtained by the fusion procedure[23)]. Section 5 outlines this method. In the last Section 6 the solutions of YBE of the 'face-model' type are discussed together with the braid representations induced by them.

Part of this paper has been written during my visit to Mathematical Sciences Research Institute in Berkeley, January 1989. I would like to thank T. Miwa for valuable comments on the manuscript.

1. The Yang-Baxter equation

1.1. Formulation. Let V be a complex vector space. Let $R(u)$ be a function of $u \in \mathbf{C}$ taking values in $\mathrm{End}_{\mathbf{C}}(V \otimes V)$. The following equation for $R(u)$ is called the *Yang-Baxter equation (YBE)*:

$$R_{12}(u)R_{13}(u+v)R_{23}(v) = R_{23}(v)R_{13}(u+v)R_{12}(u). \tag{1.1}$$

Here R_{ij} signifies the matrix on $V^{\otimes 3}$, acting as $R(u)$ on the i-th and the j-th components and as identity on the other component; e.g. $R_{23}(u) = I \otimes R(u)$. The variable u is called the spectral parameter. Often a solution of (1.1) is referred to as an R matrix.

In most cases we assume that $N = \dim V < \infty$. Upon taking a basis of V and writing

$$R(u) = \sum R_{ij}^{kl}(u) E_{ik} \otimes E_{jl},$$
$$E_{ij} = \left(\delta_{ia}\delta_{jb}\right)_{a,b=1,\cdots,N},$$

one sees that (1.1) amounts to N^6 homogeneous equations for the N^4 unknowns $R_{ij}^{kl}(u)$.

Let $P \in \mathrm{End}_{\mathbf{C}}(V \otimes V)$ denote the transposition

$$Px \otimes y = y \otimes x.$$

If $R(u)$ has the property

$$R(0) = const.P, \tag{1.2}$$

then (1.1) is identically satisfied for $u = 0$ or $v = 0$. We call (1.2) the *initial condition.*

1.2. Examples. Here are some typical examples of solutions of (1.1) in the case $V = \mathbf{C}^2$.

Example 1.1. (*McGuire*[1], *Yang*[2])

$$R(u) = \begin{pmatrix} 1+u & & & \\ & u & 1 & \\ & 1 & u & \\ & & & 1+u \end{pmatrix} = P + uI.$$

Example 1.2.

$$R(u) = \begin{pmatrix} \sin(\eta+u) & & & \\ & \sin u & \sin\eta & \\ & \sin\eta & \sin u & \\ & & & \sin(\eta+u) \end{pmatrix}.$$

Example 1.3. (Baxter[6])

$$R(u) = \begin{pmatrix} a(u) & & & d(u) \\ & b(u) & c(u) & \\ & c(u) & b(u) & \\ d(u) & & & a(u) \end{pmatrix},$$

where

$$\begin{aligned} a(u) &= \theta_0(\eta)\theta_0(u)\theta_1(\eta+u) \\ b(u) &= \theta_0(\eta)\theta_1(u)\theta_0(\eta+u) \\ c(u) &= \theta_1(\eta)\theta_0(u)\theta_0(\eta+u) \\ d(u) &= \theta_1(\eta)\theta_1(u)\theta_1(\eta+u). \end{aligned}$$

The $\theta_i(u)$ denote the elliptic theta functions

$$\begin{aligned} \theta_0(u) &= \prod_{n=1}^{\infty}(1-2p^{n-1/2}\cos 2\pi u + p^{2n-1})(1-p^n), \\ \theta_1(u) &= 2p^{1/8}\sin\pi u \prod_{n=1}^{\infty}(1-2p^n\cos 2\pi u + p^{2n})(1-p^n). \end{aligned} \tag{1.3}$$

In all cases the initial condition (1.2) is fulfilled. In examples 1.2 and 1.3, the parameters η, p are arbitrary. In fact these three are connected by specialization:

$$\text{Example 1.3} \xrightarrow{p\to 0} \text{Example 1.2} \xrightarrow{\eta\to 0} \text{Example 1.1}.$$

Let us verify that Example 1.1 solves YBE. Since both sides of (1.1) are polynomials in u of degree 2, it suffices to check (1.1) for 3 values of u.

$$\begin{aligned} u = 0 &: \text{ valid because of (1.2)}, \\ u = \infty &: P_{23} + vI = P_{23} + vI, \\ u = -v &: (P_{12} - vI)P_{13}(P_{23} + vI) = (P_{23} + vI)P_{13}(P_{12} - vI). \end{aligned}$$

The last equation reduces to the relations in the symmetric group $\mathfrak{S}_3$ $(12)(13)(23) = (23)(13)(12), (12)(13) = (13)(23)$.

Examples 1.2-1.3 can be handled in the same spirit.

1.3. Braid relations. Frequently YBE is also written in terms of the matrix

$$\check{R}(u) = PR(u). \tag{1.4}$$

Let $m \geq 2$, and define matrices on $V^{\otimes m}$ by $\check{R}_i(u) = I \otimes \cdots \otimes \check{R}(u) \otimes \cdots \otimes I$ ($\check{R}(u)$ in the $(i, i+1)$-th slot), $i = 1, \cdots, m-1$. One has then

$$\begin{aligned} \check{R}_i(u)\check{R}_j(v) &= \check{R}_j(v)\check{R}_i(u) \qquad \text{if } |i-j| > 1, \\ \check{R}_{i+1}(u)\check{R}_i(u+v)\check{R}_{i+1}(v) &= \check{R}_i(v)\check{R}_{i+1}(u+v)\check{R}_i(u). \end{aligned} \tag{1.5}$$

In the absence of the spectral parameters u, v, (1.5) is nothing other than Artin's braid relations. One notices that for the special values such that $u = u+v = v$, the $\check{R}_i(u)$ actually give rise to a representation of braid groups. The choice $u = v = 0$ usually leads to the trivial representation $\check{R}_i(0) = const.I$ (the initial condition (1.2)). In certain circumstances it makes sense to take $u = v = \infty$, leading to interesting results. See *4.3.*, *6.5.* below.

1.4. Generalizations. The above formulation of YBE admits the following extensions.

a) Instead of working with a fixed vector space V, one can equally well consider a family of vector spaces $\mathcal{F} = \{V\}$ and operators $\{R_{VV'}(u) \in \mathrm{End}_{\mathbf{C}}(V \otimes V')\}_{V,V' \in \mathcal{F}}$. YBE (1.1) is then an equation in $\mathrm{End}_{\mathbf{C}}(V_1 \otimes V_2 \otimes V_3)$, where $R_{ij}(u) = R_{V_i V_j}(u), V_i \in \mathcal{F}$. Suppose $V_1 = V_2 = V$. Regarding $\mathrm{End}_{\mathbf{C}}(V \otimes V_3) = \mathrm{End}_{\mathbf{C}}(V) \otimes \mathcal{A}, \mathcal{A} = \mathrm{End}_{\mathbf{C}}(V_3)$, let us write $R_{VV_3}(u)$ as $T(u) = \sum t_{ij}(u) E_{ij}$ with $t_{ij}(u) \in \mathcal{A}$. In this notation (1.1) becomes

$$\check{R}(u-v) T(u) \otimes T(v) = T(v) \otimes T(u) \check{R}(u-v). \tag{1.6}$$

Here $T(u) \otimes T(v) = \sum t_{ij}(u) t_{kl}(v) E_{ij} \otimes E_{kl}$, etc. (notice the ordering of the $t_{ij}(u)$ and $t_{kl}(v)$). Eq.(1.6) can be viewed as giving commutation relations among the generators $t_{ij}(u)$ of an abstract algebra $\mathcal{A}$; YBE for $\check{R}(u)$ guarantees the associativity of $\mathcal{A}$ thus defined. An important feature of (1.6) is that the map

$$\Delta : \mathcal{A} \to \mathcal{A} \otimes \mathcal{A}, \qquad \Delta(t_{ij}(u)) = \sum_k t_{ik}(u) \otimes t_{kj}(u)$$

preserves the relations (1.6) (Δ is a 'comultiplication'). The formulas (1.1) and (1.6) are the basic algebraic constituents in the quantum inverse method[9),10)].

b) One may consider YBE for a function of two variables $R(u, u')$:

$$R_{12}(u_1, u_2)R_{13}(u_1, u_3)R_{23}(u_2, u_3) = R_{23}(u_2, u_3)R_{13}(u_1, u_3)R_{12}(u_1, u_2). \quad (1.7)$$

Eq.(1.1) is a special case of (1.7) where the (u, u')-dependence enters only through the difference

$$R(u, u') \equiv R(u - u'). \quad (1.8)$$

Recently Au-Yang, Baxter, McCoy, Perk and others[24),25)] have found remarkable new solutions to (1.7) in which the spectral parameters live on curves of genus > 1. The difference property (1.8) does not hold for these solutions.

2. The classical Yang-Baxter equation

2.1. Classical limit. A solution of YBE is said to be *quasi-classical* if it contains an extra parameter $\hbar$ ('Planck constant') in such a way that as $\hbar \longrightarrow 0$ it has the expansion

$$R(u, \hbar) = (\text{scalar}) \times \left(I + \hbar r(u) + O(\hbar^2)\right). \quad (2.1)$$

The $r(u) \in \mathrm{End}_{\mathbf{C}}(V \otimes V)$ in (2.1) is called the *classical limit* of $R(u, \hbar)$. For instance, Examples 1.1-1.3 are all quasi-classical (take $u \to u/\hbar$ in Example 1.1, $\eta = \hbar$ in Examples 1.2-1.3). For quasi-classical $R(u, \hbar)$, YBE (1.1) implies the following *classical Yang-Baxter equation (CYBE)* for $r(u)$:

$$[r_{12}(u), r_{13}(u + v)] + [r_{12}(u), r_{23}(v)] + [r_{13}(u + v), r_{23}(v)] = 0. \quad (2.2)$$

As for the significance of CYBE in classical integrable systems and its algebraic/geometric meaning, see refs.12),26),27).

There are important examples of solutions of YBE which are not quasi-classical (e.g. refs.24),25),28),29)). Nevertheless quasi-classical solutions constitute an interesting class, and we shall henceforth restrict our attention to this case.

2.2. Universal solution. The characteristic feature of CYBE is that it is formulated using solely the Lie algebra structure of End(V). Let $\mathfrak{g}$ be a Lie algebra, and let $r(u)$ be a $\mathfrak{g} \otimes \mathfrak{g}$-valued function. In terms of a basis $\{X_\mu\}$ of $\mathfrak{g}$, write

$$r(u) = \sum_{\mu,\nu} r^{\mu\nu}(u) X_\mu \otimes X_\nu \quad (2.3)$$

with **C**-valued functions $r^{\mu\nu}(u)$. Let further $r_{12}(u) = \sum r^{\mu\nu}(u) X_\mu \otimes X_\nu \otimes I \in (U\mathfrak{g})^{\otimes 3}$ and so on, where $U\mathfrak{g}$ denotes the universal enveloping algebra. One has then

$$[r_{12}(u), r_{23}(v)] = \sum r^{\mu\nu}(u) r^{\rho\sigma}(u) X_\mu \otimes [X_\nu, X_\rho] \otimes X_\sigma, \ etc.,$$

so that each term in (2.2) actually lies inside $\mathfrak{g}^{\otimes 3}$. For each triplet of representations (π_i, V_i) $(i = 1, 2, 3)$ of $\mathfrak{g}$, $(\pi_i \otimes \pi_j)(r_{ij}(u))$ gives a matrix solution of CYBE in $V_1 \otimes V_2 \otimes V_3$. In this sense a $\mathfrak{g} \otimes \mathfrak{g}$-valued solution is a 'universal' solution of CYBE.

2.3. Belavin-Drinfeld theory. When $\mathfrak{g}$ is a finite dimensional complex simple Lie algebra, solutions of CYBE have been studied in detail by Belavin and Drinfeld[22)]. In the sequel we fix an orthonormal basis $\{X_\mu\}$ of $\mathfrak{g}$ with respect to a non-degenerate invariant bilinear form on $\mathfrak{g}$, and set

$$t = \sum_\mu X_\mu \otimes X_\mu.$$

By $r(u)$ we will mean a $\mathfrak{g} \otimes \mathfrak{g}$-valued meromorphic solution of (2.2) defined in a neighborhood of $0 \in \mathbf{C}$. It is said to be *non-degenerate* if $\det(r^{\mu\nu}(u)) \not\equiv 0$ in the notation of (2.3).

Theorem[22)]. *Let $r(u)$ be a non-degenerate solution of (2.2). Then*

(1) $r(u)$ extends meromorphically to the whole complex plane $\mathbf{C}$, with all its poles being simple.

(2) $\Gamma = \{$ the set of poles of $r(u)\}$ is a discrete subgroup relative to the addition of $\mathbf{C}$.

(3) As a function of u there are the following possibilities for the $r^{\mu\nu}(u)$:

rank $\Gamma = 2$:	*elliptic function,*
rank $\Gamma = 1$:	*trigonometric function*
	(i.e. a rational function in the variable $e^{const.u}$),
rank $\Gamma = 0$:	*rational function.*

Belavin-Drinfeld show further that (i) elliptic solution exists only for $\mathfrak{g} = \mathfrak{sl}(n)$, in which case it is unique (up to certain equivalence of solutions), (ii) trigonometric solutions exist for each type, and can be classified using the data from the Dynkin diagram for affine Lie algebras.

2.4. Examples. Here are two typical examples in the Belavin-Drinfeld classification.

Example 2.1. The simplest rational solution is

$$r(u) = \frac{t}{u}.$$

Example 2.2. Let $\mathfrak{g} = \mathfrak{h} \oplus (\oplus_\alpha \mathfrak{g}_\alpha)$ be the root space decomposition. Choose $X_\alpha \in \mathfrak{g}_\alpha$ so that $(X_\alpha, X_{-\alpha}) = 1$, and let

$$r = \sum_{\alpha>0} (X_\alpha \otimes X_{-\alpha} - X_{-\alpha} \otimes X_\alpha) \qquad \text{(sum over the positive roots).} \tag{2.4}$$

Then

$$r(u) = r - t + \frac{2t}{1-x}, \qquad x = e^u, \tag{2.5}$$

is a trigonometric solution.

3. The quantized universal enveloping algebra

3.1. 'Quantization'. Given a solution $r(u) \in \mathfrak{g} \otimes \mathfrak{g}$ of CYBE, one may ask whether there exists a quasi-classical $R(u, \hbar)$ having $r(u)$ as its classical limit[23)]. One has then to decide where such an object should live. A naïve candidate is $U\mathfrak{g} \otimes U\mathfrak{g}$; as it turns out, however, it is more natural to deform (or 'quantize') the algebra $U\mathfrak{g}$ according to each $r(u)$. Motivated by this 'quantization' problem, Drinfeld developed a general theory of quantum groups[17)]. In this section we shall describe a representative class of quantum groups, the quantized universal enveloping algebra $U_q\mathfrak{g}$[15)] (also called a q-analog of $U\mathfrak{g}$[16)]), which is related to the trigonometric solutions of Example 2.2. As for the case related to Example 2.1, see ref.15).

3.2. The algebra $U_q\mathfrak{g}$. Hereafter $\mathfrak{g}$ will denote a Kac-Moody Lie algebra of finite or affine type. The corresponding generalized Cartan matrix $A = (a_{ij})_{1 \le i,j \le l}$ is symmetrizable in the sense that there exist nonzero integers d_i satisfying $d_i a_{ij} = d_j a_{ji}$. Fix such $\{d_i\}$. Fix also a nonzero complex number q such that $q^{2d_i} \neq 1$. We define $U_q\mathfrak{g}$ to be the associative $\mathbf{C}$-algebra with 1, with $4l$ generators

$$X_i^+, \quad X_i^-, \quad k_i, \quad k_i^{-1} \qquad (1 \le i \le l)$$

and relations

$$
\begin{aligned}
&k_i k_j = k_j k_i, \qquad k_i k_i^{-1} = k_i^{-1} k_i = 1,\\
&k_i X_j^{\pm} k_i^{-1} = q^{\pm d_i a_{ij}/2} X_j^{\pm},\\
&[X_i^{+}, X_j^{-}] = \delta_{ij} \frac{k_i^2 - k_i^{-2}}{q^{d_i} - q^{-d_i}},\\
&\sum_{\nu=0}^{1-a_{ij}} (-)^{\nu} \begin{bmatrix} 1-a_{ij} \\ \nu \end{bmatrix}_{q^{d_i}} (X_i^{\pm})^{1-a_{ij}-\nu} (X_j^{\pm}) (X_i^{\pm})^{\nu} = 0 \qquad (i \neq j).
\end{aligned}
$$

Here we have used the notations from q-analysis

$$
\begin{bmatrix} m \\ n \end{bmatrix}_t = \frac{[m]_t!}{[n]_t![m-n]_t!}, \qquad [m]_t! = \prod_{1 \le j \le m} \frac{t^j - t^{-j}}{t - t^{-1}}.
$$

Setting formally $k_i = q^{d_i h_i/2}$ and letting $q \to 1$ one recovers the commutation relations among the Chevalley generators $\{e_i = X_i^{+}, f_i = X_i^{-}, h_i\}_{1 \le i \le l}$ of $\mathfrak{g}$.

The following comultiplication Δ, antipode S and counit ϵ endow $U_q\mathfrak{g}$ a Hopf algebra structure:

$$
\begin{aligned}
&\Delta(X_i^{\pm}) = X_i^{\pm} \otimes k_i^{-1} + k_i \otimes X_i^{\pm}, \qquad \Delta(k_i) = k_i \otimes k_i,\\
&S(X_i^{\pm}) = -q^{\mp d_i} X_i^{\pm}, \qquad S(k_i) = k_i^{-1},\\
&\epsilon(X_i^{\pm}) = 0, \qquad \epsilon(k_i) = 1.
\end{aligned}
$$

The Hopf algebra $U_q\mathfrak{g}$ was introduced by Kulish-Reshetikhin[13)] (for $\mathfrak{g} = \mathfrak{sl}(2)$), Drinfeld[15)] and the author[16)] (for a Kac-Moody algebra with symmetrizable generalized Cartan matrix). Our normalization here follows ref. 16).

3.3. Representation theory. It has been shown by Lusztig[30)] and Rosso[31)] that for generic values of q the representation theory of $U_q\mathfrak{g}$ does not change from the classical case $q = 1$.

Theorem.[30),31)] *Let* $\dim \mathfrak{g} < \infty$, *and assume that* q *is not a root of unity. Then an irreducible integrable* $\mathfrak{g}$*-module can be deformed to that of* $U_q\mathfrak{g}$. *The dimensionality of each weight space is the same as in the case* $q = 1$.

Example. Let $\mathfrak{g} = \mathfrak{sl}(2)$, and let l be a positive integer. Setting

$$\pi\left(X_1^+\right) = \begin{pmatrix} 0 & \sqrt{[l][1]} & & & & \\ & 0 & \sqrt{[l-1][2]} & & & \\ & & 0 & \sqrt{[l-2][3]} & & \\ & & & \ddots & \ddots & \\ & & & & & \sqrt{[1][l]} \\ & & & & & 0 \end{pmatrix} = {}^t\pi\left(X_1^-\right),$$

$$\pi(k_1) = \begin{pmatrix} q^{l/2} & & & \\ & q^{l/2-1} & & \\ & & \ddots & \\ & & & q^{-l/2} \end{pmatrix},$$

one gets the 'quantum deformation' of the $(l+1)$-dimensional irreducible representation of $\mathfrak{sl}(2)$. Here we have set

$$[u] = \frac{q^u - q^{-u}}{q - q^{-1}}. \tag{3.1}$$

For q a root of unity, the situation resembles the modular representation. In this case Lusztig[32)] developed a highest weight theory by modifying the definition of $U_q\mathfrak{g}$.

For affine Lie algebra $\mathfrak{g}$ of ADE type, Frenkel and Jing[33)] constructed the vertex operator representations of $U_q\mathfrak{g}$.

3.4. Universal R matrix. Drinfeld constructed a 'universal R matrix' $\mathcal{R} \in U_q\mathfrak{g} \otimes U_q\mathfrak{g}$ that enjoys the following properties[17)]:

$$\begin{aligned} \sigma \circ \Delta(a) &= \mathcal{R}\Delta(a)\mathcal{R}^{-1} \qquad \text{for } a \in U_q\mathfrak{g}, \\ (id. \otimes \Delta)\mathcal{R} &= \mathcal{R}_{13}\mathcal{R}_{12}, \\ (\Delta \otimes id.)\mathcal{R} &= \mathcal{R}_{13}\mathcal{R}_{23}, \end{aligned} \tag{3.2}$$

where $\sigma(x \otimes y) = y \otimes x$, and if $\mathcal{R} = \sum a_i \otimes b_i$ then $\mathcal{R}_{12} = \sum a_i \otimes b_i \otimes 1, \mathcal{R}_{13} = \sum a_i \otimes 1 \otimes b_i, \mathcal{R}_{23} = \sum 1 \otimes a_i \otimes b_i \in (U_q\mathfrak{g})^{\otimes 3}$. (To be precise, Drinfeld uses certain completion of $U_q\mathfrak{g}$, and $\otimes$ is to be understood in the topological sense. See ref.17) for details.)

Let $\hat{\mathfrak{g}}$ be affine, and let $\hat{\mathfrak{g}}' = [\hat{\mathfrak{g}}, \hat{\mathfrak{g}}]/(\text{the center of } \hat{\mathfrak{g}})$. Let $\mathcal{R}'$ denote the analog of $\mathcal{R}$ for $\hat{\mathfrak{g}}'$. Consider the automorphism T_x of $U_q\mathfrak{g}$ given by $T_xX_0^\pm =$

$x^{\pm 1}X_0^{\pm}, T_x X_i^{\pm} = X_i^{\pm}$ $(i \neq 0)$, where $i = 0$ is the distinguished vertex in the Dynkin diagram of $\hat{\mathfrak{g}}$. Set

$$\mathcal{R}(x) = (T_x \otimes id.)(\mathcal{R}').$$

From (3.2) it follows that $\mathcal{R}(x)$ solves YBE in the multiplicative parametrization

$$\mathcal{R}_{12}(x)\mathcal{R}_{13}(xy)\mathcal{R}_{23}(y) = \mathcal{R}_{23}(y)\mathcal{R}_{13}(xy)\mathcal{R}_{12}(x).$$

The r-matrix (2.5) is the classical limit of $\mathcal{R}(x)$ corresponding to the non-twisted loop algebra $\hat{\mathfrak{g}}' = \mathfrak{g} \otimes \mathbf{C}[\lambda, \lambda^{-1}], \dim \mathfrak{g} < \infty$. (The classical limit for twisted loop algebras can be found in ref. 22).)

Properties of $\mathcal{R}$ for finite dimensional $\mathfrak{g}$ are discussed also in ref. 34).

4. R matrix for the vector representation

4.1. Linear equations for R. Let us turn to the problem of constructing finite dimensional matrix solutions of YBE. Until the end of this paper we shall assume that q is not a root of unity, unless otherwise is stated explicitly. Retaining the notations $\hat{\mathfrak{g}}' = \mathfrak{g} \otimes \mathbf{C}[\lambda, \lambda^{-1}], \dim \mathfrak{g} < \infty$ as in *3.4.*, let $\pi : U_q\hat{\mathfrak{g}}' \to \mathrm{End}_{\mathbf{C}}(V)$ be a finite dimensional representation. Regarding $U_q\mathfrak{g}$ as a subalgebra of $U_q\hat{\mathfrak{g}}'$, we denote the restriction $\pi|_{U_q\mathfrak{g}}$ by the same letter π. We assume that at $q = 1$ the latter specializes to an irreducible representation of $\mathfrak{g}$. (An irreducible represetation of $\mathfrak{g}$ always lifts to $U_q\mathfrak{g}$ (see *3.3.*), but not always to $U_q\hat{\mathfrak{g}}'$ unless $\mathfrak{g}$ is of type A_n, cf. ref.15).)

Denoting by $P \in \mathrm{End}_{\mathbf{C}}(V \otimes V)$ the transposition, set

$$\check{R}(x) = P(\pi \otimes \pi)(\mathcal{R}(x)).$$

From (3.2) one has then

$$[\check{R}(x), (\pi \otimes \pi)(\Delta(a))] = 0 \qquad \text{for } a \in U_q\mathfrak{g}, \tag{4.1a}$$

$$\begin{aligned}\check{R}(x)(x^{\pm 1}\pi(X_0^{\pm}) \otimes \pi(k_0)^{-1} + \pi(k_0) \otimes \pi(X_0^{\pm}))\\ = (\pi(X_0^{\pm}) \otimes \pi(k_0)^{-1} + \pi(k_0) \otimes x^{\pm 1}\pi(X_0^{\pm}))\check{R}(x).\end{aligned} \tag{4.1b}$$

In fact these linear equations uniquely determine $\check{R}(x)$ up to a scalar factor[35].

Eq.(4.1a) means that $\check{R}(x)$ belongs to the centralizer of the diagonal action of $U_q\mathfrak{g}$ in $V \otimes V$. Taking a basis $\{P_k\}$ of $\mathrm{End}_{U_q\mathfrak{g}}(V \otimes V)$ one can write

$$\check{R}(x) = \sum_k \rho_k(x) P_k. \tag{4.2}$$

The coefficients ρ_k (up to an overall factor) are to be determined from (4.1b).

4.2. Vector representation. As an example let us consider the vector representation $(\pi, V_{\Lambda_1} = \mathbf{C}^N)$ of the classical Lie algebras $\mathfrak{g} = \mathfrak{sl}(N), \mathfrak{o}(N)$, or $\mathfrak{sp}(N)$ with N even. The R matrix in this representation has been calculated in refs. 36),35).

As in the classical case we have the decomposition as $U_q\mathfrak{g}$-module

$$\begin{aligned} V_{\Lambda_1} \otimes V_{\Lambda_1} &= V_{2\Lambda_1} \oplus V_{\Lambda_2} && \text{for } \mathfrak{g} = \mathfrak{sl}(N), \\ &= V_{2\Lambda_1} \oplus V_{\Lambda_2} \oplus V_0 && \text{for } \mathfrak{g} = \mathfrak{o}(N), \mathfrak{sp}(N), \end{aligned}$$

where $V_{2\Lambda_1}, V_{\Lambda_2}$ or V_0 denotes the analog of the symmetric tensor, the antisymmetric tensor or the trivial representation, respectively. Let $P_{2\Lambda_1}, P_{\Lambda_2}, P_0$ denote the corresponding orthogonal projectors relative to a $U_q\mathfrak{g}$-invariant scalar product on V_{Λ_1}. The spectral decomposition (4.2) for the $\check{R}$ matrix then reads as follows.

$\mathfrak{g} = \mathfrak{sl}(N)$:

$$\check{R}(x) = (qx - q^{-1})P_{2\Lambda_1} - (q^{-1}x - q)P_{\Lambda_2}, \tag{4.3}$$

$\mathfrak{g} = \mathfrak{o}(N), \mathfrak{sp}(N)$:

$$\begin{aligned} \check{R}(x) = &(q^{N-2\varepsilon}x - 1)(qx - q^{-1})P_{2\Lambda_1} - (q^{N-2\varepsilon}x - 1)(q^{-1}x - q)P_{\Lambda_2} \\ &+ \varepsilon(x - q^{N-2\varepsilon})(q^{-\varepsilon}x - q^{\varepsilon})P_0, \end{aligned} \tag{4.4}$$

where for convenience we have set

$$\begin{aligned} \varepsilon &= +1 && \text{for } \mathfrak{o}(N), \\ &= -1 && \text{for } \mathfrak{sp}(N). \end{aligned}$$

Together with the obvious relation $1 = \sum_k P_k$, the projectors are given by the following formulas .

$\mathfrak{g} = \mathfrak{sl}(N)$:

Set

$$T = qP_{2\Lambda_1} - q^{-1}P_{\Lambda_2}. \tag{4.5a}$$

Then

$$T^{\pm 1} = q^{\pm 1}\sum E_{ii}\otimes E_{ii} + \sum_{i\neq j} E_{ij}\otimes E_{ji} \pm (q-q^{-1})\sum_{i\lessgtr j} E_{ii}\otimes E_{jj}. \qquad (4.5b)$$

$\mathfrak{g} = \mathfrak{o}(N), \mathfrak{sp}(N)$:

Set

$$\begin{aligned} T &= qP_{2\Lambda_1} - q^{-1}P_{\Lambda_2} + \varepsilon q^{-N+\varepsilon}P_0, \\ S &= \frac{q^{N/2}-q^{-N/2}}{q-q^{-1}}\left(q^{N/2-\varepsilon}+q^{-N/2+\varepsilon}\right)P_0. \end{aligned} \qquad (4.6a)$$

Then we have

$$\begin{aligned} T^{\pm 1} &= q^{\pm 1}\sum_{i\neq i'} E_{ii}\otimes E_{ii} + \sum_{\substack{i\neq j,j' \\ or\, i=j=j'}} E_{ij}\otimes E_{ji} + q^{\mp 1}\sum_{i\neq i'} E_{ii'}\otimes E_{i'i} \\ &\pm (q-q^{-1})\sum_{i\lessgtr j}\left(E_{ii}\otimes E_{jj} - \varepsilon_i\varepsilon_j q^{\bar{i}-\bar{j}} E_{j'i}\otimes E_{ji'}\right), \qquad (4.6b) \\ S &= \varepsilon\sum \varepsilon_i\varepsilon_j q^{\bar{i}-\bar{j}} E_{j'i}\otimes E_{ji'}. \end{aligned}$$

Here $i' = N+1-i$, $\varepsilon_i = 1$ for $i < i'$, $\varepsilon_i = \varepsilon$ for $i > i'$, and

$$\begin{aligned} \bar{i} &= i + \frac{1}{2}\varepsilon \qquad (i < i') \\ &= i \qquad (i = i') \\ &= i - \frac{1}{2}\varepsilon \qquad (i > i'). \end{aligned}$$

4.3. Centralizer algebras. As we have noted in *1.3.*, if an $\check{R}$ matrix is trigonometric, i.e. a polynomial in $x = e^{cu}$ $(c \neq 0)$ up to a scalar factor, then its leading term in x gives rise to a representaion of the m string braid group B_m for any $m \geq 2$. Let T be defined by (4.5) or (4.6), and set $T_i = I\otimes\cdots\otimes T\otimes\cdots\otimes I \in \mathrm{End}_{U_q\mathfrak{g}}(V^{\otimes m})$ (T in the $(i,i+1)$-th slot). One has then

$$\begin{aligned} T_iT_j &= T_jT_i \qquad \text{if } |i-j|>1, \\ T_iT_{i+1}T_i &= T_{i+1}T_iT_{i+1}. \end{aligned} \qquad (4.7)$$

In fact, the above T is the image under $\pi\otimes\pi$ of the universal R matrix corresponding to $U_q\mathfrak{g}(\subset U_q\hat{\mathfrak{g}}')$.[34]

Let $\mathfrak{g} = \mathfrak{sl}(N)$. From (4.5a) one has in addition to (4.7)

$$(T_i - q)(T_i + q^{-1}) = 0. \qquad (4.8)$$

The relations (4.7), (4.8) mean that the braid representation factors through Iwahori's Hecke algebra[37] $H_m(q)$ for the symmetric group:

$$B_m \longrightarrow H_m(q) \xrightarrow{\rho_m} \mathrm{End}\left(V^{\otimes m}\right).$$

Moreover ρ_m commutes with the multidiagonal action of $U_q\mathfrak{g}$ given via the $(m-1)$-fold iteration of the comultiplication $\Delta^{(m)} : U_q\mathfrak{g} \rightarrow (U_q\mathfrak{g})^{\otimes m}$,

$$\Delta^{(m)}\left(X_i^{\pm}\right) = \sum_{j=1}^{m} k_i \otimes \cdots \otimes k_i \otimes \overset{j}{X_i^{\pm}} \otimes k_i^{-1} \otimes \cdots \otimes k_i^{-1}.$$

Proposition.[38] *For generic q, the two subalgebras of End$(V^{\otimes m})$*

$$\pi^{\otimes m} \circ \Delta^{(m)}\left(U_q\mathfrak{g}\right) \ \text{and} \ \rho_m\left(H_m(q)\right)$$

are commutant to each other.

This is a q-version of Weyl's reciprocity concerning the action of the symmetric group and the general linear group.

In the case $\mathfrak{g} = \mathfrak{o}(N)$ or $\mathfrak{sp}(N)$, similar statement is true[39),40)]. What replaces $H_m(q)$ is the Birman-Wenzl-Murakami algebra[39),40)], a q-analog of Brauer's centralizer algebra. Its quotient appears as the algebra generated by T_i and S_i defined similarly from (4.6).

5. The fusion procedure

5.1. Construction of R matrices. Many of the solutions of YBE known so far have been obtained by direct methods—assuming certain symmetries or guessing the functional form and solving the cubic equation (1.1) for $R(u)$. As for the quasi-classical solutions corresponding to (2.5) and its relatives, there are alternative approaches. One is to solve the linear equations (4.1). The other is the so-called *fusion procedure* initiated by Kulish, Reshetikhin and Sklyanin[23]. This method is an analog of a standard technique to get irreducible representations of Lie algebras — form a tensor product of fundamental representations and decompose it. In this section we shall describe the idea of the construction.

5.2. The fusion procedure. For later use we formulate the fusion procedure for the $\check{R}$-type matrices. Thus let $\{\check{R}_{VV'}(u) \in \mathrm{Hom}_{\mathbf{C}}(V \otimes V', V' \otimes V)\}_{V,V' \in \mathcal{F}}$ be a family of solutions of YBE written in the form

$$\begin{aligned}&\left(\check{R}_{V_2V_3}(u) \otimes I\right)\left(I \otimes \check{R}_{V_1V_3}(u+v)\right)\left(\check{R}_{V_1V_2}(v) \otimes I\right)\\ &= \left(I \otimes \check{R}_{V_1V_2}(v)\right)\left(\check{R}_{V_1V_3}(u+v) \otimes I\right)\left(I \otimes \check{R}_{V_2V_3}(u)\right).\end{aligned} \tag{5.1}$$

Here both sides map $V_1 \otimes V_2 \otimes V_3$ to $V_3 \otimes V_2 \otimes V_1$. We begin with the following observations.

(i) Fix u_1, u_2, and put

$$\check{R}_{V\otimes V'\,V''}(u) = \left(\check{R}_{VV''}(u+u_1) \otimes I\right)\left(I \otimes \check{R}_{V'V''}(u+u_2)\right). \tag{5.2a}$$

Then (5.1) remains valid by replacing $\check{R}_{V_1\,V_i}$ by $\check{R}_{V_1 \otimes V_1'\,V_i}$ $(i = 2,3)$. Likewise if we define

$$\check{R}_{V''\,V\otimes V'}(u) = \left(I \otimes \check{R}_{V''V'}(u+u_1)\right)\left(\check{R}_{V''V}(u+u_2) \otimes I\right), \tag{5.2b}$$

then (5.1) holds with $\check{R}_{V_i\,V_3 \otimes V_3'}$ in place of $\check{R}_{V_i\,V_3}$ $(i = 1,2)$.

(ii) Let $W_i \subset V_i$ be u-independent subspaces such that

$$\check{R}_{V_iV_j}(u)\,(W_i \otimes W_j) \;\subset\; W_j \otimes W_i. \tag{5.3}$$

Then (5.1) is true for the restrictions $\check{R}_{V_iV_j}(u)\big|_{W_i \otimes W_j}$.

In the notations of (5.2a), let us choose the subspace

$$W = \check{R}_{V'V}(u_2 - u_1)(V' \otimes V) \;\subset\; V \otimes V'. \tag{5.4}$$

Using YBE (5.1) one finds

$$\begin{aligned}
&\check{R}_{V\otimes V'\,V''}(u)(W\otimes V'')\\
&\quad= \left(\check{R}_{VV''}(u+u_1)\otimes I\right)\left(I\otimes \check{R}_{V'V''}(u+u_2)\right)\\
&\qquad\quad \times\left(\check{R}_{V'V}(u_2-u_1)\otimes I\right)(V'\otimes V\otimes V'')\\
&\quad= \left(I\otimes \check{R}_{V'V}(u_2-u_1)\right)\left(\check{R}_{V'V''}(u+u_2)\otimes I\right)\\
&\qquad\quad \times\left(I\otimes \check{R}_{VV''}(u+u_1)\right)(V'\otimes V\otimes V'')\\
&\quad\subset\ V''\otimes W,
\end{aligned}$$

so the condition (5.3) is satisfied for $W_1 = \check{R}_{V_1'V_1}(u_2-u_1)(V_1'\otimes V_1)\subset V_1\otimes V_1'$, $W_2 = V_2$ and $W_3 = V_3$. By the same token one has

$$\check{R}_{V''\,V\otimes V'}(u)(V''\otimes W)\ \subset\ W\otimes V''.$$

With an appropriate choice of $u_2 - u_1$ the W in (5.4) becomes a proper subspace, affording nontrivial new $\check{R}$ matrices

$$\check{R}_{W\,V''}(u) = \check{R}_{V\otimes V'\,V''}(u)\big|_{W\otimes V''}\,, \tag{5.5a}$$

$$\check{R}_{V''\,W}(u) = \check{R}_{V''\,V\otimes V'}(u)\big|_{V''\otimes W}\,. \tag{5.5b}$$

5.3. Symmetric tensors. Let us illustrate the construction above by taking the trigonometric solution $\check{R}(x)$ (4.3) for $\mathfrak{g} = \mathfrak{sl}(N)$. We shall use the multiplicative parameter $x = e^u$. There are two cases for which this $\check{R}$ matrix degenerates, namely:

$$\check{R}(q^2)\propto P_{2\Lambda_1},\qquad \check{R}(q^{-2})\propto P_{\Lambda_2}.$$

Here we consider the former case. Let $V_{\Lambda_1} = \mathbf{C}^N$, $V_{2\Lambda_1} = P_{2\Lambda_1}(V_{\Lambda_1}\otimes V_{\Lambda_1})$. Taking $V = V' = V'' = V_{\Lambda_1}$ and $W = V_{2\Lambda_1}$ in (5.5b) one has

$$\check{R}_{V_{\Lambda_1}\,V_{2\Lambda_1}}(x) = \frac{1}{x-q^{-2}}\left(I\otimes\check{R}(x)\right)\left(\check{R}(xq^2)\otimes I\right)\Big|_{V_{\Lambda_1}\otimes V_{2\Lambda_1}}$$

Likewise taking $V = V' = V_{\Lambda_1}$, $V'' = W = V_{2\Lambda_1}$ in (5.5a) one obtains

$$\check{R}_{V_{2\Lambda_1}\,V_{2\Lambda_1}}(x) = \left(\check{R}_{V_{\Lambda_1}\,V_{2\Lambda_1}}(xq^{-2})\otimes I\right)\left(I\otimes\check{R}_{V_{\Lambda_1}\,V_{2\Lambda_1}}(x)\right)\big|_{V_{2\Lambda_1}\otimes V_{2\Lambda_1}}\,. \tag{5.6}$$

By the construction, these matrices commute with the diagonal action of $U_q\mathfrak{g}$. Let

$$V_{2\Lambda_1}\otimes V_{2\Lambda_1} = V_{4\Lambda_1}\oplus V_{2\Lambda_1+\Lambda_2}\oplus V_{2\Lambda_2}$$

be the irreducible decomposition, and let $P_{4\Lambda_1}, P_{2\Lambda_1+\Lambda_2}, P_{2\Lambda_2}$ denote the corresponding orthogonal projectors. Then the spectral decomposition of (5.6) is given by

$$\begin{aligned}\check{R}_{V_{2\Lambda_1}\,V_{2\Lambda_1}}(x) =&(q^2x-1)(q^4x-1)P_{4\Lambda_1}\\ &+(q^2x-1)(q^4-x)P_{2\Lambda_1+\Lambda_2}+(q^2-x)(q^4-x)P_{2\Lambda_2}.\end{aligned}$$

In a similar manner one can construct $\check{R}$ matrices $\check{R}_{VV'}$ where V, V' are general symmetric or antisymmetric tensors (cf. refs.23),38)).

Cherednik[41)] gave a prescription to get the R matrix of type $\mathfrak{sl}(N)$ for an arbitrary pair of irreducible representations. His method applies also to the elliptic extension.

6. Face Models

6.1. Vertex vs. face models. In statistical mechanics, each solution of YBE defines a two-dimensional solvable lattice model; the matrix elements R_{ij}^{kl} of an R matrix stand for the statistical weights (Boltzmann weights) of local configurations. Usually with YBE (1.1) one associates the so-called *vertex models*, where the interaction takes place among the freedom on four edges round a lattice site, or a vertex . There are models that have dual features, in the sense that the interaction takes place among the freedom on the four sites round a face. These are called the *interaction-round-a-face models*[5)], or *face models* for short. Here YBE takes a slightly different form (though the two are mathematically equivalent). An interesting feature is that the face formulation allows to treat the case $q =$ a root of unity, by restricting the range of freedom on sites. These 'restricted' face models and the braid representations they induce play important roles in statistical mechanics, conformal field theory and operator algebras (see *6.4.*, *6.5.*).

6.2. Formulation. Let $W(u)$ be a solution of YBE (1.1) on $V \otimes V$. We say that it is of face-type if the following hold.

(i) There is a direct sum decomposition $V = \oplus_{a,b\in\mathcal{S}} V_{ab}$ into subspaces V_{ab} indexed by some (possibly infinite) set $\mathcal{S}$.

(ii) The composition of the maps

$$V_{ab}\otimes V_{b'c'} \xrightarrow{i} V\otimes V \xrightarrow{W(u)} V\otimes V \xrightarrow{p} V_{a'd'}\otimes V_{dc}$$

vanishes unless $a = a', b = b', c = c', d = d'$.

We set

$$W\begin{pmatrix} a & b \\ d & c \end{pmatrix}\!\!\left.\vphantom{\begin{pmatrix} a \\ d \end{pmatrix}}\right| u \Big) = p \circ W(u) \circ i \quad \in \mathrm{Hom}_{\mathbf{C}}(V_{ab} \otimes V_{bc}, V_{ad} \otimes V_{dc}).$$

In terms of these operators, YBE takes the form

$$\begin{aligned} &\sum_{g \in \mathcal{S}} W\left(\begin{matrix} f & g \\ e & d \end{matrix}\middle| u\right) W\left(\begin{matrix} b & c \\ g & d \end{matrix}\middle| u+v\right) W\left(\begin{matrix} a & b \\ f & g \end{matrix}\middle| v\right) \\ &\quad = \sum_{g \in \mathcal{S}} W\left(\begin{matrix} g & c \\ e & d \end{matrix}\middle| v\right) W\left(\begin{matrix} a & g \\ f & e \end{matrix}\middle| u+v\right) W\left(\begin{matrix} a & b \\ g & c \end{matrix}\middle| u\right). \end{aligned} \tag{6.1}$$

Both sides map $V_{ab} \otimes V_{bc} \otimes V_{cd}$ to $V_{af} \otimes V_{fe} \otimes V_{ed}$. In the literature the case where $\dim V_{ab} = 0$ or 1 is mainly considered; the $W\left(\begin{matrix} a & b \\ d & c \end{matrix}\middle| u\right)$ are treated as numbers subject to the relations (6.1). See ref. 5).

6.3. A vertex-face correspondence. As in section 4, let $\hat{\mathfrak{g}}' = \mathfrak{g} \otimes \mathbf{C}[\lambda, \lambda^{-1}]$ be a non-twisted loop algebra ($\dim \mathfrak{g} < \infty$), and let $\check{R}(x) = P(\pi \otimes \pi)(\mathcal{R}(x))$ be the trigonometric $\check{R}$ matrix associated with a fixed finite dimensional irreducible representation (π, V^{π}) of $U_q\hat{\mathfrak{g}}'$. Let

$$\mathcal{S} = \text{ the set of dominant integral weights of } \mathfrak{g}.$$

For each $a \in \mathcal{S}$ there exists an irreducible representation (see *3.3.*) $U_q\mathfrak{g} \to \mathrm{End}(V_a)$. The tensor module is completely reducible[31)] with respect to $U_q\mathfrak{g}$, so that one has

$$V_a \otimes V^{\pi} = \bigoplus_b V_{ab} \otimes V_b. \tag{6.2}$$

Here V_{ab} stands for the multiplicity part. We set $V_{ab} = 0$ if V_b does not appear in $V_a \otimes V^{\pi}$.

Consider now $1 \otimes \check{R}(x) \in \mathrm{End}(V_a \otimes V^{\pi} \otimes V^{\pi})$. Since this matrix commutes with the action of $U_q\mathfrak{g}$, it gives rise to a map

$$V_{ab} \otimes V_{bc} \longrightarrow \bigoplus_b V_{ab} \otimes V_{bc} \overset{1 \otimes \check{R}(x)}{\longrightarrow} \bigoplus_d V_{ad} \otimes V_{dc} \longrightarrow V_{ad} \otimes V_{dc}$$

for each $a, d \in \mathcal{S}$. We let $W\left(\begin{matrix} a & b \\ d & c \end{matrix}\middle| u\right)$ ($q^u = x$) to be the composition map. In this way, starting from a 'vertex-type' R matrix one gets a face-type solution of YBE (6.1). This construction is due to Pasquier[42)].

6.4. Vector representation. As an example, let us take again the $\check{R}$ matrix (4.3) for $\mathfrak{g} = \mathfrak{sl}(N)$, $\pi =$ the vector representation. Let $\mathcal{A} = \{\epsilon_1 - \epsilon, \cdots, \epsilon_N - \epsilon\}$ ($\epsilon = (\epsilon_1 + \cdots + \epsilon_N)/N$) denote the set of weights occurring in π, where the ϵ_i are orthonormal vectors related to the fundamental weights via $\Lambda_i = \epsilon_1 + \cdots + \epsilon_i - i\epsilon$. Defining V_{ab} by (6.2) one sees that $\dim V_{ab} \leq 1$. Clearly $\dim V_{ab} = 0$ unless $b - a$ has the form $\epsilon_i - \epsilon$. For $a \in \mathcal{S}$ we define the coordinates a_μ by

$$a_\mu = (a + \rho, \mu) \in \mathbf{Z}, \qquad \mu \in \mathcal{A},$$

where ρ signifies the half sum of the positive roots. Using the symbol $[u]$ in (3.1), one has the following expression[43] for the nonvanishing Boltzmann weights $W\left(\begin{matrix} a & b \\ d & c \end{matrix}\middle| u\right)$:

$$\begin{aligned}
W\left(\begin{matrix} a & a+\mu \\ a+\mu & a+2\mu \end{matrix}\middle| u\right) &= \frac{[1+u]}{[1]}, \\
W\left(\begin{matrix} a & a+\mu \\ a+\mu & a+\mu+\nu \end{matrix}\middle| u\right) &= \frac{[a_\mu - a_\nu - u]}{[a_\mu - a_\nu]}, \qquad (\mu \neq \nu), \\
W\left(\begin{matrix} a & a+\nu \\ a+\mu & a+\mu+\nu \end{matrix}\middle| u\right) &= \frac{[u]}{[1]} \left(\frac{[a_\mu - a_\nu + 1][a_\mu - a_\nu - 1]}{[a_\mu - a_\nu]^2} \right)^{1/2}, \qquad (\mu \neq \nu).
\end{aligned} \tag{6.3}$$

The case $\mathfrak{g} = \mathfrak{sl}(2)$ first appeared in ref.44). For the types $\mathfrak{o}(N)$ or $\mathfrak{sp}(N)$, see ref.45).

If we replace $[u]$ by the elliptic theta function $\theta_1(\pi u/L)$ in (1.3) ($L \neq 0$ being a parameter), the $W\left(\begin{matrix} a & b \\ d & c \end{matrix}\middle| u\right)$ above still solve YBE (6.1). Historically such elliptic solutions have been found by direct methods, and the above trigonometric ones came as their degenerations. The fusion procedure for (6.3) is discussed in ref.46).

Let now l be a positive integer. Consider the specialization of q to the root of unity

$$q = e^{\pi i/L}, \qquad L = l + g. \tag{6.4}$$

Here $g = N$ signifies the dual Coxeter number for $\mathfrak{sl}(N)$. Denoting by $\theta = \epsilon_1 - \epsilon_N$ the maximal root, we set

$$\mathcal{S}_l = \{ a \in \mathcal{S} \mid (a, \theta) \leq l \}.$$

It can be shown[43),45)] that, for the value (6.4) of q, YBE closes among the restricted set of Boltzmann weights $\{W\begin{pmatrix} a & b \\ d & c \end{pmatrix}\!\!\left| u\right)\}_{a,b,c,d\in\mathcal{S}_l}$. We call them *restricted face models.*

From the statistical mechanics point of view, the restricted face models are of particular interest. In the simpest case of $\mathfrak{sl}(2)$, their one point functions are known to be given in terms of the Virasoro characters in the minimal unitary series[47),48)]. Similar results have been established for a wide range of models[43),46),49)].

6.5. Fusion paths and braid representation. Let us consider the braid representation arising from (6.3). Unlike the 'vertex models' of *4.3.*, the representation space for the 'face models' does not have the tensor structure.

Let $m \geq 2$. We call a sequence of weights $p = (a_0, a_1, \cdots, a_m)$ $(a_i \in \mathcal{S})$ *fusion path* if for each i V_{a_i} appears in $V_{a_{i-1}} \otimes V^{\pi}$. For $a \in \mathcal{S}$, let $\mathcal{V}_m(a)$ be the vector space spanned by the fusion paths such that $a_0 = a$. Denoting by $W\left(\begin{matrix} a & b \\ d & c \end{matrix}\middle|\infty\right)$ the leading term of (6.3) in the variable $x = q^u$, we set for $i = 1, \cdots, m-1$

$$W_i\, p = \sum_{p'} W\left(\begin{matrix} a_{i-1} & a_i \\ a'_i & a_{i+1} \end{matrix}\middle|\infty\right) p', \qquad p \in \mathcal{V}_m(a).$$

Here the sum is over $p' = (a'_0, a'_1, \cdots, a'_m) \in \mathcal{V}_m(a)$ such that $a'_j = a_j$ for $j \neq i$. From the foregoing discussions it is clear that the W_i afford a braid representation on $\mathcal{V}_m(a)$. As in the vertex case it factorizes through the Hecke algebra.

When q is the root of unity (6.4) we define the space of restricted fusion paths $\mathcal{V}_{m\,l}(a)$ using $\mathcal{S}_l$ in place of $\mathcal{S}$. With the choice $a = 0$, the Hecke algebra representations on $\mathcal{V}_{m\,l}(0)$ coincide with the unitarizable irreducible representations of Hoefsmit[50)] and Wenzl[51)]. They also arise as the monodromy representations of N-point correlation functions in conformal field theory[19),20)]. As for the types $\mathfrak{o}(N)$ or $\mathfrak{sp}(N)$, such face type representations of the Birman-Wenzl-Murakami algebra have been studied by Murakami[52)].

References.

1) McGuire, J. B., "Study of exactly solvable one-dimensional N-body problems", J. Math. Phys. 5, 622-636 (1964).
2) Yang, C.N., "Some exact reults for the many-body problem in one dimension with repulsive delta-function interaction", Phys. Rev. Lett. 19, 1312-1314 (1967).

3) Onsager, L., "Crystal statistics I. A two dimensional model with an order-disorder transition", Phy s. Rev. 65, 117-149 (1944).
4) Lieb, E. H. and Wu, F. Y., in "Phase transitions and critical phenomena" (Domb, C. and Green, M. S. eds.), vol.1, 321-490, Academic, London, 1972.
5) Baxter, R. J., "Exactly solved models in statistical mechanics", Academic, London, 1982.
6) Baxter, R. J., "Partition function of the eight-vertex lattice model", Ann. of Phys. 70, 193-228 (1972).
7) See e.g. Zamolodchikov, A. B. and Zamolodchikov, Al. B., "Factorized S-matrices in two dimensions as the exact solution of certain relativistic quantum field theory models", Ann. of Phys. 120, 253-291 (1979).
8) Zamolodchikov, A. B., "$\mathbf{Z}_4$-symmetric factorized S-matrix in two space-time dimensions", Commun. Math. Phys. 69, 165-178 (1979).
9) Faddeev, L. D., Sklyanin, E. K. and Takhtajan, L. A., "The quantum inverse problem I", Theoret. Math. Phys. 40, 194-220 (1979).
10) Faddeev, L. D., "Integrable models in $(1+1)$-dimensional quantum field theory", Les Houches Session XXXIX, 563-608, Elsevier, Amsterdam, 1982.
11) Kulish, P. P. and Sklyanin, E. K., "Solutions of the Yang-Baxter equation", J. Soviet Math. 19, 1596-1620 (1982).
12) Kulish, P. P. and Sklyanin, E. K., "Quantum spectral transformation method. Recent developments", Lecture Notes in Physics 151, 61-119, Springer, 1982.
13) Kulish, P. P. and Reshetikhin, N. Yu., "The quantum linear problem for the sine-Gordon equation and higher representations", Zapiski nauch. LOMI 101, 101-110 (1980). The Hopf algebra structure for $U_q\mathfrak{sl}(2)$ was found in: Sklyanin, E. K., Uspekhi Math. Nauk 40, 214 (1985).
14) Sklyanin, E. K., "Some algebraic structure connected with the Yang-Baxter equation", Funct. Anal. and Appl. 16, 27-34 (1982); 17, 273-284 (1983).
15) Drinfeld, V. G., "Hopf algebars and the quantum Yang-Baxter equation", Soviet Math. Dokl. 32, 254-258 (1985).
16) Jimbo, M., "A q-difference analogue of $U_q\mathfrak{g}$ and the Yang-Baxter equation", Lett. Math. Phys. 10, 63-69 (1985).
17) Drinfeld, V. G., "Quantum Groups", ICM Proceedings, Berkeley, 798-820, 1986.
18) Jones, V. F. R., "A polynomial invariant for knots via von Neumann algebras", Bull. Amer. Math. Soc. 12, 103-111 (1985).

19) Tsuchiya, A. and Kanie, Y., "Vertex operators in conformal field theory on $\mathbf{P}^1$ and monodromy representations of braid group", Adv. Stud. Pure Math. 16, 297-372 (1988).

20) Kohno, T., "Monodromy representation of braid groups and Yang-Baxter equations", Ann. Inst. Fourier, 37, 139-160 (1987).

21) Moore, G. and Seiberg, N., "Classical and quantum conformal field theory", preprint, IASSNS-HEP-88/39, 1988.

22) Belavin, A. A. and Drinfeld, V. G., "Solutions of the classical Yang-Baxter equation for simple Lie algebras", Funct. Anal. and Appl. 16, 159-180 (1982).

23) Kulish, P. P., Reshetikhin, N. Yu. and Sklyanin, E. K., "Yang-Baxter equation and representation theory. I", Lett. Math. Phys. 5, 393-403 (1981).

24) Au-Yang, H., McCoy, B. M., Perk, J. H. H., Tang, S. and Yan, M.-L., "Commuting transfer matrices in chiral Potts models: solutions of the star-triangle equations with genus > 1", Phys. Lett. A123, 219-223 (1987).

25) Baxter, R. J., Perk, J. H. H. and Au-Yang, H., "New solutions of the star-triangle relations for the chiral Potts model", Phys. Lett. A128, 138-142 (1988).

26) Drinfeld, V. G., "Hamiltonian structures on Lie groups, Lie bialgebras and the geometric meaning of the classical Yang-Baxter equations", Soviet Math. Dokl. 27, 68-71 (1983).

27) Semenov-Tyan-Shanskii, M. A., "What is a classical r matrix ?", Funct. Anal. and Appl. 17, 259-272 (1983).

28) Fatteev V. A. and Zamolodchikov, A. B., "Self-dual solutions of the star-triangle relations on $\mathbf{Z}_N$-models", Phys. Lett. A92,37-39 (1982).

29) Kashiwara, M. and Miwa, T., "A class of elliptic solutions to the star-triangle relation", Nucl. Phys. B275 [FS17], 121-134 (1986).

30) Lusztig, G., "Quantum deformation of certain simple modules over enveloping algebras", Adv. in Math. 70, 237-249 (1988).

31) Rosso, M., "Finite dimensional representations of the qunatum analog of the enveloping algebra of a complex simple Lie algebra", Commun. Math. Phys. 117, 581-593 (1988).

32) Lusztig, G., "Modular representations and quantum groups", preprint, 1988.

33) Frenkel, I. B. and Jing, N., "Vertex representations of quantum affine algebras", preprint, 1988.

34) Reshetikhin, N. Yu., "Quantized universal enveloping algebras, the Yang-Baxter equation and invariants of links I", preprint LOMI E-4-87, 1988.

35) Jimbo, M., "Quantum R matrix for the generalized Toda system", Commun. Math Phys. 102, 537-547 (1986).

36) Bazhanov, V. V., "Trigonometric solutions of the star-triangle equation and classical Lie algebras", Phys. Lett. B159, 321-324 (1985).

37) Iwahori, N. and Matsumoto, H., "On some Bruhat decomposition and the structure of the Hecke ring of p-adic Chevalley groups", Publ. IHES 25, 5-48 (1965).

38) Jimbo, M., "A q-analogue of $U(\mathfrak{gl}(N+1))$, Hecke algebra and the Yang-Baxter equation", Lett. Math. Phys. 11, 247-252 (1986).

39) Birman, J. and Wenzl, H., "Link polynomials and a new algebra", preprint, 1986.

40) Murakami, J., "The Kaufmann polynomial of links and representation theory", Osaka J. Math. 24, 745-758 (1987).

41) Cherednik, I. V., "On 'quantum' deformations of irreducible finite dimensional representations of $\mathfrak{gl}_N$", Soviet Math. Dokl. 33, 507-510 (1986).

42) Pasquier, V., "Etiology of IRF models", Commun. Math. Phys. 118, 335-364 (1988).

43) Jimbo, M., Miwa, T. and Okado, M., "Local state probabilities of solvable lattice models: an $A_{n-1}^{(1)}$ family", Nucl. Phys. B300 [FS22], 74-108 (1988).

44) Baxter, R. J., "Eight-vertex model in lattice statistics and one-dimensional anisotropic Heisenberg chain", Ann. of Phys. 76, 1-24, 25-47, 48-71 (1973).

45) Jimbo, M., Miwa, T. and Okado, M., "Solvable lattice models related to the vector representation of classical simple Lie algebras", Commun. Math. Phys. 116, 507-525 (1988).

46) Jimbo, M., Kuniba, A., Miwa, T. and Okado, M., "An $A_{n-1}^{(1)}$ family of solvable lattice models", Commun. Math. Phys. 119, 543-565 (1988).

47) Andrews, G. E., Baxter, R. J. and Forrester, P. J., "Eight-vertex SOS model and generalized Rogers-Ramanujan-type identities", J. Stat. Phys. 35, 193-266 (1984).

48) Date, E., Jimbo, M., Kuniba, A., Miwa, T. and Okado, M., "Exactly solvable SOS models", Nucl. Phys. B290 [FS20], 231-273 (1987); Adv. Stud. Pure Math. 16, 17-122 (1988).

49) Date, E., Jimbo, M., Kuniba, A., Miwa, T. and Okado, M., "One dimensional configuration sums in vertex models and affine Lie algebra characters", preprint, 1987, to appear in Lett. Math. Phys.
50) Hoefsmit, P. N., "Representations of Hecke algebras of finite groups with BN pairs of classical type", Thesis, University of British Columbia, 1974.
51) Wenzl, H., "Representations of Hecke algebras and subfactors", Thesis, University of Pennsylvania, 1985; Invent. Math. 92, 349-383 (1988).
52) Murakami, J., "The representations of the q-analogue of Brauer's centralizer algebras and the Kaufmann polynomial of links", preprint, Osaka Univ., 1988.

INTEGRABLE SYSTEMS RELATED TO BRAID GROUPS AND YANG-BAXTER EQUATION

Toshitake KOHNO

Department of Mathematics, Nagoya University, Nagoya 464 Japan

INTRODUCTION. This note is a brief review on a recent development in the study of linear representations of the braid groups appearing as the monodromy of certain integrable connections. These connections are defined for any simple Lie algebra and its irreducible representation and appear in a natural way to describe n-point functions in the conformal field theory on the Riemann sphere with gauge symmetry due to Knizhnik and Zamolodchikov [12]. We focus the role of solutions of the *Yang-Baxter equation for the face model* to express the monodromy properties of these n-point functions. We will show that the Markov trace, which plays an important role to construct invariants of links due to Jones [10] and several other authors [1][19][22], appear as "weighted" characters of these monodromy representations. The reader may refer to [15][16] and [21] for a complete exposition on these subjects.

1. INFINITESIMAL PURE BRAID RELATIONS.

We start from a finite dimensional complex simple Lie algebra $\mathfrak{g}$ and its irreducible representation $\rho : \mathfrak{g} \longrightarrow \mathrm{End}(V)$. Let $\{I_\mu\}$ be an orthonormal basis of $\mathfrak{g}$ with respect to the Cartan-Killing form. We consider the matrices $\Omega_{\alpha\beta} \in \mathrm{End}(V^{\otimes n})$, $1 \leq \alpha < \beta \leq n$, defined by

$$(1.1) \qquad \Omega_{\alpha\beta} = \sum_{\mu} 1\otimes \ldots \otimes 1\otimes \overset{\alpha}{\rho(I_{\mu})}\otimes 1\otimes \ldots \otimes 1\otimes \overset{\beta}{\rho(I_{\mu})}\otimes 1\otimes \ldots \otimes 1$$

By using the fact that the Casimir element lies in the center of the universal enveloping algebra $U(\mathfrak{g})$ we have the relations:

$$(1.2) \qquad [\,\Omega_{\alpha\beta},\ \Omega_{\alpha\gamma}+\Omega_{\beta\gamma}] = [\,\Omega_{\alpha\beta}+\Omega_{\alpha\gamma},\ \Omega_{\beta\gamma}] = 0 \quad \text{for} \quad \alpha<\beta<\gamma$$

$$[\,\Omega_{\alpha\beta},\ \Omega_{\gamma\delta}] = 0 \quad \text{for distinct} \quad \alpha,\ \beta,\ \gamma,\ \delta\ .$$

The above relations can be considered to be a special case of the classical Yang-Baxter equation and have the following significance. Let us consider the 1-form

$$(1.3) \qquad \omega = \sum_{\alpha<\beta} \lambda\, \Omega_{\alpha\beta}\, d\log(z_{\alpha} - z_{\beta})\ .$$

with a complex parameter λ over

$$(1.4) \qquad X_n = \{(z_1, \ldots, z_n) \in \mathbb{C}^n\ ;\quad z_{\alpha} \neq z_{\beta} \quad \text{if} \quad \alpha \neq \beta\ \}$$

The relations (1.2) show that the connection ω is integrable. The fundamental group of X_n is called the pure braid group with n strings and the quadratic relations (1.2) may be consider to be an infinitesimal version of the defining relations of the pure braid group. This idea to express the relations for the fundamental group by the integrability condition goes back to Poincaré and Cartan. Following the work of Chen [4] and Sullivan [20] we can establish the precise group theoretical meaning of the relations (1.2) ([13]).

The *braid group* B_n is by definition the fundamental group of the quotient $X_n/\mathfrak{S}_n$, where the symmetric group acts as the permutation of the coordinates. Now as the monodromy of the connection ω we obtain a one parameter family of linear representations

$$(1.5) \qquad \varphi : B_n \longrightarrow \mathrm{End}(V^{\otimes n})$$

2. QUANTIZED UNIVERSAL ENVELOPING ALGEBRA AND R-MATRIX.

We present a second method to obtain linear representations of B_n. Let $U_\hbar(g)$ be the *quantized universal enveloping algebra* of g in the sense of Drinfel'd [5] and Jimbo [6]. We put $q = e^{\hbar/2}$. Let $\rho_i : U_\hbar(g) \longrightarrow \mathrm{End}(V_i)$, $i=1,2$, be the irreducible representation with the highest weight Λ_i. The tensor product $V_1 \otimes V_2$ has a structure of $U_\hbar(g)$-module by means of the comultiplication of $U_\hbar(g)$. Under the assumption that any irreducible component in $V_1 \otimes V_2$ has multiplicity one, Reshetikhin [19] obtained the following R-matrix.

$$(2.1) \qquad R^{\Lambda_1\Lambda_2} = \sum_{\Lambda} (-1)^{\varepsilon(\Lambda)} q^{\{c(\Lambda)-c(\Lambda_1)-c(\Lambda_2)\}/2} P_{\Lambda}^{\Lambda_1\Lambda_2}$$

The meaning of the notations is as follows. First, we define the q-*Clebsch-Gordan coefficient* (see Fig.1)

$$C = C_{\Lambda}^{\Lambda_1\Lambda_2}(q) \quad : \quad V_1 \otimes V_2 \longrightarrow V_\Lambda$$

Fig.1

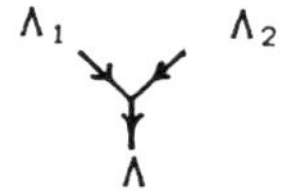

for any irreducible module V_Λ with the highest weight Λ contained in $V_1 \otimes V_2$. The row vectors of C consist of the weight vectors of V_Λ and we normalize C as $C\,{}^tC = I$. We define the projector P_Λ

by ${}^{t}C.C$. We put $c(\Lambda) = \langle\Lambda, \Lambda+2\delta\rangle$ where δ is the half sum of the positive roots of g and $\varepsilon(\Lambda)$ is the parity of V_Λ in $V_1 \otimes V_2$.

In the case g is non-exceptional and Λ_i , $i=1,2$, corresponds to the vector representation the above R-matrix is extracted from trigonometric solutions of the Yang-Baxter equation

(2.3) $$R_{12}(u)R_{23}(u+v)R_{12}(v) = R_{23}(v)R_{12}(u+v)R_{23}(u)$$

due to Jimbo [7] by tending the spectral parameter u to the infinity.

Let us suppose $\Lambda_1 = \Lambda_2$. We denote by σ_i , $1 \le i \le n-1$, the standard generators of B_n (see [2]). Then the corresponsence

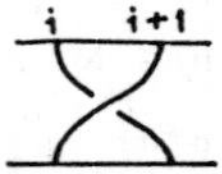

$$\sigma_i \longrightarrow R_{i,i+1} = 1\otimes \dots \otimes 1 \otimes \overset{i,i+1}{R} \otimes 1 \otimes \dots \otimes 1$$

where $R = R^{\Lambda_1\Lambda_2}$ gives a linear representation of B_n denoted by $\pi : B_n \longrightarrow \mathrm{End}(V^{\otimes n})$. This representation commutes with the diagonal action of $U_\hbar(g)$ and if we consider the classical limit $\hbar \to 0$ the above construction gives the situation due to Brauer and Weyl.

3. FUSION PATH AND NORMALIZED SOLUTIONS. In this section, we start from the vector representation $\rho : g \longrightarrow \mathrm{End}(V)$ and we suppose that $q = e^{\hbar/2}$ is not a root of unity. We denote by Π the highest weight of the vector representation. We suppose that $q = e^{\hbar/2}$ is not a root of unity. The n-fold tensor product $V^{\otimes n}$ has a decomposition $\oplus (M_\Lambda \otimes V_\Lambda)$ as a $U_\hbar(g)$-module where M_Λ stands for the multiplicity of the representation V_Λ corresponding to the highest weight Λ . We have a basis of M_Λ described in the following way.

Let $\mathcal{P}(\Lambda)$ denote the set of the sequence $(\Lambda_0, \ldots, \Lambda_n)$ of dominant integral weights of $\mathfrak{g}$ satisfying the following:

(3.1) (i) $\Lambda_0 = 0\ ,\ \ \Lambda_n = \Lambda$

(ii) $V_{\Lambda_i} \otimes V$ contains $V_{\Lambda_{i+1}}$ as a $\mathfrak{g}$-module.

An element $\not{h}$ of $\mathcal{P}(\Lambda)$ is called a *fusion path*, which corresponds to some shortest path in the decomposition diagram of $V^{\otimes n}$ as a $\mathfrak{g}$-module. We assocoate to $\not{h}$ the following composition of q-Clebsch-Gordan coefficients (see Fig.2 and 5)

(3.2) $C^{\Lambda_0 \pi}_{\Lambda_1}(q)\ :\ V^{\otimes n} \longrightarrow V_{\Lambda_1} \otimes V^{\otimes(n-2)}$

$C^{\Lambda_1 \pi}_{\Lambda_2}(q)\ :\ V_{\Lambda_2} \otimes V^{\otimes(n-2)} \longrightarrow V_{\Lambda_3} \otimes V^{\otimes(n-3)}$

.....

$C^{\Lambda_{n-1} \pi}_{\Lambda}(q)\ :\ V_{\Lambda_{n-1}} \otimes V \longrightarrow V_{\Lambda}$

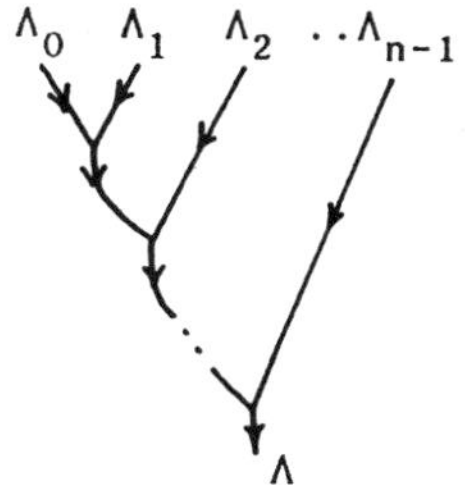

Fig.2

which defines a projector $e_{\not{h}}(q)\ :\ V^{\otimes n} \longrightarrow V_{\Lambda}$. These $e_{\not{h}}(q)$ form a basis of M_{Λ} and the action of the braid group is expressed by using W defined by

(3.3) $$W\begin{pmatrix} \Lambda_{i-1}\ , & \Lambda'_i \\ \Lambda_i\ , & \Lambda_{i+1} \end{pmatrix} = e_{\not{h}}(q) R_{i,i+1}\ {}^{t}e_{\not{h}'}(q)$$

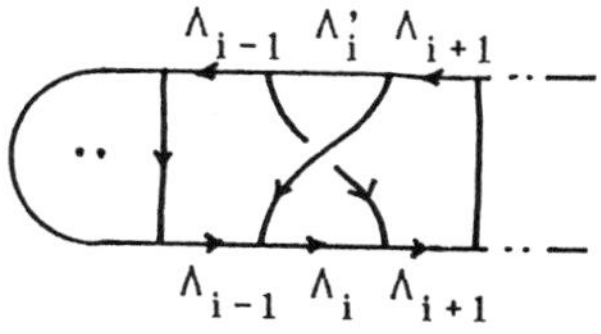

Fig.3

(see Fig.3). We have

(3.4) $$\sigma_i\ . e_{\not{h}}(q) = \sum_{\not{h}'} W\begin{pmatrix} \Lambda_{i-1}, & \Lambda'_i \\ \Lambda_i\ , & \Lambda_{i+1} \end{pmatrix} e_{\not{h}'}(q)$$

where the RHS is the sum with respect to $\lambda' = (\Lambda'_0, \ldots, \Lambda'_n) \in \mathcal{P}(\Lambda)$ satisfying $\Lambda_j \neq \Lambda'_j$ if $j \neq i$. The above coefficients W satisfies the Yang-Baxter equation for the face model

$$(3.5)\qquad \sum_g W\begin{pmatrix} f, & g \\ e, & d \end{pmatrix} W\begin{pmatrix} b, & c \\ g, & d \end{pmatrix} W\begin{pmatrix} a, & b \\ f, & g \end{pmatrix}$$

$$= \sum_g W\begin{pmatrix} a, & b \\ g, & c \end{pmatrix} W\begin{pmatrix} a, & g \\ f, & e \end{pmatrix} W\begin{pmatrix} g, & c \\ e, & d \end{pmatrix}$$

and they are extracted from the Boltzmann weights for the IRF model due to Jimbo, Miwa and Okado [8] by taking the critical limit and tending the spectral parameter to the infinity.

example. Let us consider the case $\mathfrak{g} = sl(N,\mathbb{C})$. We suppose that Λ_{i-1} corresponds to the Young diagram of type $(d_1, \ldots, d_m)$, $d_1 \geq \ldots \geq d_m \geq 0$ and Λ_i and Λ'_i are obtained by adjoining one node to the r-th and s-th row $(r \neq s)$ respectively (see Fig.4). We put $d = (d_r - r) - (d_s - s)$. In this case W is given by

$$(3.6)\qquad W\begin{pmatrix} \Lambda_{i-1} & , \Lambda'_i \\ \Lambda_i & , \Lambda_{i+1} \end{pmatrix} = \sqrt{[d-1][d+1]/[d]^2}$$

where $[k]$ stands for $(q^k - q^{-k})/(q - q^{-1})$.

Fig.4

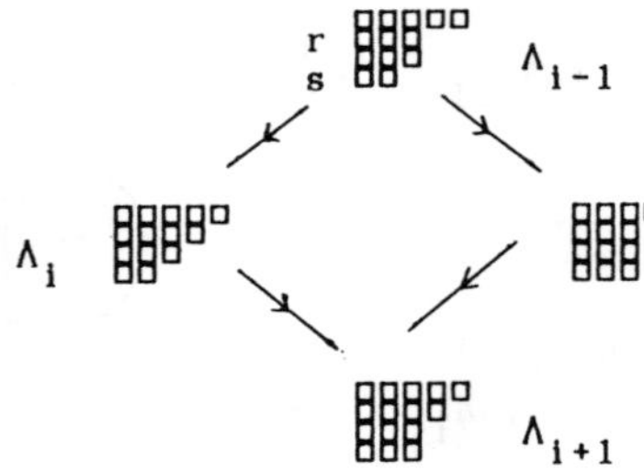

The above description of the action of the braid group by means of the face language can be used effectively to describe the monodromy of ω defined in (1.3) with respect to the *normalized solutions associated with the fusion paths* in the following sense. Let $w_1, \ldots, w_{n-1}$ be the blowing up coordinates of X_n such that $w_k=0$ corresponds to $z_1= \ldots = z_{k+1}$. The residue of ω along $w_k=0$ is given by $\sum_{1\le\alpha<\beta\le k+1} \Omega_{\alpha\beta}$. These elements are diagonalized simultaneouly with respect to the basis $e_{\hbar} = \lim_{q\to 1} e_{\hbar}(q)$ with the eigenvalues

$$(3.7) \qquad \mu_k = \frac{1}{2}\lambda\ (c(\Lambda_{k+1}) - (k+1)c(\Pi))\ , \quad 1\le k\le n-1\ ,$$

Let us suppose that $q = \exp(\pi i\lambda)$ is not a root of unity. Then the total differential equation $d\Phi = \omega\Phi$ has solutions associated with the fusion paths given by

$$(3.8) \qquad \phi_{\hbar}(z)= w_1^{\mu_1} w_2^{\mu_2} \cdots w_{n-1}^{\mu_{n-1}} \{ e_{\hbar} + \text{(higher order terms)} \}$$

We can show (see [16]) that after a certain normalization $\phi^{\sim}_{\hbar}(z) = \alpha_{\hbar}(\lambda)\phi_{\hbar}(z)$ the monodoromy of the braid group is expressed as

$$(3.9) \qquad \sigma_i^* \phi^{\sim}_{\hbar}(z) = \sum_{\hbar'} W\begin{pmatrix} \Lambda_{i-1}, & \Lambda_i' \\ \Lambda_i & , \Lambda_{i+1} \end{pmatrix} \phi^{\sim}_{\hbar'}(z)$$

To show this we used the following expansion

$$(3.10) \qquad (R_{12}R_{23} \cdots R_{k-1,k})^k = 1 + \hbar \sum_{1\le\alpha<\beta\le k} \Omega_{\alpha\beta} + O(\hbar^2)$$

together with a description of the Riemann-Hilbert correspondence for the pure braid group obtained by investigating the group theoretical meaning of the infinitesimal pure braid relations (see [14]).

4. FUSION METHODS. The following priciple to compute the monodromy by localizing the situation to the case of four variables was discovered by Tsuchiya and Kanie [21]. We start from $\mathfrak{g} = \mathfrak{sl}(N,\mathbb{C})$ and its vector representation. We consider the fusion paths connecting Λ_{i-1} and Λ_{i+1} (see Fig.4). Such a fusion path $\hbar$ defines a $\mathfrak{g}$-homomorphism $e_\hbar : V_{i-1}\otimes V\otimes V \longrightarrow V_{i+1}$ where V_j denotes the irreducible representation with the highest weight Λ_j. We have

$$(4.1) \qquad \Omega_{i-1,i}\, e_\hbar = \frac{1}{2}\,\{c(\Lambda_i) - c(\Pi) - c(\Lambda_{i-1})\}\, e_\hbar$$

$$(\Omega_{i-1,i} + \Omega_{i-1,i+1} + \Omega_{i,i+1})\, e_\hbar = \frac{1}{2}\,\{c(\Lambda_{i+1}) - 2c(\Pi) - c(\Lambda_{i-1})\}\, e_\hbar$$

Let us denote by $\Delta\, e_\hbar$ the RHS of the second equation. Let $\omega\hat{}$ be the connection defined by

$$(4.2) \qquad \omega\hat{} = \sum_{i-1\le\alpha<\beta\le i+1} \lambda\, \Omega_{\alpha\beta}\, d\log(z_\alpha - z_\beta)$$

We put $z_{i-1} = 0$. The total differential equation $d\Phi = \omega\hat{}\,\Phi$ can be written in the form

$$(4.3) \qquad \frac{d}{d\zeta}\Psi_0(\zeta) = \lambda\,\{\,\Omega_{i-1,i}/\,\zeta + \Omega_{i,i+1}/(\zeta-1)\,\}\,\Psi_0(\zeta)$$

Here we put $\Phi(z_i, z_{i+1}) = z_{i+1}^{\Delta}\,\Psi_0(\zeta)$, $\zeta = z_i/z_{i+1}$.
In our case this is essentially the Gauss hypergeometric differential equation and by means of the classical methods we can compute the

matrix relating the solution Ψ_0 normalized at 0 and the solution Ψ_∞ normalized at the infinity. This method enables us to express the normalizing factor $\alpha_\hbar(\lambda)$ appiearing in the previous section by means of the Γ functions in the following way. In the situation of Fig.4 we define $\gamma_i(\hbar)$ to be

$$\gamma_i(\hbar) = \Gamma(\lambda d)/\sqrt{\Gamma(\lambda(d-1))\ \Gamma(\lambda(d+1))}$$

If there is no such $\Lambda_i' \neq \Lambda_i$, we put $\gamma_i(\hbar)=1$. Then the gamma factor $\alpha_\hbar(\lambda)$ is given by the product $\gamma_1(\hbar) \ldots \gamma_n(\hbar)$.

This principle can be also applied to higher representations of $\mathfrak{g}$. We have a formula analogous to (3.9) where W is computed from the R-matrix associated with higher representaitions. Let us note that the linear representations of the braid groups defined by this R-matrix were used by Akutsu and Wadati [1] and Murakami [17] to construct invariants of links.

5. ALGEBRAS FACTORING THROUGH THE MONODROMY. We suppose that $\mathfrak{g}$ is a non-exceptional simple Lie algebra and $\rho : \mathfrak{g} \longrightarrow \mathrm{End}(V)$ its vector representation. In the case of $\mathfrak{g}$ is of type A the monodromy representation φ is equivalent to the higher order Temperley-Lieb representation and it factors through the Iwahori's Hecke algebra. In the other cases φ factors through a speciali-zation of the algebra with two parameter $\mathcal{C}_n(\alpha,\beta)$ discovered by Birman-Wenzl [3] and Murakami [18]. These algebras are denoted by $\mathcal{C}_n(\mathfrak{g},q)$ and may be considered to be a q-analogue of Brauer's centra-lizer algebras.

The following Markov trace was related to invariants of links by Jones [10] and Turaev [22].

(5.1) $\tau(x) = \chi^{-n} \operatorname{Tr}((q^{-\delta}|V)^{\otimes n}.\varphi(x))$ for $x \in B_n$

where δ is the half sum of the positive coroots and χ is $\operatorname{Trace}(q^{-\delta}|V)$. The above τ gives a functional on $\mathscr{C}_\infty(g,q)$.

We will see in the next section that the case q is a root of unity is important from the viewpoint of the conformal field theory. In this case the algebra $\mathscr{C}_n(g,q)$ is not semi-simple but the above Markov trace gives us a method to construct its semi-simple quotient. Let us suppose that $q = \exp(\pi i/(\ell+g))$ where ℓ is a positive integer called a level and g is the corresponding dual Coxeter number (see [11]). We consider

(5.2) $J_n = \{ x \in \mathscr{C}_n(g,q) \; ; \; \tau(xy) = 0 \text{ for any } y \in \mathscr{C}_n(g,q) \}$

Then it turns out that the quotient algebra $\overline{\mathscr{C}}_n = \mathscr{C}_n(g,q)/J_n$ is semi-simple. The irreducible representations of this algebra are described in the following way. Let $\mathscr{P}_\ell(\Lambda)$ be the subset of $\mathscr{P}(\Lambda)$ consisting of $\hbar = (\Lambda_0, \ldots, \Lambda_n)$ such that $\langle\Lambda_i,\theta\rangle \leq \ell$, for any i, where θ denotes the highest root and the Cartan-Killing form is normalized as $\langle\theta,\theta\rangle = 2$. For $\hbar$, $\hbar' \in \mathscr{P}_\ell(\Lambda)$, we put

(5.3) $w_{\hbar'\hbar} = \lim_{q\to\xi} W\begin{pmatrix} \Lambda_{i-1} & \Lambda'_i \\ \Lambda_i & \Lambda_{i+1} \end{pmatrix}$

where $\xi = \exp(\pi i/(\ell+g))$. It turns out that the above limit is a non-zero finite number. Then the representations of B_n given by

(5.4) $\sigma_i \cdot e_{\mathfrak{h}} = \sum_{\mathfrak{h}'} w_{\mathfrak{h}'\mathfrak{h}}\, e_{\mathfrak{h}'}$

give all irreducible representations of $\overline{\mathscr{C}}_n$. The above construction corresponds to the *restricted model* in the terminology of the solvable lattice models (see [8]).

Moreover, we can show that the Markov trace τ defines a positive definite bilinear form on the algebra $\overline{\mathscr{C}}_n$.

example. We illustrate some examles of the decomposition of the algebra $\overline{\mathscr{C}}_n$ in the following figure.

Fig.5 $g=sl(2,\mathbb{C})$, $\ell=2$

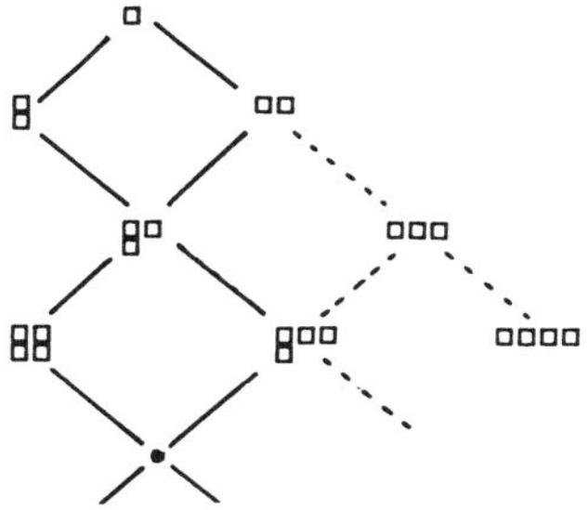

6. MONODROMY OF n-POINT FUNCTIONS. We discuss how the framework described in the previous sections can be applied to illustrate the monodromy properties of the n-point functions in the conformal field theory on the Riemann sphere with gauge symmetry of affine Lie algebras. We refer to [12] and [21] for the operator formalism in this theory, which we shall review briefly.

Let $g^{\wedge}$ denote the affine Lie algebra associated with g, which is defined to be the canonical central extension of the loop algebra $g\otimes\mathbb{C}[t, t^{-1}]$. Starting from a finite dimensional irreducible g-module V whose highest weight Λ satisfies $\langle\Lambda,\theta\rangle\leq\ell$ it is known by Kac [11] that we can associate an irreducible $g^{\wedge}$-module generated by

$$X_1(-n_1)\ X_2(-n_2)\ \dots\ X_k(-n_k).\ v\ ,\quad X_i\in g\ ,\ n_i>0\ ,\quad 1\leq i\leq k\ ,$$

for $v\in V_\Lambda$ on which the central element $\hat{c}$ of $g^{\wedge}$ acts as $\ell\times id$. Here $X(n)$ stands for $X\otimes t^n$. This is called the *integrable highest weight module of level* ℓ with the highest weight Λ and is denoted by $\mathcal{H}_\Lambda$. The *Sugawara form*

$$(6.1)\qquad L_n = \frac{1}{2(\ell+g)}\sum_\mu \sum_{k\in\mathbb{Z}} :I_\mu(-k)\ I_\mu(n+k):$$

satisfies the relation of the *Virasoro Lie algebra*

$$(6.2)\qquad [\,L_m\ ,\ L_n\,] = (m-n)\,L_{m+n} + \frac{m^3-m}{12}\,\delta_{m+n,0}\,c$$

with the central charge $c = \ell\dim g/(\ell+g)$. A *primary field* $\Phi(u,z)$ is an operator on $\oplus_{\langle\Lambda,\theta\rangle\leq\ell}\mathcal{H}_\Lambda$ depending linearly on $u\in V_\pi$ with some fixed π, depending holomorphically on $u\in\mathbb{C}-\{0\}$ and satisfying the following conditions.

$$(6.3)\qquad [\,L_m\ ,\ \Phi(u,z)\,] = z^m\{z(\partial/\partial z) + (m+1)\Delta_\pi\}\,\Phi(u,z)$$

with $\Delta_\pi = c(\pi)/(\ell+g)$,

(6.4) $[\, X(m),\ \Phi(u,z)\,] = z^m\, \Phi(Xu,z)$ for $X \in g$.

For simplicity we suppose that V_π is the vector representation. Assocoated with an ℓ-contraint fusion path $\not{h}=(\Lambda_0, \ldots, \Lambda_n) \in \mathcal{P}_\ell(\Lambda)$ Tsuchiya and Kanie constructed a vertex operator Φ_i for each i which is a primary field sending $\mathcal{H}_{\Lambda_i}$ to $\mathcal{H}_{\Lambda_{i+1}}$.

It was shown by Knizhnik and Zamolodchikov [12] that the n-point function

(6.5) $\phi_{\not{h}}(z) = \langle u|\Phi_n\Phi_{n-1} \cdots \Phi_1|vac\rangle \quad , \ u \in V_\Lambda^\dagger$

is a solution of the total differential $d\Phi = \omega\Phi$ where ω is defined in (1.3) with the parameter $\lambda = 1/(\ell+g)$. It turns out that the above n-point function is the normalized sulution associated with the fusion path $\not{h}$ in the sense of Section 3 . Now the monodromy of the above n-pont functions is described in the following way. We have a non-zero constant $\alpha_{\not{h}}$ such that the monodromy is expressed by using $w_{\not{h}',\not{h}}$ defined in (5.3) as

(6.6) $\sigma_i^*\ \alpha_{\not{h}}\phi_{\not{h}}(z) = \sum_{\not{h}'} w_{\not{h}',\not{h}}\ \alpha_{\not{h}'}\phi_{\not{h}'}(z)$

in the case g is non-exceptional. Consequently the monodromy of n-point functions factors through the semi-simple algebra $\overline{\mathcal{C}}_n$ defined in the previous section carrying a positive Markov trace. This algebra coincides with the Jones algebra of index $4\cos^2(\frac{1}{\ell+2})$ in the case $g = sl(2,\mathbb{C})$ (see [9],[21] and [23]).

References

[1] Y. Akutsu and M. Wadati, Knot invariants and the critical statistical systems, J. Phys. Soc. Japan 56 (1987), 839-842.

[2] J. Birman, Braids, links, and mapping class groups, Ann. Math. Stud. 82 (1974).

[3] J. Birman and H. Wenzl, Link polynomials and a new algebra, preprint, 1986.

[4] K.T. Chen, Iterated path integrals, Bull. Amer. Math. Soc. 83 (1977), 831-879.

[5] V.G. Drinfel'd, Quantum groups, Proc. of ICM, Berkley 1986, 798-820.

[6] M. Jimbo, A q-difference analogue of U(g) and Yang-Baxter equation, Lett. in Math. Phys. 10 (1985), 63-69.

[7] M. Jimbo, Quantum R matrix for the generalized Toda system, Comm. Math. Phys. 102 (1986), 537-547.

[8] M. Jimbo, T. Miwa and M. Okado, Solvable lattice models related to the vector representation of classsical simple Lie algebras, Commun. Math. Phys. 116 (1988) 507-525.

[9] V. Jones, Index of subfactors, Invent. Math. 72 (1983), 1-25.

[10] V. Jones, Hecke algebra representations of braid groups and link polynomials, Ann. of Math. 126 (1987), 335-388.

[11] V.G. Kac, Infinite dimensional Lie algebras, Progress in Math. 44, Birkhäuser (1983).

[12] V.G. Knizhnik and A.B. Zamolodchikov, Current algebra and Wess-Zumino models in two dimensions, Nucl. Phys. B247 (1984), 83-103.

[13] T. Kohno, Série de Poincaré-Koszul associée aux groupes de tresses pures, Invent. Math. 82 (1985), 57-75.

[14] T. Kohno, Linear representations of braid groups and classical Yang-Baxter equations, Contemp. Math. 78 (1988), "Braids, Santa Cruz, 1986", 339-363.

[15] T. Kohno, Monodromy representations of braid groups and Yang-Baxter equations, Ann. Inst. Fourier, 37, 4 (1987) 139-160.

[16] T. Kohno, Quantized universal enveloping algebras and monodromy of braid groups, preprint 1988.

[17] J. Murakami, On the Jones invariant of paralleled links and linear representations of braid groups, preprint, (1986).

[18] J. Murakami, The Kauffman polynomial of links and representation theory, Osaka J. Math. 24 (1987), 745-758.

[19] N.Y. Reshetikhin, Quantized universal enveloping algebras, the Yang-Baxter equation and invariants of links I,II, LOMI preprint, 1987.

[20] D. Sullivan, Infinitesimal computations in topology, Publ. IHES 47 (1977), 269-331.

[21] A. Tsuchiya and Y. Kanie, Vertex operators in two dimensional conformal field theory on $\mathbb{P}^1$ and monodromy representations of braid groups, Advanced Studies in Pure Math. 16 (1988) 297-372.

[22] V.G. Turaev, The Yang-Baxter equation and invariants of links, Invent. math. 92 (1988), 527-553.

[23] H. Wenzl, Representations of Hecke algebras and subfactors, Thesis, Univ. of Pensylvenia (1985).

The Yang-Baxter Relation: A New Tool for Knot Theory

Yasuhiro AKUTSU, Tetsuo DEGUCHI* and Miki WADATI*

Institute of Physics, Kanagawa University, Rokkakubashi, Kanagawa-ku, Yokohama 221, Japan
**Institute of Physics, College of Arts and Sciences, University of Tokyo, Komaba, Meguro-ku, Tokyo 153, Japan*

Abstract

We present a general theory for construction of link polynomial, topological invariant for knots and links, from an exactly solvable model satisfying the Yang-Baxter relation. First, we present a method to make braid group representation from the Yang-Baxter operator, the constituent of the diagonal-to-diagonal transfer matrix, for the model at criticality. Second, we construct the Markov trace whose existence is a consequence of the crossing symmetry or the second inversion relation satisfied by the solvable model. Third, the general theory is applied to various models, by which a list of new link polynomials are constructed. Lastly, some extensions of the theory are presented. We conclude that a new and powerful approach to knot theory based on the theory of exactly solvable models has been established.

1 Introduction

Recent development in the theory of quantum completely integrable systems provides us a unified treatment of various exactly solvable models in 1 + 1 dimensional field theory and in two-dimensional classical statistical mechanics[1,2]. The key point is that to each model we can associate a commuting family of transfer matrices which are the generators of an infinite number of conserved quantities. The commutability condition is called the *Yang-Baxter relation.*

It was two dacades ago when Yang-Baxter relation appeared in physics. It was the applicability condition of the *Bethe-ansatz* for one-dimensional δ-function gases and is the factorizability condition of many-body scattering matrices[3]. How could one imagine that the same theory provides a totally novel approach to knot theory? In this article we report this unexpected close connection between physics and mathematics. We present a general method to construct topological invariants for knots and links by applying the theory of exactly solvable models.

The Yang-Baxter relation takes several forms depending on the types of models under consideration. For the 1 + 1 dimensional field theory, the Yang-Baxter relation is the factorization condition[3,4] for the many-body S-matrices and is called the *factorization equation.* Let us denote the scattering amplitude for the process $(i,j) \to (k,l)$ by $S_{jl}^{ik}(u)$, where the indices i,j,k and l specify inner degrees of freedom such as charge, spin (z-component), flavor etc.. The parameter u, called spectral parameter, is the rapidity difference between the two colliding particles. The factorization equation reads as

$$\sum_{\alpha\beta\gamma} S_{\gamma r}^{\beta q}(v) S_{k\gamma}^{\alpha p}(u+v) S_{j\beta}^{i\alpha}(u) = \sum_{\alpha\beta\gamma} S_{\beta q}^{\alpha p}(u) S_{\gamma r}^{i\alpha}(u+v) S_{k\gamma}^{j\beta}(v) \quad (1.1)$$

which is schematically explained in Fig.1. For two-dimensional statistical mechanics, we have two types of models, the vertex models and the IRF (Interaction Round a Face) models[5] (Fig.2). As was pointed out by Zamolodchikov[6], any factorized S-matrix can be interpreted as the Boltzmann weight of a solvable vertex model where the inner degrees of freedom of the former correspond to the egde

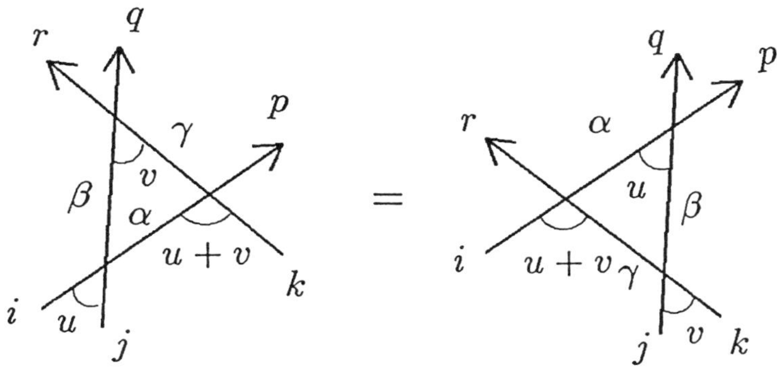

Figure 1: Factorization equation.

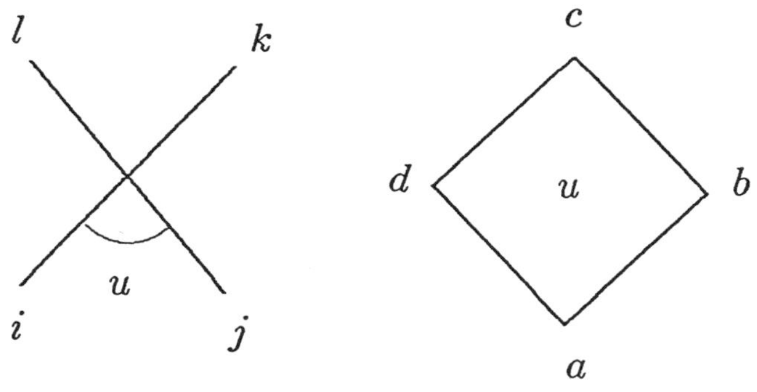

Figure 2: Left: Boltzmann weight $S_{jl}^{ik}(u)$ of vertex model. Right: Boltzmann weight $w(a,b,c,d;u)$ of IRF model.

variables of the latter. The factorization equation (1.1) is then the commutability condition of the transfer matrices of the vertex model. The IRF model is the one where the 'spin variables' are located on the sites of the square lattice. The Boltzmann weight is assigned to each spin configuration round a unit square (face). For the IRF models, the commutability condition is called the *star-triangle relation*[5] which reads as

$$\sum_c w(b,d,c,a;u)w(a,c,f,g;u+v)w(c,d,e,f;v)$$
$$= \sum_c w(a,b,c,g;v)w(b,d,e,c;u+v)w(c,e,f,g;u). \qquad (1.2)$$

Here $w(a,b,c,d;u)$ denotes the Boltzmann weight for the spin configuration (a,b,c,d) round a face.

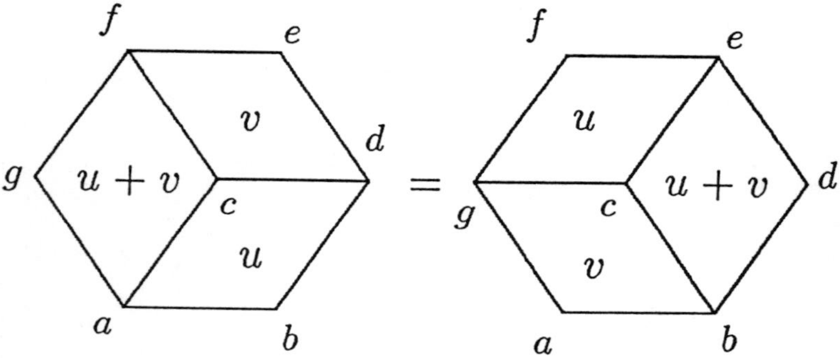

Figure 3: Star-triangle relation for the IRF model.

Some of the IRF models have their equivalent vertex models through the Wu-Kadanoff-Wegner transformation[7] but many have not. In this sence, IRF models form a larger class of solvable models than that of vertex models.

Before going into the knots and links, we should mention the *braid* and the *braid group*[8,9] which lie at the starting point of our study. Braids are formed when n points on a horizontal line are connected by n strings to n points on another horizontal line directly below the first n points. Trivial n-braid is a configuration where no cross between the strings is present (Fig.4).

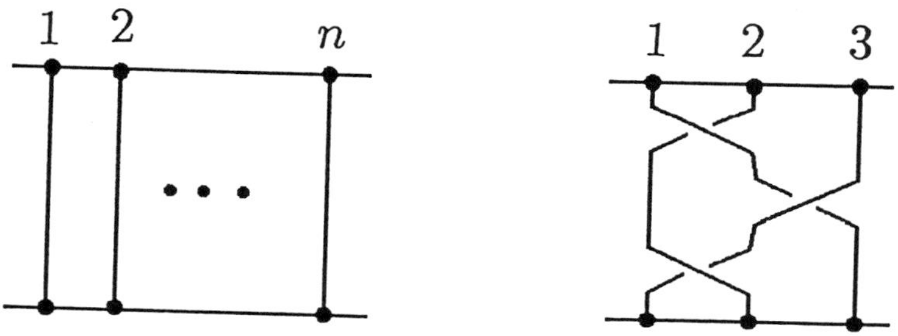

Figure 4: Example of braids: A trivial n-braid (left) and a non-trivial 3-braid.

A general n-braid is constructed from the trivial n-braid by successive applications of operations $b_i, i = 1, 2, \cdots, n-1$. The operation b_i and its inverse b_i^{-1} are best understood by the graphs (Fig.5).

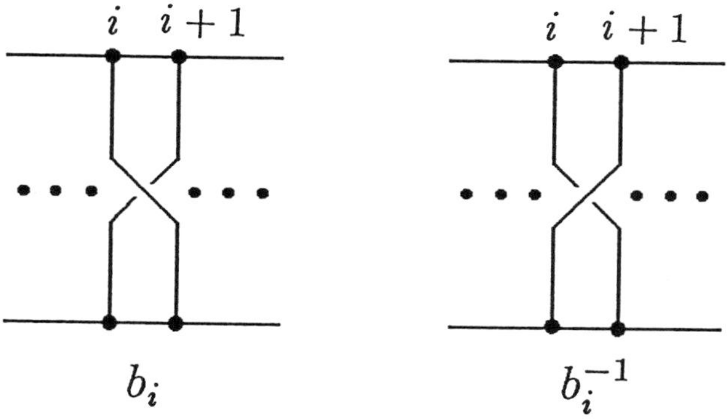

Figure 5: Elementary operations b_i and b_i^{-1}.

A set of generators, $b_1, b_2, \cdots, b_{n-1}$, defines the braid group B_n. By regarding the trivial n-braid as the identity operation in B_n, we can identify any element in B_n as an n-braid. To guarantee the topological equivalence between different expressions of a braid in terms of braid group elements, Artin[8] proved that the following conditions are necessary and sufficient (Fig.6):

$$b_i b_j = b_j b_i, \qquad |i - j| \geq 2,$$

$$b_i b_{i+1} b_i = b_{i+1} b_i b_{i+1}. \tag{1.3}$$

We call them *defining relations* of B_n.

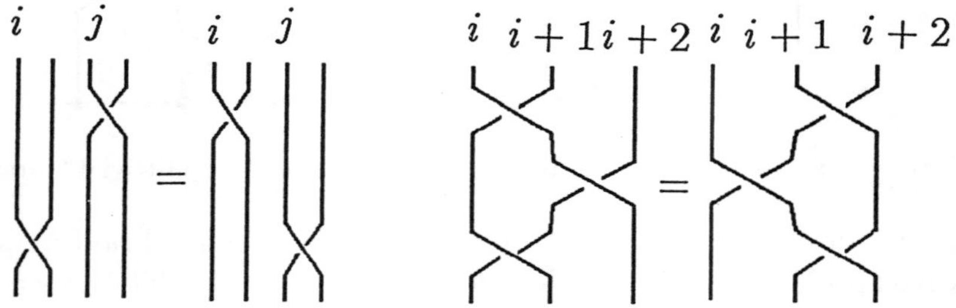

Figure 6: Defining relations of the braid group.

Then, each topologically equivalent class (isotopy class) of the braids is identified with an element in B_n. The braid goup appeared in physics as an explanation for the fractional statistics in two-dimensional quantum systems[10]. Recently, it also appears in the two-dimensional conformal field theory as the monodromy of the differential equation satisfied by the N-point functions[11]. Note here the graphical similarity between Fig. 1 and Fig. 6 (right). Moreover as will be given in §2, eq.(1.3) is, leaving the spectral parameter aside, nothing but the equation satisfied by the elementary diagonal-to-diagonal transfer matrices[5]. These similarities are the motivation of our study.

The braid group B_n has a significance for knot theory in the following sense. Given a braid, one may form a link by tying opposite ends (Fig.7).

What is important is that *the converse is true*: According to Alexander's theorem[12], *any* link is represented by a *closed braid*. This fact gives the braid group a fundamental role in the knot theory. One can, however, find that the representation of a link as a closed braid is highly non-unique: infinitely many braids give the same link when they are closed. Therefore, we need another important theorem

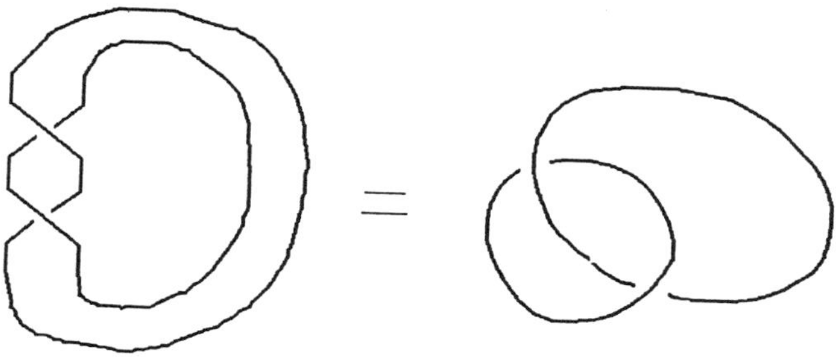

Figure 7: Example of a link as a closed braid.

due to A.A. Markov[13]. The equivalent braids expressing the same link are mutually transformed by successive applications of two types of operations, *type* I *and type* II *Markov moves* (Fig.8):

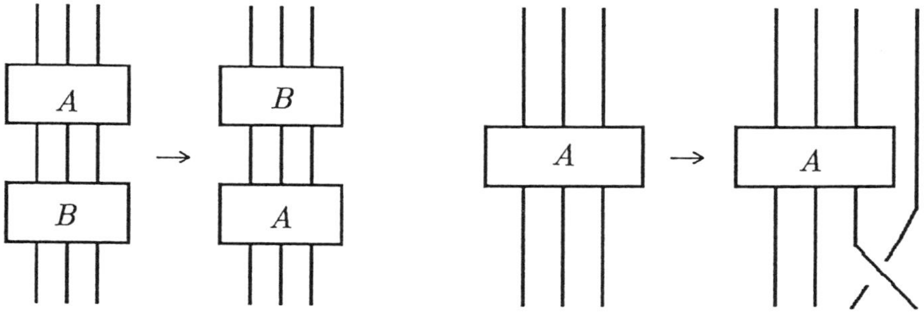

Figure 8: Markov moves I (left) and II (right).

$$\text{I.} \qquad AB \to BA, \qquad (A, B \in B_n), \tag{1.4}$$
$$\text{II.} \qquad A \to Ab_n, A \to Ab_n^{-1}, \quad (A \in B_n, b_n \in B_{n+1}). \tag{1.5}$$

The above consideration leads to a strategy to construct topological invariant of knots and links called *link polynomial.* We first make explicit representation of B_n, and then find a quantity defined on B_n which is invariant under the Markov move hence is a link invariant. The present article is based on this line of thinking.

The outline of this article is as follows. In §2 we present a general method to construct a braid group representation from a solvable model. Also, basic properties and symmetries of solvable models are summarized. In §3 we give the Markov trace and the link polynomial associated with the braid group representation. It is shown that the crossing symmetry assures the existence of the Markov trace. In §4, the general theory described in §2 and §3 is applied to various models. In §5, extensions of our theory are presented. The last section is devoted to summary and discussion.

2 Yang-Baxter Algebra and the Braid Group

2.1 Symmetries and basic properties of solvable models

Let us summarize here basic properties posessed by the models satisfying the Yang-Baxter relation.

(a) **overall normalization**

It is simple to see that the Yang-Baxter relations (1.1) and (1.2) are invariant under transformations

$$\begin{aligned} S_{jl}^{ik}(u) &\rightarrow N(u)\cdot S_{jl}^{ik}(u), \\ w(a,b,c,d;u) &\rightarrow N(u)\cdot w(a,b,c,d;u), \end{aligned} \tag{2.1}$$

where $N(u)$ is arbitrary and index-independent. We can freely choose overall normalization of the Boltzmann weights.

(b) **standard initial condition**

$$\begin{aligned} S_{jl}^{ik}(0) &= C\cdot\delta_{il}\delta_{jk}, \\ w(a,b,c,d;0) &= C\cdot\delta_{ac}, \end{aligned} \tag{2.2}$$

where C is a constant depending on the overall normalization of the Boltzmann weights.

(c) **unitarity condition or the first inversion relation**

$$\sum_{p,q} S_{pk}^{ql}(-u) S_{jq}^{ip}(u) = \rho(u)\rho(-u)\delta_{ik}\delta_{jl},$$
$$\sum_{e} w(e,c,d,a;-u) w(b,c,e,a;u) = \rho(u)\rho(-u)\delta_{bd}. \tag{2.3}$$

The function $\rho(u)$ is sometimes called *unitarity function.* We can always choose normalization where $\rho(u) \equiv 1$. Putting $u = 0$ in (2.3) and recalling (2.2), we have $C^2 = \rho(0)^2$. We choose a normalization of $\rho(u)$ such that

$$\rho(0) = C. \tag{2.4}$$

(d) **crossing symmetry**

$$S_{jl}^{ik}(u) = [\frac{r(i)r(l)}{r(j)r(k)}]^{1/2} F(u) S_{\bar{k}\bar{i}}^{jl}(\lambda - u),$$
$$w(a,b,c,d;u) = [\frac{\psi(a)\psi(c)}{\psi(d)\psi(b)}]^{1/2} F(u) \cdot w(d,a,b,c;\lambda - u). \tag{2.5}$$

In the above, λ is called crossing point of the spectral parameter or *crossing parameter* for short. For the vertex model, the *bar*-operation on the indices is usually interpreted as 'charge conjugation'

$$\bar{i} = -i. \tag{2.6}$$

Different interpretations of the bar-operation are also possible. We call the factors $\{r(i)\}$ and $\{\psi(a)\}$ *crossing multipliers.* In relation with (2.6),

$$r(\bar{i}) = \frac{1}{r(i)} \tag{2.7}$$

is assumed. The function $F(u)$ depends on the overall normalization but always satisfies

$$F(u)F(\lambda - u) = 1. \tag{2.8}$$

We can normalize the Boltzmann weights so that $F(u) \equiv 1$.

(e) **second inversion relation**

$$\sum_{p,q} S_{ql}^{kp}(\lambda+u)S_{pi}^{jq}(\lambda-u)\cdot\frac{r(q)r(p)}{r(j)r(k)} = \rho(u)\rho(-u)\delta_{ik}\delta_{jl},$$

$$\sum_{e} w(c,b,a,e;\lambda-u)w(a,b,c,e;\lambda+u)\cdot\frac{\psi(e)\psi(b)}{\psi(a)\psi(c)} = \rho(u)\rho(-u)\delta_{bd}. \tag{2.9}$$

For models with crossing symmetry, the second inversion relation is equivalent to the unitarity relation. There are, however, solvable models satisfying the Yang-Baxter relation *without* the crossing symmetry, whose example is the $A_m^{(1)}$ IRF model (see §4). Nevertheless, such models satisfy the second inversion relation. The factors $r(\cdot)$ and $\psi(\cdot)$ appearing in (2.9), for those models, are also called crossing multipliers.

(f) **CPT invariances or reflection symmetry**

$$\begin{aligned} S_{jl}^{ik}(u) &= S_{\bar{j}\bar{l}}^{\bar{i}\bar{k}}(u) \quad \text{(C invariance)} \\ &= S_{ik}^{jl}(u) \quad \text{(P invariance)} \\ &= S_{lj}^{ki}(u) \quad \text{(T invariance)}, \end{aligned} \tag{2.10}$$

$$\begin{aligned} w(a,b,c,d;u) &= w(a,d,c,b;u) \\ &= w(c,b,a,d;u) \\ &= w(c,d,a,b;u). \end{aligned} \tag{2.11}$$

(g) **charge conservation condition**

$$S_{jl}^{ik}(u) = 0 \qquad \text{unless } i+j=k+l. \tag{2.12}$$

Not all of the solvable vertex models satisfy the charge conservation condition. However, as will be seen later, the existence of nontrivial link polynomial associated with the model requires the charge conservation condition. The IRF analogue of this condition is the single-valuedness of the spin-value round a face, which is automatically satisfied for all IRF models.

(h) **symmetry-breaking transformations**

(vertex model)

$$S_{jl}^{ik}(u) \to \tilde{S}_{jl}^{ik}(u) = \alpha_{ij,kl}(u) \cdot \beta_{ij,kl} \cdot \gamma_{ij,kl} \cdot S_{jl}^{ik}(u), \tag{2.13}$$

where each constituent transformation is given by

$$\alpha_{ij,kl}(u) = \exp[\mu(k-i-l+j)u], \tag{2.14}$$

$$\beta_{ij,kl} = \exp[\nu(l-i-k+j)], \tag{2.15}$$

$$\gamma_{ij,kl} = \exp[\omega(kl-ij)]. \tag{2.16}$$

In the above, μ, ν and ω are free parameters. From a symmetric model satisfying the Yang-Baxter relation *and* the charge conservation condition, we can produce asymmetric model with Boltzmann weights $\{\tilde{S}_{jl}^{ik}(u)\}$ which also satisfy the Yang-Baxter relation, by applying the symmetry-breaking transformation.

(IRF model)

$$\begin{aligned} w(a,b,c,d;u) &\to \tilde{w}(a,b,c,d;u) \\ &= \alpha_{abcd}(u) \cdot \beta_{abcd} \cdot \gamma_{abcd} \cdot w(a,b,c,d;u), \end{aligned} \tag{2.17}$$

where

$$\alpha_{abcd}(u) = \exp\{[-p(a)+p(b)-p(c)+p(d)]u\}, \tag{2.18}$$

$$\beta_{abcd} = \frac{q(a)}{q(c)}, \tag{2.19}$$

$$\gamma_{abcd} = \exp\{\omega[(b-c)(c-d)-(a-d)(b-a)]\}, \tag{2.20}$$

with arbitrary functions $p(\cdot)$ and $q(\cdot)$, and arbitrary parameter ω. Unlike the case of vertex model, the symmetry-breaking transformation is applicable to *any* IRF model because the 'charge conservation condition' is trivial for IRF models. We should note that the asymmetrized weights also satisfy both of the first- and second-inversion relation with the unitarity function $\rho(u)$ left unchanged. The symmetry-breaking transformations, both for vertex models and for IRF models, are sometimes useful in constructing an interesting braid group representation from a given model.

2.2 Yang-Baxter operator

The Boltzmann weights satisfying (1.1) or (1.2) define a sequence of operators $X_1(u), X_2(u), \cdots, X_i(u)$, ..., which satisfy the relations:

$$X_i(u)X_j(v) = X_j(v)X_i(u), \qquad \text{for} \quad |i-j| \geq 2,$$
$$X_i(u)X_{i+1}(u+v)X_i(v) = X_{i+1}(v)X_i(u+v)X_{i+1}(u). \qquad (2.21)$$

The operators $\{X_i(u)\}$, which we call *Yang-Baxter operators*, are constituents of the diagonal-to-diagonal transfer matrix[5]. The relation (2.21) is the Yang-Baxter relation (1.1) or (1.2) in disguise. For vertex models, $\{X_i(u)\}$ is expressed as a sum of matrix tensor products:

$$X_i(u) = \sum_{klmp} \quad S_{lp}^{km}(u) \cdot I^{(1)} \otimes I^{(2)} \otimes \cdots$$
$$\otimes E_{pk}^{(i)} \otimes E_{ml}^{(i+1)} \otimes I^{(i+2)} \otimes \cdots \otimes I^{(n)}, \qquad (2.22)$$

where $I^{(j)}$ is the identity acting at the j-th position and E_{pk} is a matrix such that $(E_{jk})_{pq} = \delta_{jp}\delta_{kq}$. For IRF models, we have

$$[X_i(u)]_{l_0 l_1 l_2 \cdots l_n}^{l'_0 l'_1 l'_2 \cdots l'_n} = \delta_{l_0 l'_0}\delta_{l_1 l'_1} \cdots \delta_{l_{i-1} l'_{i-1}} \times$$
$$\times w(l_i, l_{i+1}, l'_i, l_{i-1}; u) \cdot \delta_{l_{i+1} l'_{i+1}} \cdots \delta_{l_n l'_n}, \qquad (2.23)$$

where n corresponds to the horizontal size of the system (Fig.9).

2.3 Braid group representation

We can see close similarity between (1.3) and (2.21). The only difference is that in (2.21) we have spectral parameters as the arguments. If we can get rid of the spectral parameters, we then have a representation of $\{b_i\}$. A way to do so is to set all the spectral parameters equal:

$$u = u + v = v. \qquad (2.24)$$

An obvious solution is

$$u = v = 0. \qquad (2.25)$$

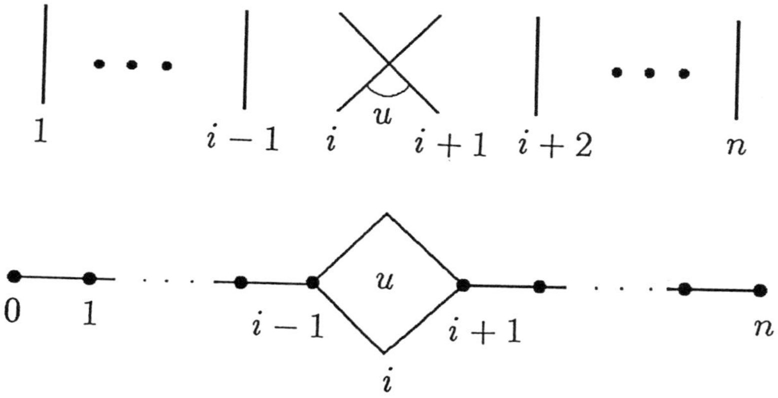

Figure 9: The Yang-Baxter operator $X_i(u)$ for vertex model (upper) and for IRF model (lower).

The resulting representation is, however, trivial because the braid group generator b_i given by an identification

$$b_i \to X_i(0) \tag{2.26}$$

is a constant multiple of unity. In fact, due to the standard initial condition we have

$$X_i(0) = C \cdot I, \tag{2.27}$$

where C is a constant and I is the identity operator. It is important to notice that (2.24) has another 'possible solution'

$$u = u + v = v = \infty. \tag{2.28}$$

This solution is meaningful if, after suitable normalization, the following limit exists:

$$\lim_{u\to\infty} X_i(u) \equiv x_i. \tag{2.29}$$

It is simple to see that, by taking the $u \to \infty$ limit of both sides in (2.21), these operators $\{x_1, x_2, ..., x_{n-1}\}$ satisfy the defining relations (1.3) of B_n. We may then have a non-trivial representation of B_n through the identification

$$b_i = x_i = \lim_{u\to\infty} X_i(u). \tag{2.30}$$

We have shown[14]~[22] that there actually exist solvable models which have the limit (2.29). The 6-vertex model[23] is one of the simplest examples. These models have a common feature that they are parametrized by *trigonometric* or *hyperbolic* functions. For such models, we can always choose a direction in the complex u-plane along which $u \to \infty$ gives a definite limit. Recall that a model with trigonometric/hyperbolic parametrization corresponds to the *critical limit* of a more 'general' model (e.g., the 8-vertex model[24] for the 6-vertex model) with elliptic parametrization which does not have the well-defined limit (2.29) due to the double (quasi) periodicities of elliptic functions. We can, therefore, say that the braid group have a close connection with a *critical statistical system.* This considerations naturally explain why the Temperley-Lieb algebra[25] (see §2.4) which describes the 6-vertex model and the critical (self-dual) Potts model, appeared in the original construction of the Jones polynomial[26].

Given a solvable model parametrized by trigonometric or hyperbolic functions, we must fix the overall normalization of the Boltzmann weights so that (2.29) does not diverge to infinity nor shrink to zero but has actually a finite limit. A convenient choice is the one where the unitarity function $\rho(u)$ is unity. As such we use the Boltzmann weights normalized by the unitarity function. In terms of the Yang-Baxter operator, this means the use of a renormalized operator

$$\tilde{X}_i(u) = X_i(u)/\rho(u) \tag{2.31}$$

instead of $X_i(u)$ in the formula (2.29). By taking account of the unitarity condition (2.3) we then have the following representation of generators[14,15,17]:

$$b_i = \lim_{u\to\infty} \tilde{X}_i(u) = \lim_{u\to\infty} X_i(u)/\rho(u), \tag{2.32}$$

$$b_i^{-1} = \lim_{u\to\infty} \tilde{X}_i(-u) = \lim_{u\to\infty} X_i(-u)/\rho(-u). \tag{2.33}$$

It should be remarked that making the inverse operator is equivalent to the substitution $u \to -u$. The identity operator $I \in B_n$ is also expressed in terms of the Yang-Baxter operator. From (2.2) and (2.4) we have

$$I = \tilde{X}_i(0) = [X_i(u)/\rho(u)]_{u=0} \qquad \text{for all } i. \tag{2.34}$$

Before proceed further, we give some comments on the formula (2.29) when applied to the IRF models. In most of the solvable IRF models, there imposed constraints on the spin configurations round a face. Hence $\{X_i(u)\}$ or $\{b_i\}$ for IRF models should be regarded as operators acting on a 'constrained Hilbert space'. Accordingly, even the identity operator $I \in B_n$

$$[I]_{l_0 l_1 l_2 \cdots l_n}^{l'_0 l'_1 l'_2 \cdots l'_n} = \delta_{l_0 l'_0} \delta_{l_1 l'_1} \cdots \delta_{l_{n-1} l'_{n-1}} \delta_{l_n l'_n} \tag{2.35}$$

is subject to the constraint on the neighboring indices. The constraint depends on the model. In the case of eight-vertex SOS model (see §4.2), only the configurations with $|l_i - l_{i-1}| = 1 (i = 1, 2, ..., n)$ are allowed. We should note that the constraint can be manifestly taken into account by the identification

$$I = \tilde{X}_1(0) \cdot \tilde{X}_2(0) \cdots \tilde{X}_n(0), \tag{2.36}$$

which is, in effect, equivalent to (2.34). Similarly, we can write

$$b_i = I \cdot [\lim_{u \to \infty} \tilde{X}_i(u)] = [\lim_{u \to \infty} \tilde{X}_i(u)] \cdot I, \tag{2.37}$$

with I given by (2.36). Here, in the last equality, we have used the commutability

$$[X_i(u), X_j(0)] = 0, \qquad \text{for any } i \text{ and } j. \tag{2.38}$$

2.4 Crossing symmetry: Temperley-Lieb algebra and monoid diagram

For models with the crossing symmetry, the Yang-Baxter operators $X_i(u)$ constructed from them become the Temperley-Lieb operators[25] at the crossing point $u = \lambda$. For definiteness, let us consider the case of vertex model or factorized S-matrix. Assume that the Boltzmann weights are normalized so that they satisfy the standard initial condition (2.2) with $C = 1$ and have the crossing symmetry (2.5) with $F(u) = 1$. At $u = \lambda$, we have

$$S_{jl}^{ik}(\lambda) = \delta_{j\bar{i}}\delta_{\bar{k}l} r(i) r(l). \tag{2.39}$$

Putting this into (2.22) and writing

$$U_i \equiv X_i(\lambda), \tag{2.40}$$

we can verify the following relations:

$$U_i U_j = U_j U_i, \quad \text{for} \quad |i-j| \geq 2. \tag{2.41}$$

$$U_i U_{i\pm 1} U_i = U_i, \tag{2.42}$$

$$U_i^2 = \sqrt{q} U_i, \tag{2.43}$$

where

$$\sqrt{q} = \sum_k r(k)^2. \tag{2.44}$$

Relations (2.41)-(2.43) are the defining relations of the Temperley-Lieb algebra. From the Temperley-Lieb algebra we can construct a representation of B_n. Let us introduce a parameter t by

$$q = t + \frac{1}{t} + 2. \tag{2.45}$$

Using (2.41)-(2.43), we can show that the operators $\{b_i\}$ given by

$$b_i = 1 - \sqrt{t} U_i, \qquad (i = 1, 2, \ldots, n-1) \tag{2.46}$$

satisfy the defining relations of B_n. This representation of B_n is essentially the one utilized by Jones[26].

We can associate the Temperley-Lieb operator U_i with the *monoid diagram*[27] (Fig.10). We then sometimes call the Temperley-Lieb operator as monoid operator.

In terms of the monoid diagram, the defining relation of the Temperley-Lieb algebra are schematically shown in Fig.11.

We have the following physical interpretation of monoids[20]. The crossing symmetry can be considered as a transformation from the scattering channel to the crossing channel. This corresponds to the observation of the scattering diagram from a 90°-rotated direction (Fig.12). Then, the scattering with $u = \lambda$ in the scattering channel

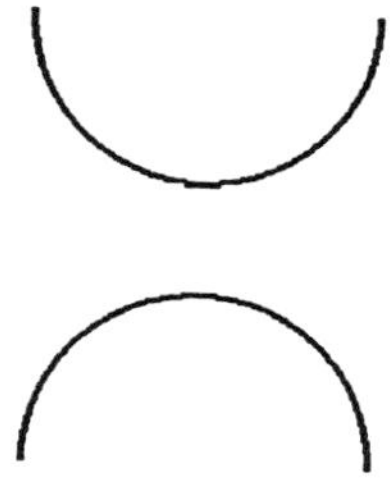

Figure 10: Monoid diagram.

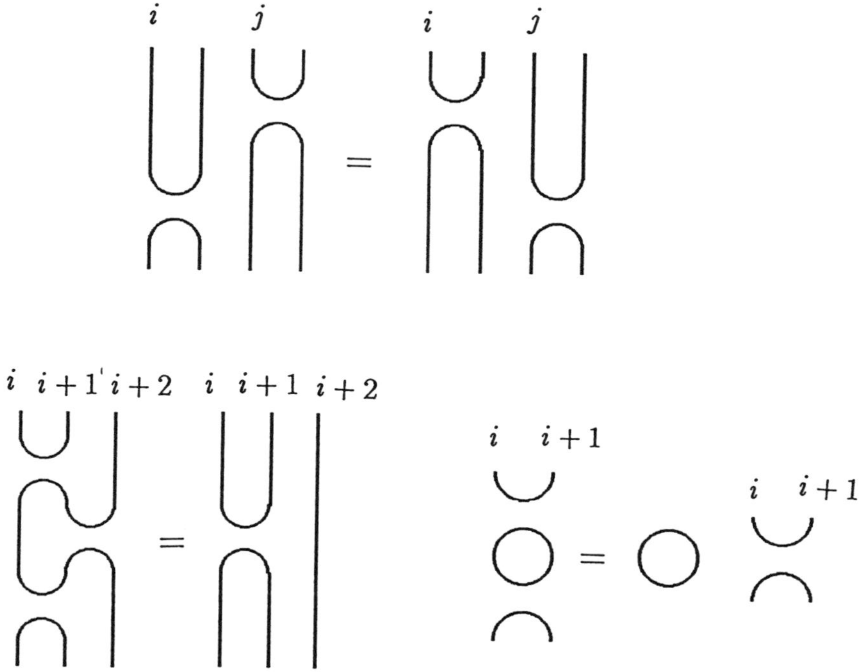

Figure 11: Temperley-Lieb algebra.

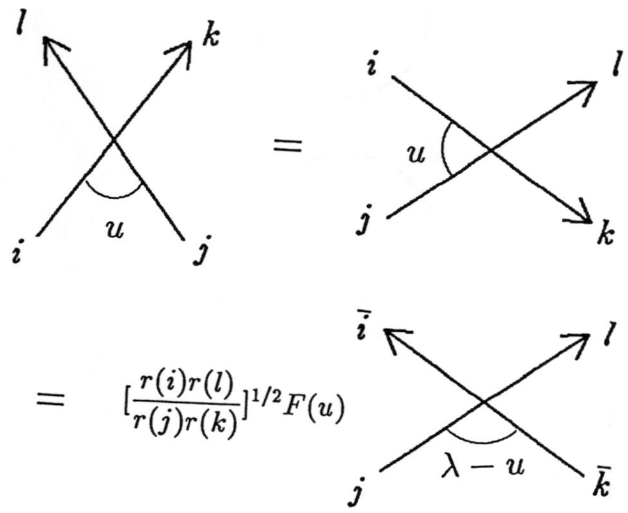

Figure 12: Crossing transformation.

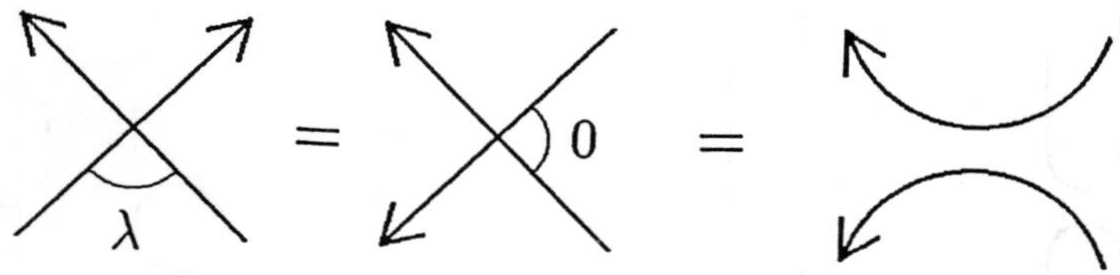

Figure 13: Crossing transformation and the monoid operator.

is regarded as the scattering with $u = 0$ in the crossing channel (Fig. 13).

To summarize, while the braid operator corresponds to S-matrix at $u = \infty$ (high energy limit), the monoid operator corresponds to S-matrix at $u = \lambda$ describing creation-annihilation of a particle-antiparticle pair. The monoid diagram and monoid operator are helpful for algebraic formulation of link polynomials[20] (see §5.2).

3 Solvable Models and Markov Traces

3.1 Markov trace and link polynomial

As was briefly mentioned in §1, construction of link polynomial reduces to finding of Markov-move invariant quantity defined on B_n. For this purpose, let us introduce a functional $\phi(\cdot)(\phi : B_n \to \mathbf{C})$ defined on B_n. We call $\phi(\cdot)$ *Markov trace* if it has the following properties

type I *Markov property*

$$\phi(AB) = \phi(BA), \quad \text{for} \quad A, B \in B_n, \tag{3.1}$$

type II *Markov property*

$$\begin{aligned} &\phi(Ab_n) = \tau \cdot \phi(A), \\ &\phi(Ab_n^{-1}) = \bar{\tau} \cdot \phi(A), \quad \text{for} \quad A \in B_n, b_n \in B_{n+1}. \end{aligned} \tag{3.2}$$

The constants τ and $\bar{\tau}$ are given by

$$\tau = \phi(b_i), \quad \bar{\tau} = \phi(b_i^{-1}) \quad (\text{for all } i). \tag{3.3}$$

With the Markov trace $\phi(\cdot)$, link polynomial $\alpha(\cdot)$ is given by a general formula[14,15]

$$\alpha(A) = (\frac{1}{\tau\bar{\tau}})^{\frac{n-1}{2}} \cdot (\frac{\bar{\tau}}{\tau})^{\frac{e(A)}{2}} \cdot \phi(A) \qquad (A \in B_n), \tag{3.4}$$

where $e(A)$ is the exponent sum of the generators $\{b_i\}$ appearing in the braid A (for example, if $A = b_2^3 b_1^{-2}$ then $e(A) = 3 - 2 = 1$).

The quantity $\alpha(\cdot)$ is indeed a link polynomial because it is invariant under the the Markov moves:

$$\alpha(AB) = \alpha(BA) \quad (A, B \in B_n), \tag{3.5}$$

$$\alpha(Ab_n) = \alpha(Ab_n^{-1}) = \alpha(A) \quad (A \in B_n). \tag{3.6}$$

The property (3.5) directly follows from (3.1). The property (3.6) is easily verified as follows:

$$\begin{aligned}
\alpha(Ab_n) &= (\tau \cdot \bar{\tau})^{-(n+1-1)/2} (\frac{\bar{\tau}}{\tau})^{e(Ab_n)/2} \cdot \phi(Ab_n) \\
&= (\tau \cdot \bar{\tau})^{-(n+1-1)/2} (\frac{\bar{\tau}}{\tau})^{e(Ab_n)/2} \cdot \tau\phi(A) \\
&= [(\tau \cdot \bar{\tau})^{-1/2} (\frac{\bar{\tau}}{\tau})^{1/2} \cdot \tau](\tau \cdot \bar{\tau})^{-(n-1)/2} (\frac{\bar{\tau}}{\tau})^{e(A)/2} \cdot \phi(A) \\
&= (\tau \cdot \bar{\tau})^{-(n-1)/2} (\frac{\bar{\tau}}{\tau})^{e(A)/2} \cdot \phi(A) = \alpha(A).
\end{aligned} \tag{3.7}$$

Note that (3.2) and $e(Ab_n) = e(A) + 1$ have been used in the second and the third lines, respectively.

3.2 Crossing symmetry and extended Markov property

Let us show that the existence of the Markov trace mentioned above is a consequence of the *crossing symmetry* of the Bolztmann weight. Recall that the crossing symmetry with $F(u) = 1$ reads as follows: (vertex model)

$$S_{jl}^{ik}(u) = [\frac{r(i)r(l)}{r(j)r(k)}]^{1/2} \cdot S_{\bar{k}\bar{i}}^{jl}(\lambda - u), \tag{3.8}$$

(IRF model)

$$w(a, b, c, d; u) = [\frac{\psi(a)\psi(c)}{\psi(d)\psi(b)}]^{1/2} \cdot w(d, a, b, c; \lambda - u). \tag{3.9}$$

Using the crossing multipliers $r(\cdot)$ and $\psi(\cdot)$ we can show that the following property which we call *extended Markov property* holds[19]: (vertex model)

$$\sum_k S_{kl}^{lk}(u) \cdot r(k)^2 = \tilde{H}(u) \quad \text{(independent of } k\text{)} \tag{3.10}$$

(IRF model)

$$\sum_b w(a,b,a,c;u)\frac{\psi(b)}{\psi(a)} = \tilde{H}(u) \quad \text{(independent of } a \text{ and } c\text{)}. \tag{3.11}$$

Under the assumption $[X_i(u), X_i(\lambda)] = 0$, (3.10) is proved as follows. The standard initial condition and the crossing symmetry give[20]

$$S_{jl}^{ik}(\lambda) = r(i)r(l)\delta_{\bar{k}l}\delta_{j\bar{i}}. \tag{3.12}$$

Let us introduce a matrix E such that

$$(E)_{pq,ij} = S_{jp}^{iq}(\lambda) = r(p)r(i)\delta_{\bar{q}p}\delta_{\bar{j}i} \tag{3.13}$$

which corresponds to the local transfer matrix at $u = \lambda$. It is important to note that E is essentially a one-dimensional (un-normalized) projector:

$$E = |e><e|, \tag{3.14}$$

where $|e>$ (resp. $<e|$) is a column (resp. row) vector with double-index entries

$$(|e>)_{ij} = r(i)\delta_{j\bar{i}}. \tag{3.15}$$

From the property of the one-dimensional projector we have, for any matrix A commuting with E,

$$EA = \lambda_a E, \tag{3.16}$$

with λ_a being a constant given by

$$\lambda_a = <e|A|e> / <e|e>. \tag{3.17}$$

For $A = X(\lambda - u)$ where $[X(\lambda - u)]_{pq,ij} = S_{jp}^{iq}(\lambda - u)$, (3.16) reads

$$EX(\lambda - u) = \tilde{H}(u)E, \tag{3.18}$$

with

$$\tilde{H}(u) = <e|X(\lambda - u)|e> / <e|e>. \tag{3.19}$$

The LHS of (3.18) is explicitly written as

$$\begin{aligned}
[EX(\lambda-u)]_{pq,ij} &= \sum_{\alpha,\beta} r(p)r(\alpha)\delta_{\bar{q}p}\delta_{\bar{\beta}\alpha}[X(\lambda-u)]_{\alpha\beta,ij} \\
&= r(p)r(i)\delta_{\bar{q}p}\sum_{\alpha} r(\alpha)[X(\lambda-u)]_{\alpha\bar{\alpha},ij} \\
&= r(p)r(i)\delta_{\bar{q}p}\sum_{\alpha} r(\alpha)^2[\frac{r(i)}{r(j)}]^{1/2}[X(u)]_{\bar{i}\alpha,j\alpha} \\
&= \delta_{\bar{q}p}\delta_{j\bar{i}}[\frac{r(p)r(i)}{r(q)r(j)}]^{1/2}\sum_{\alpha} r(\alpha)^2[X(u)]_{j\alpha,j\alpha} \qquad (3.20)
\end{aligned}$$

In the above, we have used the crossing symmetry and charge conservation condition. Comparing this expression with the RHS of (3.18), we obtain that the quantity

$$\begin{aligned}
&\sum_{\alpha} r(\alpha)^2[X(u)]_{j\alpha,j\alpha} \\
&\quad = \sum_{\alpha} r(\alpha)^2 S^{j\alpha}_{\alpha j}(u) \qquad (3.21)
\end{aligned}$$

is independent of j, proving the extended Markov property (3.10). Proof for the IRF model can be done in a similar fashion by showing that the Yang-Baxter operator at $u = \lambda$ is a one-dimensional projector.

We should mention that there exist models which, in a naive sense, do not have the crossing symmetry but have the *second inversion relation*. By explicit calculations, we have verified the extended Markov properties also for these models, where the 'crossing multipliers' are understood as those appearing in the second inversion relation[20,22].

3.3 Markov trace for vertex models

Now we are in a position to write down the explicit form of the Markov trace $\phi(\cdot)$. For vertex models, $\phi(\cdot)$ is given as [14,15,17]:

$$\phi(A) = \mathrm{Tr}(HA). \qquad (3.22)$$

Here Tr stands for the ordinary trace (sum of diagonal elements) and the matrix H is given by a tensor product of a diagonal matrix h:

$$H = \overbrace{h \otimes h \otimes \cdots \otimes h}^{n \text{ times}}, \tag{3.23}$$

with

$$\begin{aligned}(h)_{pq} &= r(p)^2\delta_{pq}/(\sum_p r(p)^2) \\ &\equiv h_p\delta_{pq}.\end{aligned} \tag{3.24}$$

The denominator in (3.24) gives the proper normalization

$$\phi(I) = 1 \qquad (I : \text{identity}). \tag{3.25}$$

The Markov trace defined by (3.22)-(3.24) is a generalization[14,15,17] of Powers state[28]. To show that $\phi(\cdot)$ indeed has the Markov property, we first note that the type I Markov property (3.1) reads

$$\mathrm{Tr}(HAB) = \mathrm{Tr}(HBA). \tag{3.26}$$

To prove this, it is sufficient to show that

$$[H, X_i(u)] = 0. \tag{3.27}$$

In terms of the elements of the weight matrix $[X(u)]_{pq,ij} = S^{iq}_{jp}(u)$, (3.27) is

$$r(k)^2 r(l)^2 [X(u)]_{kl,ij} = [X(u)]_{kl,ij} r(i)^2 r(j)^2, \tag{3.28}$$

which follows from the charge conservation condition, $i + j = k + l$. The type II Markov property (3.2) reads

$$\begin{aligned}&\sum_l x^{(+)}_{kl,kl} h_l = \tau, \qquad \text{(independent of } k),\\ &\sum_l x^{(-)}_{kl,kl} h_l = \bar{\tau}, \qquad \text{(independent of } k),\\ &x^{\pm}_{kl,ij} = \lim_{u\to\infty} [X(\pm u)]_{kl,ij}/\rho(\pm u),\end{aligned} \tag{3.29}$$

which is easily verified from the extended Markov property by taking the $u \to \infty$ limit. The τ-factors are given by

$$\begin{aligned}\tau &= \lim_{u\to\infty} \frac{\tilde{H}(u)}{\tilde{H}(0)} \cdot \frac{\rho(0)}{\rho(u)},\\ \bar{\tau} &= \lim_{u\to\infty} \frac{\tilde{H}(-u)}{\tilde{H}(0)} \cdot \frac{\rho(0)}{\rho(-u)}.\end{aligned} \tag{3.30}$$

3.4 Markov trace for IRF models

For any element $A \in B_n$ which has a matrix representation

$$A = (A^{l'_0 l'_1 l'_2 \cdots l'_n}_{l_0 l_1 l_2 \cdots l_n}), \tag{3.31}$$

we define 'constrained trace' $\tilde{\mathrm{Tr}}(A)$ by

$$\tilde{\mathrm{Tr}}(A) = \sum_{l_1 l_2 \cdots l_n ; l_0 : \text{fixed}}^{\sim} A^{l_0 l_1 l_2 \cdots l_n}_{l_0 l_1 l_2 \cdots l_n} \frac{\psi(l_n)}{\psi(l_0)}. \tag{3.32}$$

In the above, the symbol $\tilde{\Sigma}$ represents the summation under the constraint imposed on the IRF model. The Markov trace $\phi(\cdot)$ is then given by[19]

$$\phi(A) = \tilde{\mathrm{Tr}}(A)/\tilde{\mathrm{Tr}}(I) \quad (A, I \in B_n). \tag{3.33}$$

The type I Markov property (3.1) of $\phi(\cdot)$ is equivalent to the condition

$$\tilde{\mathrm{Tr}}(AB) = \tilde{\mathrm{Tr}}(BA) \qquad (A, B \in B_n), \tag{3.34}$$

which immediately follows from the definition (3.32). It is easy to see[19] that $\phi(\cdot)$ has the type II Markov property if the following conditions are satisfied:

$$\sum_{b \sim a} \sigma^{\pm}(a,b,a,c) \frac{\psi(b)}{\psi(a)} = \chi^{\pm} \quad \text{(independent of } a \text{ and } c)$$
$$\sigma^{\pm}(a,b,c,d) = \lim_{u \to \infty} w(a,b,c,d; \pm u) \tag{3.35}$$

and

$$\sum_{b \sim a} \frac{\psi(b)}{\psi(a)} = \xi \quad \text{(independent of } a). \tag{3.36}$$

Here $b \sim a$ means that the summation is over all values of spin (or height) b admissible to the nearest neighbor spin (or height) a. The constants $\chi^{\pm}$ and ξ are given by

$$\chi^{\pm} = \lim_{u \to \infty} \tilde{H}(\pm u)/\rho(\pm u),$$
$$\xi = \tilde{H}(0)/\rho(0). \tag{3.37}$$

We should note that (3.35) and (3.36) are obtained as special limits, $u \to \pm\infty$ (for the former) and $u = 0$ (for the latter), of the extended Markov property. The τ-factors are given

$$\tau = \chi^+/\xi = \lim_{u\to\infty} \frac{\tilde{H}(u)}{\tilde{H}(0)} \cdot \frac{\rho(0)}{\rho(u)},$$
$$\bar{\tau} = \chi^-/\xi = \lim_{u\to\infty} \frac{\tilde{H}(-u)}{\tilde{H}(0)} \cdot \frac{\rho(0)}{\rho(-u)}. \tag{3.38}$$

4 Application to Various Models

4.1 N-state vertex models

Among many solvable models with hyperbolic (or equivalently trigonometric) parametrization, we first consider a series of vertex models (the N-state vertex model) proposed by Sogo, Akutsu and Abe[29]. This series includes the 6-vertex model as N=2 case and the 19-vertex model[30] as N=3 case. By applying the general theory presented in the preceding section, we obtain a series of new link polynomials[14,15,16,17].

The edge variables i,j,k and l of the Boltzmann weights $\{S_{jl}^{ik}(u)\}$ for the N-state vertex model take the following values:

$$i, j, k, l = -s, -s+1, \cdots, s-1, s, \tag{4.1}$$

where 'spin' or 'charge' s is related to the state number N by

$$N = 2s + 1. \tag{4.2}$$

The model satisfies the charge conservation condition

$$S_{jl}^{ik}(u) = 0, \qquad \text{unless} \quad i + j = k + l. \tag{4.3}$$

Maximally symmetric Boltzmann weights satisfy the CPT invariances and the crossing symmetry with trivial crossing multipliers,

$$S_{jl}^{ik}(u) = S_{\bar{k}\bar{i}}^{jl}(\lambda - u). \tag{4.4}$$

The charge conservation condition allows us to asymmetrize the maximally symmetric Boltzmann weights by the symmetry-breaking transformation mentioned in §2.1. In doing so, only the factor $\alpha_{ij,kl}(u)$ plays an essential role. By $\tilde{S}_{jl}^{ik}(u)$ we denote the transformed asymmetric Boltzmann weight:

$$\tilde{S}_{jl}^{ik}(u) = \alpha_{ij,kl}(u) \cdot S_{jl}^{ik}(u), \tag{4.5}$$

with

$$\alpha_{ij,kl}(u) = \exp[\mu(k-i-l+j)u]. \tag{4.6}$$

Then, $\{\tilde{S}_{jl}^{ik}(u)\}$ satisfy the crossing symmetry with non-trivial crossing multiplier:

$$\tilde{S}_{jl}^{ik}(u) = [\frac{r(i)r(l)}{r(j)r(k)}]^{1/2} \cdot \tilde{S}_{\bar{k}\bar{i}}^{jl}(\lambda - u), \tag{4.7}$$

with

$$r(p) = \exp(-2\mu\lambda p). \tag{4.8}$$

In the below, we explicitly write down the maximally symmetric Boltzmann weights for N =2, 3 and 4. Only a part of the Boltzmann weights are shown; all other weights are obtained by taking account of the crossing symmetry and the CPT invariances. For notational simplicity, the argument u of the S-matrices is suppressed.

$N = 2$ (s=1/2) case (6-vertex model)

$$\begin{aligned} S_{1/2\,1/2}^{1/2\,1/2} &= \sinh(\lambda - u), \\ S_{-1/2\ \ 1/2}^{\ \ 1/2\,-1/2} &= \sinh\lambda. \end{aligned} \tag{4.9}$$

$N = 3$ (s=1) case

$$\begin{aligned} S_{1\,1}^{1\,1} &= \sinh(\lambda - u)\sinh(2\lambda - u), \\ S_{-1\ \ 1}^{\ \ 1\,-1} &= \sinh\lambda\sinh 2\lambda, \\ S_{0\,0}^{1\,1} &= \sinh u \sinh(\lambda - u), \\ S_{0\,1}^{1\,0} &= \sinh 2\lambda \sinh(\lambda - u), \\ S_{0\,0}^{0\,0} &= \sinh\lambda\sinh 2\lambda - \sinh u\sinh(\lambda - u). \end{aligned} \tag{4.10}$$

$N = 4$ (s=3/2) case

$$
\begin{aligned}
S_{3/2\,3/2}^{3/2\,3/2} &= \sinh(\lambda - u)\sinh(2\lambda - u)\sinh(3\lambda - u),\\
S_{-3/2\ \ 3/2}^{\ 3/2\,-3/2} &= \sinh\lambda\sinh 2\lambda\sinh 3\lambda,\\
S_{1/2\,1/2}^{3/2\,3/2} &= \sinh u\sinh(\lambda - u)\sinh(2\lambda - u),\\
S_{1/2\,3/2}^{3/2\,1/2} &= \sinh(\lambda - u)\sinh(2\lambda - u)\sinh 3\lambda,\\
S_{-1/2\ \ 1/2}^{\ 3/2\,1/2} &= 2\sqrt{\sinh\lambda\sinh 3\lambda}\cosh\lambda\sinh u\sinh(\lambda - u),\\
S_{-1/2\ \ 3/2}^{\ 3/2\,-1/2} &= \sinh(\lambda - u)\sinh 2\lambda\sinh 3\lambda,\\
S_{1/2\,1/2}^{1/2\,1/2} &= \sinh(\lambda - u)[\sinh 2\lambda\sinh 3\lambda - \sinh u\sinh(\lambda - u)],\\
S_{-1/2\ \ 1/2}^{\ 1/2\,-1/2} &= \sinh\lambda[\sinh 2\lambda\sinh 3\lambda - 2\cosh\lambda\sinh u\sinh(\lambda - u)].
\end{aligned}
\tag{4.11}
$$

To obtain the Boltzmann weights for general N, we have the inductive method[29] or the fusion method[2, 20]. The N-state vertex model satisfies the unitarity condition with the unitarity function $\rho(u)$ given by

$$
\begin{aligned}
\rho(u) &= \prod_{l=1}^{N-1}\sinh(l\lambda - u)\\
&= S_{ss}^{ss}(u).
\end{aligned}
\tag{4.12}
$$

As was shown in §3, interesting link polynomial is the one constructed from a model with non-trivial crossing multiplier; the τ-factors in (3.2) and (3.4) are determined by the crossing multiplier. Hence we make a symmetry-breaking transformation by $\alpha_{ij,kl}(u)$ to obtain an asymmetric model. We have some choice of μ in (4.6). Existence of the well-defined braid-group representation requires

$$
|\mu| \leq 1/2. \tag{4.13}
$$

We have shown[14, 15, 17] that the 'edge' of the region, $\mu = \pm 1/2$, gives 'interesting' representation. For $\mu = 1/2$, the crossing factor is

$$
r(p) = e^{-p\lambda}. \tag{4.14}
$$

The N-state model has the extended Markov property (3.10) with the $\tilde{H}(u)$-function given by

$$\tilde{H}(u) = \prod_{l=2}^{N} \sinh(l\lambda - u), \tag{4.15}$$

which lead to the Markov trace with τ-factors

$$\tau = \tau(t) = \frac{1}{1 + t + t^2 + \cdots + t^{N-1}}, \tag{4.16}$$

$$\bar{\tau} = \bar{\tau}(t) = \frac{t^{N-1}}{1 + t + t^2 + \cdots + t^{N-1}} = \tau(1/t). \tag{4.17}$$

In the above, the parameter t has been introduced as

$$t = e^{2\lambda}. \tag{4.18}$$

From the explicit matrix form of the generator b_i, we obtain the following N-th order algebraic relation satisfied by b_i:

$$(b_i - c_1)(b_i - c_2) \cdots (b_i - c_N) = 0, \tag{4.19}$$

where the i-th root c_i is given by

$$c_i = (-1)^{i+1} t^{[N(N-1)-(N-i+1)(N-i)]/2}. \tag{4.20}$$

We call (4.19) *reduction relation* because it reduces a high power of b_i to lower powers of b_i. In paricular for $N = 2$, relation (4.19) is expanded into

$$b_i^2 = (1 - t)b_i + t, \tag{4.21}$$

which has been the characteristic relation for the Jones polynomial. The quadratic relation (4.21) is the source of the Alexander-Conway relation[31] or *skein relation* for the Jones polynomial[26]:

$$\alpha(L_+) = (1 - t)\sqrt{t}\,\alpha(L_0) + t^2\alpha(L_-). \tag{4.22}$$

In (4.22), L_+, L_0 and L_- are links which have the configurations of b_i, b_i^0 ($=I$: identity) and b_i^{-1}, respectively, at an intersection of the link projection. Since the reduction relation satisfied by $N \geq 3$ representation is not quadratic, the polynomials constructed thereby *do not* satisfy the skein relation (4.22). Instead, they satisfy what we call *generalized skein relations*[14,15,17]. For N=3 it reads

$$\alpha(L_{2+}) = t(1-t^2+t^3)\alpha(L_+)+t^2(t^2-t^3+t^5)\alpha(L_0)-t^8\alpha(L_-). \quad (4.23)$$

and for N=4 it reads

$$\begin{aligned}\alpha(L_{3+}) &= t^{3/2}(1-t^3+t^5-t^6)\alpha(L_{2+}) \\ &+t^6(1-t^2+t^3+t^5-t^6+t^8)\alpha(L_1) \\ &-t^{9/2}(1-t+t^3-t^6)\alpha(L_0)-t^{20}\alpha(L_-). \end{aligned} \quad (4.24)$$

The generalized skein relation is useful for actual calculation of polynomial of a given link. We have explicitely calculated the N=3 polynomials for 3-braids[16]. There are inifinitely many links which are different but have the identical Jones polynomial (and its two-variable extension, HOMFLY polynomial[32,33]). Examples are given by Birman[34] and Kanenobu[35]. It should be noted that, for the Birman's example, the N=3 polynomial detects the difference[16]. We suppose that the powerfulness of $N \geq 3$ polynomial comes from the higher-order reduction relation which reflects the property of higher-spin strings.

4.2 Graph-state IRF models in the Temperley-Lieb class

Constraint imposed on the IRF spin configurations can be expressed as a graph[19,21,36]. For instance, the unrestricted eight-vertex solid-on-solid (8VSOS, for short) model[37] corresponds to a one-dimensional lattice of infinite length. For any graph we can always obtain a solvable IRF model by constructing the Temperley-Lieb operators[19,21]. Possible values of the spin variable correspond to the lattice (not necessarily one-dimensional) points of the graph and the bond connecting the points means that the pair of spin-values can be neigbours on the 'real' lattice. The crossing multipliers $\{\psi(a)\}$ are eigenfunctions of the difference equation defined on the graph:

$$\sum_{c\sim a} \psi(c) = \sqrt{q}\psi(a). \quad (4.25)$$

Here the 'eigenvalue' q which corresponds to the state number of the Potts model is related to the crossing parameter λ as

$$q = 4\cos^2\lambda \quad (4.26)$$

In this class (the Temperley-Lieb class) of models, the Boltzmann weights have the form

$$w(a,b,c,d;u) = \delta_{ac}\sin(\lambda-u) + \delta_{bd}\cdot[\frac{\psi(a)\psi(c)}{\psi(d)\psi(b)}]^{1/2}\sin u. \quad (4.27)$$

The hyperbolic parametrization is also possible by changes; $u \to iu, \lambda \to i\lambda$. Unitarity function $\rho(u)$ is given by

$$\rho(u) = \sin(\lambda - u). \quad (4.28)$$

In the case of the unrestricted 8VSOS model, the spin variables are heights of an interface. The height variable l_i takes

$$l_i = 0, \pm 1, \pm 2, ..., \pm\infty, \quad (4.29)$$

under the constraint

$$|l_i - l_j| = 1, \quad \text{for nearest neighbor sites } i \text{ and } j. \quad (4.30)$$

Solution of (4.25) for this model is given by a 'plane wave'

$$\psi(a) = \sin(a\lambda + \omega_0). \quad (4.31)$$

The crossing point λ, corresponding to the wave number, is an arbitrary parameter. The parameter ω_0 is also arbitrary expressing the translational invariance of the graph. Originally, the Boltzmann weights of the unrestricted 8VSOS model are parametrized in terms of the elliptic theta functions[37]. The expression (4.27) is the critical limit where the *nome* of the theta functions is set to be zero.

From any graph-state IRF model given by (4.27), we obtain the well-defined operator b_i by the limit $u \to i\infty$ following the prescription described in §2. The models have the extended Markov property (3.11) with

$$\tilde{H}(u) = \sin(2\lambda - u). \quad (4.32)$$

Then from (3.38), the τ factors in the Markov trace are given by

$$\begin{aligned} \tau &= \lim_{u\to+i\infty}[\frac{\sin(2\lambda-u)}{\sin(\lambda-u)}]/[\frac{\sin\lambda}{\sin 2\lambda}] = \frac{1}{1+t}, \\ \bar{\tau} &= \lim_{u\to-i\infty}[\frac{\sin(2\lambda-u)}{\sin(\lambda-u)}]/[\frac{\sin\lambda}{\sin 2\lambda}] = \frac{t}{1+t}, \end{aligned} \quad (4.33)$$

with

$$t = e^{-2i\lambda}. \tag{4.34}$$

This class of models gives the Jones polynomial with t given by (4.34). For the case of the unrestricted 8VSOS model, this is manifest because in $\omega_0 \to \pm\infty$ limit the critical unrestricted 8VSOS model is equivalent to the 6-vertex model[19,21], N=2 case in the preceding subsection, through the Wu-Kadanoff-Wegner transformation[7].

4.3 ABCD IRF models: HOMFLY polynomial and Kauffman polynomial

We consider the unrestricted IRF models related to the vector representations of affine Lie algebras[38], $A^{(1)}_{m-1}$, $B^{(1)}_m$, $C^{(1)}_m$ and $D^{(1)}_m$. The model corresponding to algebra $A^{(1)}_{m-1}$ ($B^{(1)}_m$, $C^{(1)}_m$ and $D^{(1)}_m$)[39] is called the $A^{(1)}_{m-1}$ ($B^{(1)}_m$, $C^{(1)}_m$ and $D^{(1)}_m$) model. In those models vector-valued 'spin' located on the square planar lattice takes weight vectors in the weight space associated with the algebra. The Boltzmann weight $w(\vec{a}, \vec{b}, \vec{c}, \vec{d}; u)$ is assigned to the configuration $(\vec{a}, \vec{b}, \vec{c}, \vec{d})$ round a face (Fig. 14).

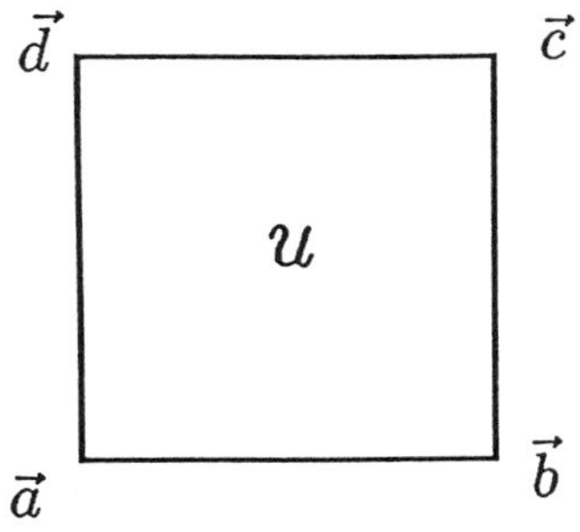

Figure 14: ABCD IRF model.

There are restrictions on nearest neighboring spin pairs. Let Σ be a given set of weight vectors. For a configuration $(\vec{a}, \vec{b}, \vec{c}, \vec{d})$ with fixed $\vec{d}$, we assume that $\vec{a}$, $\vec{b}$ and $\vec{c}$ satisfy the following conditions:

$$\vec{c} - \vec{d},\ \ \vec{a} - \vec{d},\ \ \vec{b} - \vec{c},\ \ \vec{b} - \vec{a} \in \Sigma. \tag{4.35}$$

The spin $\vec{a}$ is said to be allowable or admissible to $\vec{d}$ when $\vec{a}-\vec{d}\in\Sigma$. We represent this relation as

$$\vec{a}\sim\vec{d}. \tag{4.36}$$

The restriction (4.35) means that all nearest-neighbor spin pairs should be admissible. In terms of the orthonormal weight vectors $\{\vec{e}_i\}$ satisfying

$$(\vec{e}_i,\vec{e}_j)=\delta_{ij}, \tag{4.37}$$

where $(\cdot,\cdot)$ is the symmetric bilinear form on the weight space, the set Σ is given as follows:

$$\begin{aligned}
\Sigma &= \{\vec{e}_1-\frac{1}{m}(\vec{e}_1+\cdots+\vec{e}_m),\cdots,\vec{e}_m-\frac{1}{m}(\vec{e}_1+\cdots+\vec{e}_m)\}, \\
&\qquad\qquad \text{for } A^{(1)}_{m-1}, \\
\Sigma &= \{0,\pm\vec{e}_1,\cdots,\pm\vec{e}_m\}, \quad \text{for } B^{(1)}_m, \\
\Sigma &= \{\pm\vec{e}_1,\cdots,\pm\vec{e}_m\}, \quad \text{for } C^{(1)}_m \text{ and } D^{(1)}_m.
\end{aligned} \tag{4.38}$$

The Boltzmann weights of these models are expressed in terms of the elliptic theta function (we omit some factor) with *nome* p ($|p|\le 1$):

$$h(u)=\sin u\prod_{n=1}^{\infty}(1-2p^n\cos(2u)+p^{2n})(1-p^n). \tag{4.39}$$

To write down the explicit forms of Boltzmann weights let us introduce some notations:

$$\begin{aligned}
d_\mu &= \omega\cdot(\vec{d}+\vec{\rho},\vec{\mu}), \quad \vec{\mu}\in\Sigma, \quad \vec{\mu}\neq 0, \\
d_0 &= -\frac{\omega}{2}, \\
d_{\mu\nu} &= d_\mu-d_\nu, \qquad \vec{\mu},\vec{\nu}\in\Sigma,
\end{aligned} \tag{4.40}$$

where $\vec{d}$ is a weight vector, $\vec{\rho}$ a fixed weight vector and ω a free parameter. The following is remarked:

$$\begin{aligned}
(\vec{d}+\vec{\mu})_{\nu\kappa} &= d_{\nu\kappa}+\omega\delta_{\mu\nu}-\omega\delta_{\mu\kappa}, \quad \text{for } A^{(1)}_{m-1}, B^{(1)}_m, C^{(1)}_m \text{ and } D^{(1)}_m \\
(\vec{d}+\vec{\mu})_\nu &= d_\nu+\omega\delta_{\mu\nu}, \qquad \text{for } B^{(1)}_m, C^{(1)}_m \text{ and } D^{(1)}_m.
\end{aligned} \tag{4.41}$$

The Boltzmann weights of the $A^{(1)}_{m-1}$ model are given by[22,39]

$$\begin{aligned}
w(\vec{d}+\vec{\mu},\vec{d}+2\vec{\mu},\vec{d}+\vec{\mu},\vec{d}:u) &= \frac{h(\omega-u)}{h(\omega)},\\
w(\vec{d}+\vec{\mu},\vec{d}+\vec{\mu}+\vec{\nu},\vec{d}+\vec{\mu},\vec{d}:u) &= \frac{h(d_{\mu\nu}+u)}{h(d_{\mu\nu})},\\
w(\vec{d}+\vec{\mu},\vec{d}+\vec{\mu}+\vec{\nu},\vec{d}+\vec{\nu},\vec{d}:u) &= \frac{h(u)}{h(\omega)}[\frac{h(d_{\mu\nu}+\omega)h(d_{\mu\nu}-\omega)}{h(d_{\mu\nu})^2}]^{1/2},
\end{aligned} \tag{4.42}$$

where $\vec{\mu},\vec{\nu}\in\Sigma$ and $\vec{\mu}\neq\vec{\nu}$. The Boltzmann weights of the $B^{(1)}_m$, $C^{(1)}_m$ and $D^{(1)}_m$ models are given by

$$w(\vec{d}+\vec{\mu},\vec{d}+2\vec{\mu},\vec{d}+\vec{\mu},\vec{d}:u) = \frac{h(\lambda-u)h(\omega-u)}{h(\lambda)h(\omega)}, \quad \text{for} \quad \vec{\mu}\neq 0,$$

$$w(\vec{d}+\vec{\mu},\vec{d}+\vec{\mu}+\vec{\nu},\vec{d}+\vec{\mu},\vec{d}:u) = \frac{h(\lambda-u)h(d_{\mu\nu}+u)}{h(\lambda)h(d_{\mu\nu})}, \quad \text{for } \vec{\mu}\neq\pm\vec{\nu},$$

$$\begin{aligned}
&w(\vec{d}+\vec{\mu},\vec{d}+\vec{\mu}+\vec{\nu},\vec{d}+\vec{\nu},\vec{d}:u)\\
&\quad = \frac{h(\lambda-u)h(u)}{h(\lambda)h(\omega)}[\frac{h(d_{\mu\nu}+\omega)h(d_{\mu\nu}-\omega)}{h(d_{\mu\nu})^2}]^{1/2}, \quad \text{for } \vec{\mu}\neq\pm\vec{\nu},
\end{aligned}$$

$$\begin{aligned}
w(\vec{d}+\vec{\mu},\vec{d},\vec{d}+\vec{\nu},\vec{d}:u) &= \frac{h(u)h(d_{\mu-\nu}+\omega-\lambda+u)}{h(\lambda)h(d_{\mu-\nu}+\omega)}(g_{d\mu}g_{d\nu})^{1/2}\\
&\quad +\delta_{\mu\nu}\frac{h(\lambda-u)h(d_{\mu-\nu}+\omega+u)}{h(\lambda)h(d_{\mu-\nu}+\omega)}, \quad \text{for } \vec{\mu}\neq 0,
\end{aligned}$$

$$w(\vec{d},\vec{d},\vec{d},\vec{d};u) = \frac{h(\lambda+u)h(2\lambda-u)}{h(\lambda)h(2\lambda)} - \frac{h(u)h(\lambda-u)}{h(\lambda)h(2\lambda)}\cdot J_{d0}, \tag{4.43}$$

where $\vec{\mu},\vec{\nu}\in\Sigma$ and

$$g_{d\mu} = \sigma\frac{s(d_\mu+\omega)}{s(d_\mu)}\prod_{\kappa\neq\pm\mu,0}\frac{h(d_{\mu\kappa}+\omega)}{h(d_{\mu\kappa})}, \quad \text{for } \vec{\mu}\neq 0, \tag{4.44}$$

$$g_{d0} = 1, \tag{4.45}$$

$$J_{d0} = \sum_{\kappa\neq 0}\frac{h(d_\kappa+\omega/2-2\lambda)}{h(d_\kappa+\omega/2)}g_{d\kappa}. \tag{4.46}$$

The sign factor σ, crossing point λ and the function $s(z)$ is summarized in Table 1.

Table 1.

	$A^{(1)}_{m-1}$	$B^{(1)}_m$	$C^{(1)}_m$	$D^{(1)}_m$
λ	$m\omega/2$	$(2m-1)\omega/2$	$(m+1)\omega$	$(m-1)\omega$
σ	1	1	-1	1
$s(z)$	1	$h(z)$	$h(2z)$	1

The $B^{(1)}_m$, $C^{(1)}_m$ and $D^{(1)}_m$ models have the crossing symmetry but $A^{(1)}_{m-1}$ model does not. Nevertheless, all these models satisfy the second inversion relation with the 'crossing multiplier' $\{\psi(\vec{d})\}$ given by

$$\begin{aligned}\psi(\vec{d}) &= \prod_{1\leq i<j\leq m} h(d_{ij}), \quad \text{for } A^{(1)}_{m-1}, \\ &= \epsilon(\vec{d})\prod_{k=1}^{m} h(d_k) \prod_{1\leq i<j\leq m} h(d_{ij})h(d_{i-j}), \\ &\qquad \text{for } B^{(1)}_m, C^{(1)}_m \text{ and } D^{(1)}_m, \end{aligned} \tag{4.47}$$

with $\epsilon(\vec{d})$ being a sign factor such that $\epsilon(\vec{d}+\vec{\mu})/\epsilon(\vec{d}) = \sigma$. Note that

$$g_{d\mu} = \frac{\psi(\vec{d}+\vec{\mu})}{\psi(\vec{d})}. \tag{4.48}$$

Given the Boltzmann weight we can apply the general procedure presented in §2 and §3 to construct link polynomial. In doing so, we put $|p|$=0 (model is then at criticality) to have a well-defined representation of B_n in the $u \to i\infty$ limit. For BCD-type models, the extended Markov property holds due to the crossing symmetry[22]. For A-type model we have verified the extended Markov property by explicit calculation[22]. In the below we summarize the results.

(a) reduction relations

$$(b_i - 1)(b_i + \gamma^2) = 0 \quad \text{for } A^{(1)}_{m-1}, \tag{4.49}$$

$$(b_i - 1)(b_i - \beta)(b_i + \gamma^2) = 0 \quad \text{for } B^{(1)}_m, C^{(1)}_m \text{ and } D^{(1)}_m, \tag{4.50}$$

with

$$\gamma = e^{-i\omega}, \qquad \text{for } A^{(1)}_{m-1}, B^{(1)}_m, C^{(1)}_m \text{ and } D^{(1)}_m, \tag{4.51}$$

$$\beta = \sigma e^{-i[2\lambda+\omega(1+\sigma)]}, \quad \text{for } B^{(1)}_m, C^{(1)}_m and D^{(1)}_m. \tag{4.52}$$

(b) τ-factors

$$\begin{aligned}\tau &= e^{i(m-1)\omega}\frac{\sin\omega}{\sin(m\omega)},\\ \bar{\tau} &= e^{-i(m-1)\omega}\frac{\sin\omega}{\sin(m\omega)}, \quad \text{for } A^{(1)}_{m-1},\end{aligned}\tag{4.53}$$

$$\begin{aligned}\tau &= e^{i[2\lambda+\omega(\sigma-1)]}\frac{\sin\lambda\sin\omega}{\sin 2\lambda\sin(\sigma\omega+\lambda)},\\ \bar{\tau} &= e^{-i[2\lambda+\omega(\sigma-1)]}\frac{\sin\lambda\sin\omega}{\sin 2\lambda\sin(\sigma\omega+\lambda)}, \quad \text{for} B^{(1)}_m, C^{(1)}_m, D^{(1)}_m.\end{aligned}\tag{4.54}$$

(c) generalized skein relations

$$\alpha(L_{2+}) = (1-\gamma^2)\gamma^{m-1}\alpha(L_+) + \gamma^{2m}\alpha(L_0), \quad \text{for } A^{(1)}_{m-1},\tag{4.55}$$

$$\begin{aligned}\alpha(L_{3+}) &= (1-\gamma^2+\beta)e^{-i[2\lambda+\omega(\sigma-1)]}\alpha(L_{2+})\\ &\quad +(\gamma^2+\beta\gamma^2-\beta)e^{-2i[2\lambda+\omega(\sigma-1)]}\alpha(L_+)\\ &\quad -\gamma^2\beta e^{-3i[2\lambda+\omega(\sigma-1)]}\alpha(L_0), \quad \text{for } B^{(1)}_m, C^{(1)}_m, D^{(1)}_m.\end{aligned}\tag{4.56}$$

The above polynomials are one-variable ones for each fixed m. But if we regard m as a continuous parameter or make ‘analytic continuation’ with respect to m, we have a two-variable polynomial. The A-type model corresponds to the HOMFLY polynomial[32,33], the two-variable extention of the Jones polynomial, with

$$\begin{aligned}t &= e^{-2i\omega},\\ z &= e^{i(m-1)\omega}\frac{\sin\omega}{\sin(m\omega)},\\ \bar{z} &= e^{-i(m-1)\omega}\frac{\sin\omega}{\sin(m\omega)},\end{aligned}\tag{4.57}$$

in the notation of Ocneanu[32]. The BCD-type polynomial corresponds to the Kauffman polynomial[27] with

$$\begin{aligned}\ell &= \frac{\sqrt{-1}\gamma}{\beta},\\ m &= \sqrt{-1}(\gamma-\gamma^{-1}).\end{aligned}\tag{4.58}$$

We thus have explicit model-realizations of the HOMFLY and Kauffman polynomials.

4.4 Other models

We have many other solvable models satisfying the Yang-Baxter relation, to which the general construction procedure of link polynomials is applied. For example, we have the $SU(m)$ vertex model[40,41] and its extension[42]. When applied to these models, we reproduce the recent work of Turaev[43] in a straightforward manner. Another example is the 'fusion hierarchy' of the IRF models[44]. The link polynomials constructed from them are essentially the ones constructed from the N-state vertex models given in §4.1.

We should remark here that the charge conservation condition, in the case of vertex models, is not the prerequisite for existence of a link polynomial. There actually exist many solvable vertex models without the charge conservation with trigonometric/hyperbolic parametrization assuring a well-defined representation of B_n. But most of those models have the crossing symmetry with trivial crossing multiplier, leading to a 'trivial' Markov trace which means the ordinary trace (=sum of diagonal elements). The resulting link polynomials may be simple but not necessarily be trivial. There may be a possibility that we have simple but useful link polynomials constructed from those models[45,46].

In the construction of the N-state polynomial, we obtained the braid-group representation with $\mu = \pm 1/2$. It is natural to ask how is for the case $|\mu| < 1/2$. There exists braid group representation for this case. Moreover, the representation can be extended into a multivariable one. We can construct a multivariable link polynomial from the representation[45]. It is an interesting problem to explore further the nature of this multivariable polynomial.

5 Extensions of the Theory

5.1 Composite string representations and two-variable polynomials

The N-state vertex model can be considered to describe the scattering of spin $(N-1)/2$ particles which has the factorization property. In particular, the 6-vertex model (N=2) corresponds to spin 1/2 factorized S-matrix. We know that a mutliplet of spin 1/2 particles contains higher spin particles. For example, from a pair of spin 1/2 particles, we can make two 'composite particles'; one with spin 1 and the other with spin 0. The N=3 vertex model can be interpreted as the factorized S-matrix for the composite spin 1 particles[2]. Extension of this idea leads us to higher spin representations of the braid group, which we call *composite string representation* [17,18,20].

Suppose that a set of generators $\{g_i\}_{i=1,\dots n-1}$ satisfies the followings:

$$g_i g_j = g_j g_i, \qquad |i-j| \geq 2, \tag{5.1}$$

$$g_i g_{i+1} g_i = g_{i+1} g_i g_{i+1}, \tag{5.2}$$

$$g_i^2 = (1-t) g_i + t, \tag{5.3}$$

The relations (5.1)-(5.3) define the Hecke algebra[47] $H(t,n)$. The braid group representation made from the Temperley-Lieb algebra, through (2.46), gives a represention of $H(t,n)$. Note, however, that there exist representations of $H(t,n)$ which do not have the Temperley-Lieb algebra behind them, whose example are the one constructed from the $SU(m)$ vertex model and $A_{m-1}^{(1)}$ IRF model mentioned in §4.4.

We form a 'composite string' by combining $(N-1)$ strings and attaching a projector $P^{(N)}$ at each end. We first make a multiplet of $(N-1)$ spin 1/2 particles, and then extract the spin $(N-1)/2$ component by the projectors. Explicit form of the projector $P_i^{(N)}$ can be derived through a recursion formula[18]

$$P_i^{(N)} = P_i^{(N-1)} h_{i+N-3}^{(N)} P_i^{(N-1)}, \quad P_i^{(2)} \equiv 1, \tag{5.4}$$

where

$$h_j^{(N)} = \frac{\tau_{N-2}}{\tau_{N-1}}(\frac{t^{N-2}}{\tau_{N-2}} + g_i) \tag{5.5}$$

with

$$\tau_m = 1 + t + t^2 + \cdots + t^{m-1}. \tag{5.6}$$

We have another method to construct the projector from the composite Yang-Baxter operator[20]. Hereafter, we sometimes write P_i instead of $P_i^{(N)}$. The projector P_i defined by (5.4) satisfies the relations

$$\text{(a)} \quad P_i^2 = P_i, \tag{5.7}$$

$$\text{(b)} \quad P_i(g_{i+N-2}g_{i+N-3}\cdots g_i) = (g_{i+N-2}g_{i+N-3}\cdots g_i)P_{i+1},$$
$$P_i(g_{i+N-2}^{-1}g_{i+N-3}^{-1}\cdots g_i^{-1}) = (g_{i+N-2}^{-1}g_{i+N-3}^{-1}\cdots g_i^{-1})P_{i+1}, \tag{5.8}$$

$$\text{(c)} \quad P_i g_i = P_i, \tag{5.9}$$

$$\text{(d)} \quad P_i \Delta_i^2 = P_i, \tag{5.10}$$

where

$$\Delta_i = (g_i g_{i+1}\cdots g_{i+N-3})(g_i g_{i+1}\cdots g_{i+N-4})\cdots(g_i). \tag{5.11}$$

The operator Δ_i is called a half-twist[9].

Let us denote spin-s representation of B_n by $B_n^{[s]}$. Using the composite string, we define generators $\{G_i; i = 1, 2, \cdots, n-1\}$ of $B_n^{[s]}$ as follows. For notational simplicity, we use

$$k \equiv N - 1 = 2s. \tag{5.12}$$

We prepare $n-1$ sets of k strings and combine a set of k strings into a composite string with projectors at both ends. The generator G_i corresponds to an operation in Fig.15.

To describe this, we introduce an operator $\bar{G}_i^{(N)}$ by

$$\bar{G}_i^{(N)} = g_i^{(1)} g_i^{(2)} \cdots g_i^{(N-1)}, \tag{5.13}$$

where

$$g_i^{(\ell)} = g_{ik+1-\ell}\, g_{ik+2-\ell} \cdots g_{(i+1)k-\ell} \quad (\ell = 1, 2, \cdots, N-1). \tag{5.14}$$

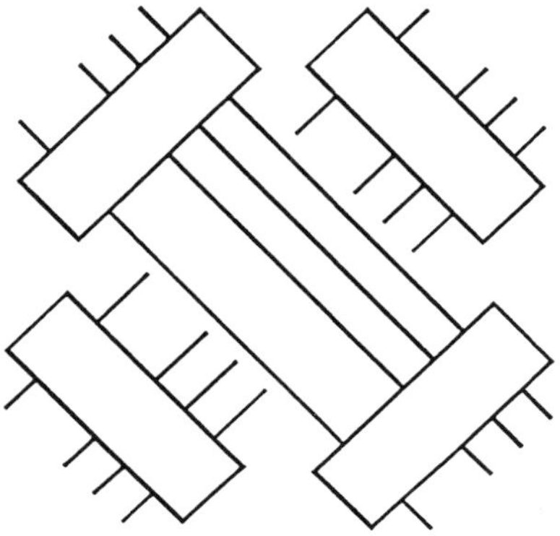

Figure 15: Braid group generator in the composite string representation. The boxes correspond to the projectors.

Then, the generator G_i of $B_n^{[s]}$ is expressed as

$$G_i = P^{(N)}_{(i-1)k+1} P^{(N)}_{ik+1} \bar{G}^{(N)}_i P^{(N)}_{(i-1)k+1} P^{(N)}_{ik+1} \tag{5.15}$$

Using (5.1)-(5.3) and (5.7)-(5.10), we can show that the generators G_i, G_2, $\cdots$, G_{n-1} satisfy the defining relation of the braid group:

$$G_i G_j = G_j G_i, \quad |i-j| \geq 2, \tag{5.16}$$

$$G_i G_{i+1} G_i = G_{i+1} G_i G_{i+1}. \tag{5.17}$$

In the composite string representation, the identity operator I and the inverse generator G_i^{-1} are given by

$$I = P^{(N)}_1 P^{(N)}_{k+1} \cdots P^{(N)}_{(i-1)k+1} \cdots P^{(N)}_{(n-1)k+1}, \tag{5.18}$$

$$G_i^{-1} = P^{(N)}_{(i-1)k+1} P^{(N)}_{ik+1} [\bar{G}^{(N)}_i]^{-1} P^{(N)}_{(i-1)k+1} P^{(N)}_{ik+1}. \tag{5.19}$$

Appearance of the projectors in the above means that we are considering only the 'highest spin' space for each set of strings.

Let us construct the Markov trace associated with $B_n^{[s]}$. By $\psi(\cdot)$ we denote the Markov trace associated with a starting representation of $B_n^{[1/2]}$ (=$H(t,n)$). It satisfies the normalization condition

$$\psi(I) = 1, \quad I\text{: identity in } B_n^{[1/2]}, \tag{5.20}$$

and the Markov properties

$$\begin{aligned} &\text{I.} \quad \psi(AB) = \psi(BA), \qquad (A, B \in B_n^{[1/2]}). && (5.21)\\ &\text{II.} \quad \psi(Ag_n) = z\psi(A), \\ &\qquad \psi(Ag_n^{-1}) = \bar{z}\psi(A), \qquad (A \in B_n^{[1/2]}, g_n \in B_{n+1}^{[1/2]}), && (5.22) \end{aligned}$$

where

$$z = \psi(g_i), \quad \bar{z} = \psi(g_i^{-1}) \quad \text{for all } i. \tag{5.23}$$

From $\psi(\cdot)$ we can make a trace functional $\psi^{[s]}(\cdot)$ which acts on $B_n^{[s]}$:

$$\psi^{[s]}(A) = \frac{\psi(A)}{\psi(P^{(N)})^n}, \quad (A \in B_n^{[s]}), \tag{5.24}$$

with

$$\psi(P_j^{(N)}) \equiv \psi(P^{(N)}) \quad \text{(independent of } j). \tag{5.25}$$

Note that we have regarded $B_n^{[s]}$ as a sub group-algebra in $B_n^{[1/2]}$. It is straightforward to see that $\psi^{[s]}(\cdot)$ satisfy the Markov properties

$$\begin{aligned} &\text{I.} \quad \psi^{[s]}(AB) = \psi^{[s]}(BA), \qquad (A, B \in B_n^{[s]}). && (5.26)\\ &\text{II.} \quad \psi^{[s]}(AG_n) = Z\psi^{[s]}(A), && (5.27)\\ &\qquad \psi^{[s]}(AG_n^{-1}) = \bar{Z}\psi^{[s]}(A), \qquad (A \in B_n^{[s]}, G_n \in B_{n+1}^{[s]}), && (5.28) \end{aligned}$$

with

$$\begin{aligned} Z &= \psi^{[s]}(G_j) = \frac{z^{N-1}}{\psi(P^{(N)})}, \\ \bar{Z} &= \psi^{[s]}(G_j^{-1}) = \frac{\bar{z}^{N-1}}{\psi(P^{(N)})} \end{aligned} \tag{5.29}$$

We should note that the above composite string representatio of B_n and associated Markov trace are always applicable so long as the 'starting' representation $B_n^{[1/2]}$ satisfies the defining relation of the Hecke algebra (5.1)-(5.3). In the Ocneanu's formulation of HOMFLY polynomial[32], value of z characterizing the Markov trace $\psi(\cdot)$ is independent of the parameter t appearing in the representation $H(t,n)$; the pair of parameter (t,z) enters into the two-variable polynomial. Hence, starting from the Ocneanu's representation of $H(t,n)$ and $\psi(\cdot)$, the above procedure leads us to 'higher spin' generalization of the HOMFLY polynomial. Let us make a change of variables from (t,z) to (t,ω) where

$$\omega = \frac{\bar{z}}{z}. \tag{5.30}$$

Then the factors Z and $\bar{Z}$ are written explicitly as

$$\begin{aligned} Z &= \frac{(1-t)(1-t^2)\cdots(1-t^{N-1})}{(1-\omega t)(1-\omega t^2)\cdots(1-\omega t^{N-1})}, \\ \bar{Z} &= \frac{\omega^{N-1}(1-t)(1-t^2)\cdots(1-t^{N-1})}{(1-\omega t)(1-\omega t^2)\cdots(1-\omega t^{N-1})}. \end{aligned} \tag{5.31}$$

The two-variable link polynomial $\alpha_\omega^{[s]}(\cdot)$ is given by[17,18,20]

$$\alpha_\omega^{[s]}(A) = (Z\bar{Z})^{-(n-1)/2}(\frac{\bar{Z}}{Z})^{e(A)/2}\psi^{[s]}(A), \quad (A \in B_n^{[s]}), \tag{5.32}$$

where $e(A)$ is the exponent sum of G_i's appearing in the braid A. We should remark here that these new two-variable link polynomials $\alpha_\omega^{[s]}(\cdot)$ reduce to the one-variable link polynomials constructed from the N-state vertex models when we set $\omega = t$.

5.2 Braid-monoid algebra

Let us discuss here an algebra associated with the link polynomials constructed from solvable models[17,18,20]. As was mentioned in §2, the Yang-Baxter algebra generated by $\{X_i(u)\}$ becomes the Temperley-Lieb algebra at $u = \lambda$ and, the Temperley-Lieb operators correspond to monoid diagrams. What we want to discuss here is an algebra generated by the monoids and the braids which lies behind our new link polynomials discussed in §4. In the algebraic formulation of the Kauffman polynomial, Birman-Wenzl[48] and Murakami[49] devised an algebra $C_n(\ell, m)$ which is generated by $\{1\}, \{E_i\}_{i=1}^{n-1}$ (monoid operators) and $\{G_i\}_{i=1}^{n-1}$ (braid operators). In addition to the definig relation of the braid group for $\{G_i\}_{i=1}^{n-1}$, the braid and monoid operators satisfy the following:

$$G_i + G_i^{-1} = m(1 + E_i), \tag{5.33}$$

$$E_i E_{i\pm1} E_i = E_i, \tag{5.34}$$

$$G_{i\pm1} G_i E_{i\pm1} = E_i G_{i\pm1} G_i = E_i E_{i\pm1}, \tag{5.35}$$

$$G_{i\pm1} E_i G_{i\pm1} = G_i^{-1} E_{i\pm1} G_i^{-1}, \tag{5.36}$$

$$G_{i\pm1} E_i E_{i\pm1} = G_i^{-1} E_{i\pm1}, \tag{5.37}$$

$$E_{i\pm1} E_i G_{i\pm1} = E_{i\pm1} G_i^{-1}, \tag{5.38}$$

$$G_i E_i = E_i G_i = \ell^{-1} E_i, \tag{5.39}$$

$$E_i G_{i\pm1} E_i = \ell E_i, \tag{5.40}$$

$$E_i E_j = E_j E_i \quad \text{if } |i - j| \geq 2, \tag{5.41}$$

$$E_i^2 = [m^{-1}(\ell + \ell^{-1}) - 1] E_i, \tag{5.42}$$

$$G_i^2 = m(G_i + \ell^{-1} E_i) - 1. \tag{5.43}$$

These relations have diagramatical meanings as partly shown in Fig.16. Let us call the algebra $C_n(\ell, m)$ as Birman-Wenzl-Murakami (BWM, for short) algebra.

We should note that the BWM algebra characterizes a class of link polynomial with cubic reduction relation; combining (5.33) and (5.43) we see that the generator G_i satisfies a cubic relation

$$G_i^3 = (m + \frac{1}{\ell}) G_i^2 - (1 + \frac{m}{\ell}) G_i + \frac{1}{\ell}. \tag{5.44}$$

As is expected from the cubic relation for G_i, the N=3 polynomial and the BCD polynomial both satisfy the above relations (5.33)-(5.43) with specified ℓ and m. The operators G_i and E_i are given by constant multiple of b_i and $X_i(\lambda)$ respectively. To be precise, we have for the N=3 polynomial

$$G_i = \frac{b_i}{t\sqrt{-1}}, \tag{5.45}$$

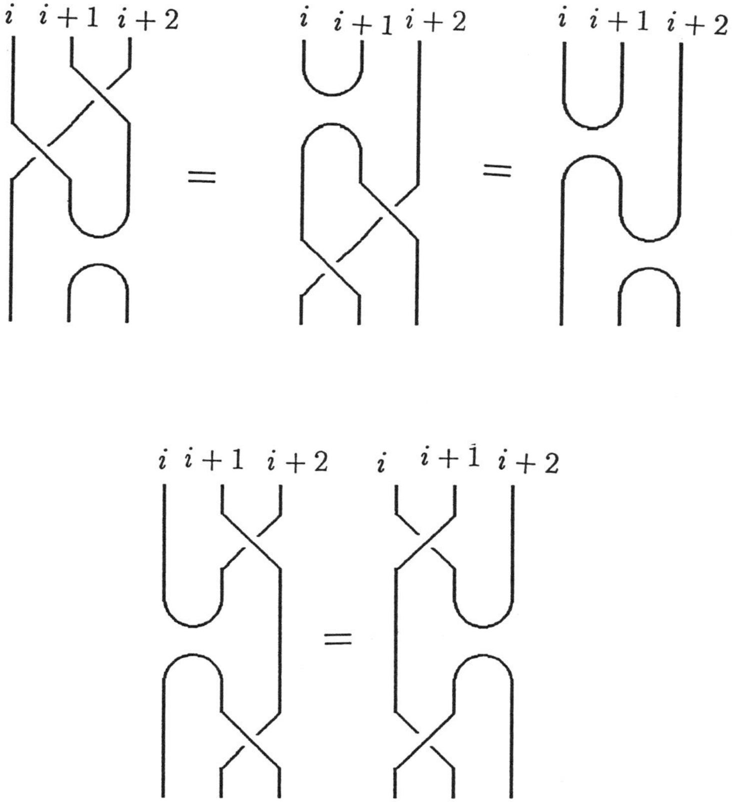

Figure 16: Two of the relations of Birman-Wenzl-Murakami algebra. Upper: (5.35). Lower: (5.36)

$$E_i = \frac{1}{t^3(t-1)(t+1)}(b_i - 1)(b_i + t^2), \tag{5.46}$$

$$m = \frac{1-t^2}{t\sqrt{-1}}, \tag{5.47}$$

$$\ell = \frac{\sqrt{-1}}{t^2}, \tag{5.48}$$

and for the BCD polynomial

$$G_i = \frac{b_i}{\gamma\sqrt{-1}}, \tag{5.49}$$

$$E_i = \frac{1}{\beta(\gamma^2-1)}(b_i - 1)(b_i + \gamma^2), \tag{5.50}$$

$$m = \sqrt{-1}(\gamma - \gamma^{-1}), \tag{5.51}$$

$$\ell = \frac{\gamma\sqrt{-1}}{\beta}. \tag{5.52}$$

The constants β and γ have been defined in (4.51) and (4.52). We thus see that the $N = 3$ polynomial (one-variable) corresponds to a specialization of the Kauffman polynomial (two variable) with ℓ and m substituted by (5.47) and (5.48) respectively. Note, however, that the two-variable extension of the $N = 3$ polynomial presented in §5.1 is different from the Kauffman polynomial, which comes from the difference in the Markov trace.

We have an analogus but new relations for general N-state polynomial:

$$(b_i - C_1)(b_i - C_2)\cdots(b_i - C_N) = 0 \tag{5.53}$$

$$E_i = \frac{t^{(N-1)/2} + \cdots + t^{-(N-1)/2}}{(C_N - C_1)\cdots(C_N - C_{N-1})}(b_i - C_1)\cdots(b_i - C_{N-1}), \tag{5.54}$$

$$E_i^2 = (t^{(N-1)/2} + \cdots + t^{-(N-1)/2})E_i, \tag{5.55}$$

$$b_i E_i = C_N E_i, \tag{5.56}$$

$$b_i b_{i+1} b_i = b_{i+1} b_i b_{i+1}, \tag{5.57}$$

$$E_i E_{i\pm1} E_i = E_i, \tag{5.58}$$

$$E_i b_{i\pm1} b_i = b_{i\pm1} b_i E_{i\pm1} = (-t^{1/2})^{(N-1)^2} E_i E_{i\pm1}, \tag{5.59}$$

$$b_{i\pm1}E_ib_{i\pm1} = t^{(N-1)^2}b_i^{-1}E_{i\pm1}b_i^{-1}, \tag{5.60}$$
$$b_{i\pm1}E_iE_{i\pm1} = (-t^{1/2})^{(N-1)^2}b_i^{-1}E_{i\pm1}, \tag{5.61}$$
$$E_{i\pm1}E_ib_{i\pm1} = (-t^{1/2})^{(N-1)^2}E_{i\pm1}b_i^{-1}, \tag{5.62}$$
$$E_ib_{i\pm1}E_i = C_N^{-1}(-t^{1/2})^{(N-1)^2}E_i = t^{-(N-1)/2}E_i, \tag{5.63}$$
$$b_ib_j = b_jb_i, \quad \text{for } |i-j| \geq 2, \tag{5.64}$$
$$E_iE_j = E_jE_i, \quad \text{for } |i-j| \geq 2, \tag{5.65}$$
$$C_r = (-1)^{r+1}t^{[N(N-1)-(N-r)(N-r+1)]/2}, \quad (r = 1, 2, \ldots, N). \tag{5.66}$$

These have been proved by taking a full advantage of the composite string representation[17,18] and the composite Yang-Baxter operator[20]. The above relations (5.53)∼(5.66) will be helpful for the 'graphical calculation'[27,50,20] of higher N link polynomials.

6 Summary and Discussion

In this article, we have presented a general theory for construction of link polynomials from exactly solvable models satisfying the Yang-Baxter relation. Most important point is that the Yang-Baxter operators $\{X_i(u)\}$ and the *Yang-Baxter algebra* generated by them are very close to, but far more general than, the braid group generators $\{b_i\}$ and the braid group B_n. A representation of the braid group is obtained in a straightforward manner by tending the spectral parameter u to infinity. To change words, the braid group is a 'high energy limit' of the Yang-Baxter algebra. The existence of the well-defined limit requires the model to be at criticality, having the trigonometric/hyperbolic parametrization.

Construction of the Markov trace $\phi(\cdot)$ is also straightforward in our theory. The extended Markov property, which holds for finite spectral parameter u, assures the existence of $\phi(\cdot)$. The explicite form of the Markov trace is immeditely given from the extended Markov property, with the crossing multiplier and a function characterizing the extended Markov property. Here we again need the 'criticality condition': Non-trivial crossing multipliers can be associated only for models with charge conservation which have, in general, trigonometric/hyperbolic parametrization hence are 'critical'. The extended Markov property is also the one for the critical models. It seems that the 'criticality condition' is vital to our approach where the braid group, only *a limit* of 'general' Yang-Baxter algebra, lies at its basis. There may remain a possibility that 'full' Yang-Baxter algebra can be utilized leading to super-multivariable link invariant.

In our approach, regarding any link as a closed braid, we utilize the braid group and its asociated Markov trace to have a link invariant based on the Markov's theorem. For calculational convenience, a method which allows one to work directly on the link diagram, not necessarily a closed braid, may be helpful. Our theory presented in this article also includes a 'unbraiding' approach[20, 51, 52, 53] which is based on the Reidemister moves[54] instead of the Markov moves. Using the $u \to \infty$ limit of the Boltzmann weights of solvable model, we can assign a 'statistical weights' to a given link diagram. Link polynomial is, then, given by a weighted sum over 'possible configurations', or 'partition function', of the link diagram.

We have many applications of link polynomials to physics. The most classic and most popular link invariant in physics is the Gauss linking number. What is special for the Gauss linking number is that it has an 'integral representation': It is expressed as integral along space curves forming a link, corresponding to Biot-Savart law and Ampere's law of classical electrodynamics. Such an explicit 'integral representation' has not been known for link polynomials (Jones, HOMFLY, N-state hierarchy, Kauffman,...) discussed in this article. It is one of the most challenging problems left for future study. There have been some attempts for such direction[55, 56]. We hope that the Yang-Baxter relation, which has physical origin such as the factorized S-matrix, may again play an important role for this purpose.

In conclusion, we like to emphasize that the theory of exactly solvable models offers an extremely powerful method to construct new link polynomials. A key is the Yang-Baxter relation proving that a new concept in physics accompanies a new mathematics.

Acknowledgements

The authors thank Professor C. N. Yang and Professor K. Murasugi for continuous encouragements.

References

[1] L. D. Faddeev: Sov. Sci. Rev. Math. Phys. **C1** (1981) 107.
H. B. Thacker: Rev. Mod. Phys. **53** (1981) 253.
M. Wadati: in *Dynamical Problems in Soliton Systems*, ed. S. Takeno (Springer-Verlag, 1985) p.68.

[2] P. P. Kulish and E. K. Sklyanin: Lecture Notes in Physics (Springer-Verlag, New York, 1982) vol.151, p.61.

[3] C.N. Yang: Phys. Rev. Lett. **19** (1967) 1312.

[4] M. Karowski, H.J. Thun, T.T. Truong and P.H. Weisz: Phys. Lett. **67B** (1977) 321.
A. B. Zamolodchikov and A. B. Zamolodchikov: Ann. of Phys. (N.Y.) **120** (1979) 253.
K. Sogo, M. Uchinami, A. Nakamura and M. Wadati: Prog. Theor. Phys. **66** (1981) 1284.
K. Sogo, M. Uchinami, Y. Akutsu and M. Wadati: Prog. Theor. Phys. **68** (1982) 508.

[5] R. J. Baxter: *Exactly Solved Models in Statistical Mechanics* (Academic Press, 1982).

[6] A. B. Zamolodchikov: Commun. Math. Phys. **69** (1979) 165.

[7] F. Y. Wu: Phys. Rev. **B4** (1971) 2312.
L. P. Kadanoff and F. J. Wegner: Phys. Rev. **B4** (1971) 3989.

[8] E. Artin: Ann. of Math. **48** (1947) 101.

[9] J. S. Birman: *Braids, Links and Mapping Class Group* (Princeton University Press, 1974).

[10] Y. S. Wu: Phys. Rev. Lett. **52** (1984) 2103.

[11] A. Tsuchiya and Y. Kanie: Adv. Stud. in Pure Math. **16** (1988) 297.
M. Wadati, Y. Yamada and T. Deguchi: J. Phys. Soc. Jpn. (to appear).

[12] J. W. Alexander: Proc. Nat. Acad. **9** (1923) 93.

[13] A. A. Markov: Recuell Math. Moscov (1935) 73.

[14] Y. Akutsu and M. Wadati: J. Phys. Soc. Jpn. **56** (1987) 839.

[15] Y. Akutsu and M. Wadati: J. Phys. Soc. Jpn. **56** (1987) 3039.

[16] Y. Akutsu, T. Deguchi and M. Wadati: J. Phys. Soc. Jpn. **56** (1987) 3464.

[17] Y. Akutsu and M. Wadati: Commun. Math. Phys. **117** (1988) 243.

[18] T. Deguchi, Y. Akutsu and M. Wadati: J. Phys. Soc. Jpn. **57** (1988) 757.

[19] Y. Akutsu, T. Deguchi and M. Wadati: J. Phys. Soc. Jpn. **57** (1988) 1173.

[20] T. Deguchi, M. Wadati and Y. Akutsu: J. Phys. Soc. Jpn.**57** (1988) 1905.

[21] M. Wadati and Y. Akutsu: Prog. Theor. Phys. Suppl. **94** (1988) 1.

[22] T. Deguchi, M. Wadati and Y. Akutsu: J. Phys. Soc. Jpn. **57** (1988) 2921.

[23] E. H. Lieb and F. Y. Wu: in *Phase Transitions and Critical Phenomena*, ed. C. Domb and M. S. Green, vol. 1 (Academic Press, London, 1972) p.331.

[24] R. J. Baxter: Ann. Phys. **70** (1972) 193.

[25] H. N. V. Temperley and E.H. Lieb: Proc. R. Soc. London **A322** (1971) 251.

[26] V. F. R. Jones: Bull. Amer. Math. Soc. **12** (1985) 103.

[27] L. Kauffman: Topology **26** (1987) 395.

[28] R. T. Powers: Ann. of Math. **86** (1967) 138.
M. Pimsner and S. Popa: Ann. Scient. Ec. Norm. Sup., 4^e serie t. **19** (1986) 57.

[29] K. Sogo and Y. Akutsu and T. Abe: Prog. Theor. Phys. **70** (1983) 730 and 739.

[30] A. B. Zamolodchikov and V. A. Fateev: Sov. J. Nucl. Phys. **32** (1980) 298.

[31] J. W. Alexander: Trans. Amer. Math. Soc. **30** (1928) 275.
J. H. Conway: in *Computational Problems in Abstract Algebra*, ed. J. Leech (Pergamon Press, London, 1969) p.329.

[32] P. Freyd, D. Yetter, J. Hoste, W.B.R. Lickorish, K. Millett and A. Ocneanu: Bull. Amer. Math. Soc. **12** (1985) 239.

[33] J. H. Przytycki and K. P. Traczyk: Kobe J. Math. **4** (1987) 115.

[34] J. S. Birman: Invent. Math. **81** (1985) 287.

[35] T. Kanenobu: Math. Ann. **275** (1986) 555.

[36] V. Pasquire: J. Phys. A. Math. Gen. **20** (1986) L217 and L221.
A. Kuniba and T. Yajima: J. Phys. A. Math. Gen. **21** (1988) 519.
P. A. Pearce and K. A. Seaton: Phys. Rev. Lett. **60** (1988) 1347.

[37] G. E. Andrews, R.J. Baxter and P.J. Forrester: J. Stat. Phys. **35** (1984) 193.

[38] V. G. Kac and D.H. Peterson: Adv. Math. **5** (1984), 125.

[39] M. Jimbo, T. Miwa and M. Okado: Commun. Math. Phys. **116** (1988) 353.

[40] O. Babelon, H.J. de Vega and C.M. Viallet: Nucl. Phys. **B190** (1981) 542.

[41] J. H. H. Perk and C. L. Schultz: Phys. Lett. **84A** (1981) 407.

[42] M. Jimbo: Commun. Math. Phys. **102** (1986) 537.

[43] V. G. Turaev: Invent. Math. **92** (1988) 527.

[44] A. Kuniba, Y. Akutsu and M. Wadati: J. Phys. Soc. Jpn. **55** (1986) 1092, 2170 and 3338.
A. Kuniba, Y. Akutsu and M. Wadati: Phys. Lett. **116A** (1986) 382 and **117A** (1986) 358.
R. J. Baxter and G. E. Andrews: J. Stat. Phys. **44** (1986) 249.
G. E. Andrews and R. J. Baxter: J. Stat. Phys. **44** (1986) 713.
Y. Akutsu, A. Kuniba and M. Wadati: J. Phys. Soc. Jpn. **55** (1986) 2907.
E. Date, M. Jimbo, T. Miwa and M. Okado: Lett. Math. Phys. **12** (1986) 209.
E. Date, M. Jimbo, T. Miwa and M. Okado: Phys. Rev. **B35** (1987) 2105.
E. Date, M. Jimbo, A. Kuniba, T. Miwa and M. Okado: Nucl. Phys. **B290 [FS20]** (1987) 231.

[45] T. Deguchi, M. Wadati and Y. Akutsu: preprint, 1988.

[46] Z. Q. Ma and B. H. Zhao: preprint, 1988.

[47] N. Bourbaki: *Groupes et algebras de Lie* (Hermann, Paris, 1968) Chap. 4.

[48] J. S. Birman and H. Wenzl: preprint, 1987.

[49] J. Murakami: Osaka J. Math. **24** (1987) 745.

[50] L. H. Kauffman and P. Vogel: preprint, 1987.
L. H. Kauffman: *On Knots* (Princeton University Press, 1987).

[51] L. H. Kauffman: preprint, 1987.

[52] V. F. R. Jones: preprint, 1988.

[53] M. Wadati, T. Deguchi and Y. Akutsu: preprint, 1989.

[54] K. Reidemeister: *Knotentheorie* (Springer-Verlag, reprinted 1974).

[55] A. M. Polyakov: Mod. Phys. Letters **3A** (1988) 325.

[56] E. Witten: preprint, 1988.

AKutzu-Wadati Link Polynomials from Feynman-Kauffman Diagrams

Mo-Lin Ge, Lu-Yu Wang and Kang Xue
Theoretical Physics Division, Nankai Institute of Mathematics
Nankai University, Tianjin, P.R. China

Yong-Shi Wu
Department of Physics, University of Utah
Salt Lake City, Utah 84112

Abstract

By employing techniques familiar to particle physicists, we develop Kauffman's state model for the Jones polynomial, which uses diagrams looking like Feynman diagrams for scattering, into a systematic, diagrammatic approach to new link polynomials. We systematize the ansatz for S-matrix by symmetry considerations and find a natural interpretation for CPT symmetry in the context of knot theory. The invariance under Reidemeister moves of type III, II and I can be imposed diagrammatically step by step, and one obtains successively braid group representations, regular isotopy and ambient isotopy invariants from Kauffman's bracket polynomials. This procedure is explicitly carried out for the N = 3 and 4 cases, N being the number of particle labels (or charges). With appropriate symmetry ansatz and with annihilation and creation included in the S-matrix, we have obtained link polynomials which generalize the definition of the Akutzu-Wadati polynomials from closed braids to any oriented knots or links with explicit invariance under Reidemeister moves.

I. Introduction

Recently, the braid group and knot/link polynomials have received much attention from theoretical and mathematical physicists working in various fields. These fields include exotic quantum statistics in either 2+1[1-2] or 1+1[3-4] dimensions, exactly soluable models in statistical mechanics[5-10], conformal field theories in 2 dimensions[11-15,3], Chern-Simons theories in 3 dimensions[16] and even quantum gravity in 3+1 dimensions[17]. All these concerns are, in this or that way, related to the Yang-Baxter equations[18], which seems to serve as an organizing principle for the recent developments in statistical mechanics and conformal field theories. Mathematically, the special Yang-Baxter equations, in which the Yang-Baxter matrix has no usual parameter dependence, are nothing but the non-trivial part of the defining relations for the braid group. So a solution to the special Yang-Baxter equations gives rise to a representation of the braid group. Since one obtains knots or links by closing the ends of braids (see Fig. 1), invariant polynomials can be obtained from the representations of the braid group satisfying certain extra conditions, which express the desired topological invariance under continuous deformations for knots and links in three-space. What is interesting to physicists is that these extra conditions coincide with what we need or obtain on physical grounds in various fields in physics as mentioned above. Thus not only the representations of the braid group but also the knot/link polynomials become interesting objects to construct in physics.

However, the relationship between mathematics and physics is not unidirectional from the former to the latter. The progress in physics, either conceptual or technical, may inspire important developments in mathematics. Recently there are two examples of this sort, dealing with the problem of constructing link polynomials by methods familiar to physicists. Here we are referring to the work of Akutzu and Wadati[7,8] and of Kauffman[6, 19]:

1) In Akutzu and Wadati's approach, they have exploited the well-known fact that the Boltzmann weights or elements of the diagonal-to-diagonal transfer matrix of a soluable model in d=2 statistical mechanics satisfy the Yang-Baxter equations. When the starting model is critical and parametrized by trigonometric or hyperbolic functions, the Boltzmann weights, after appropriate parameter-dependent phase transformations, have well-defined limits as the spectral parameter is sent to $+\infty$, which satisfy the special Yang-Baxter equations. In this way, a representation of the braid group is obtained from the already-known solution to a soluable vertex or IRF model. With the Markov trace constructed by generalizing the Powers state, a sequence of new link polynomials were constructed by Akutzu and Wadati explicitly in, e.g., the N-state vertex models with N=3 and N=4 and the unrestricted eight-vertex SOS model.
2) In Kauffman's diagrammatic approach, he has exploited the well-known fact that the special Yang-Baxter equations can be viewed as the factorizable condition for an S-matrix, which depends only on the spin (or charge) of the particles but nothing else. The key observation is that the conditions expressing the topological invariance for obtaining link polynomials from braid group representations can also be formulated as conditions on the S-matrix, such as the unitarity and cross-channel unitarity. All these conditions, as have been familiar to physicists, can be conveniently put into the form of Feynman diagrams for scatterings. As solutions to these conditions, the well-known Jones polynomial[20] and a two-variable generalization of it, the so-called HOMFLY polynomial[21],

were derived in Kauffman's original papers[6,19] with some simple ansatz for the S-matrix.

We like the diagrammatic approach, because Feynman diagrams are what physicists, in particular particle physicists, have been familiar with. However, we notice that Kauffman has made use of only scattering diagrams. To obtain more new link polynomials, one may need to include creation and annihilation diagrams. This observation motivates the present investigation. With appropriate symmetry ansatz for the S-matrix and with annihilation and creation included, we develop a general and systematic procedure for new link polynomials, with a new form of "cross-channel unitarity" needed for more complicated ansatz than Kauffman's. In the cases of N=3 or 4, we are able to derive the Akutzu-Wadati link polynomials for any oriented knot or link with explicit invariance under all Reidemeister moves. This shows the strength and effectiveness of the diagrammatic approach, and we believe that more new link polynomials can be found by appropriately modifying the ansatz for the S-matrix.

This paper is organized as follows. In Sec. II we review Kauffman's state model and establish our notations and conventions. In Sec. III we discuss the discrete C, P, T symmetries for S-matrix and find that CPT symmetry has a natural interpretation in the context of knot theory and, therefore, should be considered as indispensable. To narrow the possible choices we impose CP- and T-symmetries and charge conservation, and write down a quite general ansatz for the S-matrix for the N=3 case (N = number of particle charges) which contains several parameters and incorporates annihilation-creation terms. Sec. IV is devoted to the diagrammatic techniques for finding out the constraints on the parameters in the ansatz from the "unitarity", Yang-Baxter equations and "cross-channel unitarity". This reduces the number of independent parameters and leads to regular isotopy invariants, i.e. invariants under Reidemeister moves of type

II and III. In Sec. V we finally impose the invariance under Reidemeister moves of type I and obtain an (ambient isotopy) invariant polynomial with only one variable, which is invariant under all Reidemeister moves. Then we derive in Sec. VI the reduction relation and the generalized Alexander-Conway relation for our link polynomials and show that they are identical to what obtained by Akutzu and Wadati for N=3. This tells us that our polynomial, which is defined for any orient knot or link, is actually the generalization of the N=3 Akutzu and Wadati polunomial which is defined only for closed braids. In Sec. IV, we present the results for N=4 without details, again generalizing the Akutzu-Wadati N=4 link polynomial to any oriented knots or links. The final section, Sec. VII, is devoted to a summary of our diagrammatic approach and some discussions.

For an introduction to the braid group and knot theory see, for example, the books [22,23] and also refs. [3,6,7,19].

II. The Diagrammatic Approach to the Jones Polynomial

Let us first briefly review Kauffman's diagrammatic approach[6,19]. As in all other approaches, one start with the construction of a representation of the braid group, and then use it to construct a link polynomial by imposing additional conditions for the topological invariance for knots or links. The key observation for the diagrammatic approach is that all these conditions can be expressed in terms of diagrams, which are analogous to Feynman diagrams for S-matrix in quantum field theories. The use of diagrams also makes the bookkeeping of the indices for the S-matrix elements (or for the braid-group representation) an easy job.

To start, we notice that the generators of the braid group are the exchange of two neighboring (say, the i-th and (i+1)-th) strings, denoted by σ_i and σ_i^{-1} for the two senses of exchange. For a linear representation of the braid group, one associates a vector space to each string, say V_i to the i-th string, and each generator σ_i is represented by a linear transformation in the direct-product space $V_i \otimes V_{i+1}$. For simplicity, from now on we assume that all V_i are copies of one and the same vector space V of dimension N and all σ_i are represented by copies of one and the same linear transformation. Choosing a basis $\{e^a\}$ in V, we have a basis $\{e^a \otimes e^b\}$ in $V_i \otimes V_{i+1}$, in which each σ_i is represented by one and the same $N^2 \times N^2$ matrix S_{ab}^{cd}. Here a, b, c, d take N values, labelling the base vectors e^a of V. The representation of the generator σ_i^{-1} is, of course, the inverse $(S^{-1})_{ab}^{cd}$.

Following Kauffman[6], we think of S_{ab}^{cd} as the S-matrix elements of some quantum field theory, which depend not on momenta of the particles but only on the particle labels a, b, c and d. Kauffman called these labels as "spins"; instead, we will call them as "isospin variables" and their values as "charge." Diagrammatically we represent S_{ab}^{cd} and its inverse by

S_{ab}^{cd} : 

and $\left(S^{-1}\right)_{ab}^{cd}$: 

(1)

in which the arrows on the lines indicates the direction of time pointing upward (i.e. a,b are the initial states and c,d the final states). For a braid standing upright, this means that we have assigned an orientation by having the arrows on the strings all pointing upward.

It is well-known that in order for S_{ab}^{cd} and $\left(S^{-1}\right)_{ab}^{cd}$ to provide a braid-group representation, they have to satisfy the so-called unitarity condition

$$S_{ab}^{ef} \cdot \left(S^{-1}\right)_{ef}^{cd} = \delta_a^c \, \delta_b^d \qquad (2)$$

and the special Yang-Baxter equations

$$S_{ab}^{de} \, S_{ec}^{fk} \, S_{df}^{ij} = S_{bc}^{de} \, S_{ad}^{if} \, S_{fe}^{jk} \qquad (3)$$

which represents nothing but the nontrivial defining relations for the braid group $\sigma_i \, \sigma_{i+1} \, \sigma_i = \sigma_{i+1} \, \sigma_i \, \sigma_{i+1}$. The other set of defining relations $\sigma_i \sigma_j = \sigma_j \sigma_i$ with $|i-j| \geq 2$ are automatically satisfied by representing σ_i in the space $V_i \otimes V_{i+1}$. In the S-matrix language, the special Yang-Baxter equations are just the "factorization condition" in (1+1)-dimensional integrable quantum field theories[24]. Using the diagrammatic representation (1), these conditions can be depicted as

"unitarity" condition:

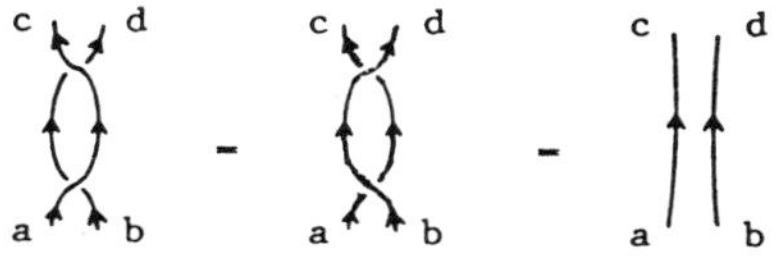

(4)

"factorization" condition:

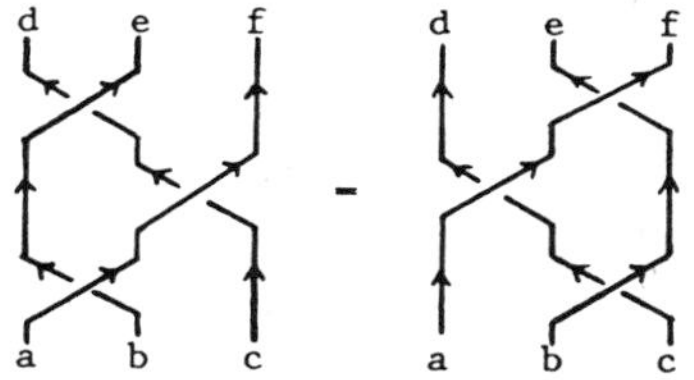

(5)

Here a simple vertical line denotes the Kronecker delta and a line connecting legs of two crossings implies the summation over the repeated indices on the legs.

To see the advantages of the diagrammatic approach, let us consider the following simple example first considered by Kauffman[6,19]:

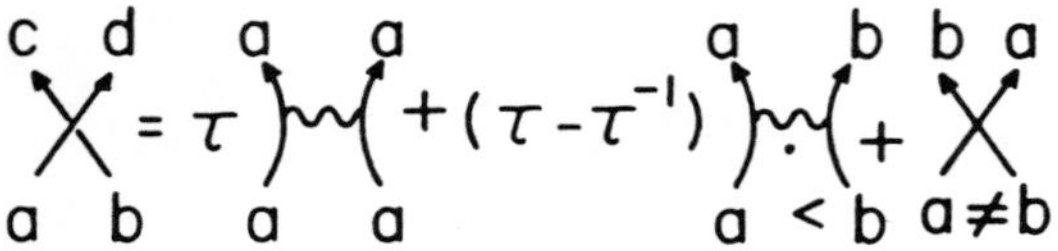
c d
a b
= τ
a a
a a
+ (τ - τ⁻¹)
a b
a < b
+
b a
a≠b

(6)

which corresponds to the following Ansatz for the S-matrix:

$$S_{ab}^{cd} = \begin{cases} \tau - \tau^{-1} & \text{for } a=c < b=d \\ \tau & \text{for } a=c = b=d \\ 1 & \text{for } a=d \neq b=c \\ 0 & \text{otherwise} \end{cases} \qquad (6')$$

Here we have used a straight solid line to denote δ_a^c, a wavy line to stand for a = b and a wavy line with a dot below for a < b. Thus, the inverse matrix is given by

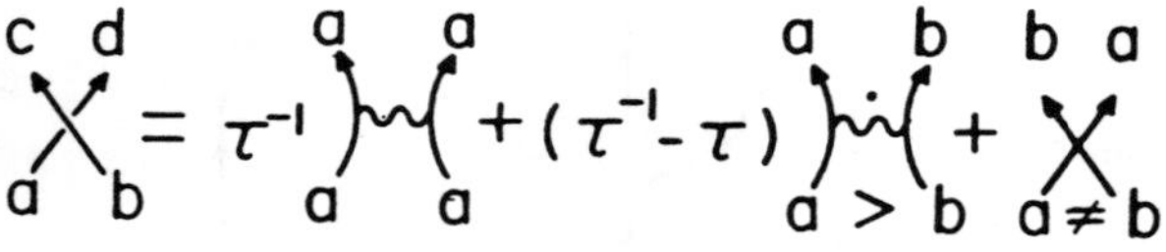
c d
a b
= τ⁻¹
a a
a a
+ (τ⁻¹ - τ)
a b
a > b
+
b a
a≠b

(7)

which corresponds to

$$\begin{cases} \tau^{-1} - \tau & \text{for } a=c > b=d \\ \tau^{-1} & \text{for } a=c = b=d \end{cases}$$

$$\left\{S^{-1}\right\}_{ab}^{cd} = \begin{cases} -1 & \text{for } a=d \neq b=c \\ 0 & \text{otherwise} \end{cases} \qquad (7')$$

Substitute the right-hand sides of eqs. (6) and (6') into eqs. (4) and (5), it is easy to check that these equations are indeed satisfied. For example, for a typical case of indices we have

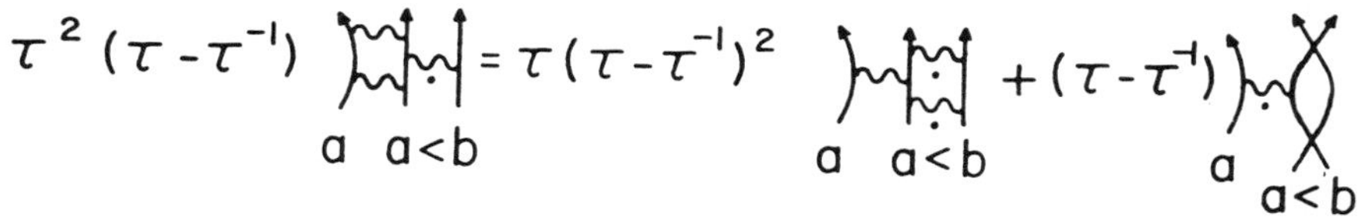

$$\tau^2(\tau-\tau^{-1}) \underset{a \quad a<b}{\text{⟩⟨}} = \tau(\tau-\tau^{-1})^2 \underset{a \quad a<b}{\text{⟩⟨}} + (\tau-\tau^{-1}) \underset{a \quad a<b}{\text{⟩⟨}} \qquad (8)$$

for the left-hand and right-hand sides of eq. (5). The equality holds true since $\tau^2(\tau-\tau^{-1}) = \tau(\tau-\tau^{-1})^2 + (\tau-\tau^{-1})$. Here we have only drawn non-zero terms in the expansion for the indicated choice of indices. Note that two-fold crossing of the second pair of lines in the right-hand side is allowed since $a \neq b$. A similar check can be done for all other choices of indices. Therefore, eqs. (6) and (6') give us a representation of the braid group.

To construct a link polynomial from a braid-group representation, it is noted that by connecting the top and bottom ends of a braid (see Fig. 1), one always obtains a knot or link, which is called the closure of the braid. By the Alexander theorem[25] (also see [22]), any oriented knot or link is equivalent, by the (three types of) Reidemeister moves, to the closure of some braid. Kauffman has noticed that the diagrammatic equivalences (4) and (5) just express the topological invariance under the Reidemeister moves of type II and III involving strings with parallel arrows. For a general knot or link (see Fig. 1c), we may also need the invariance under the type-II moves involving strings with opposite arrows:

"cross-channel unitarity": 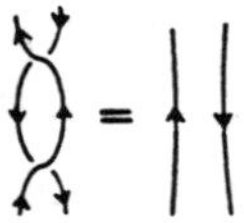 (9)

However, in a knot or link the strings must form closed loops, so it is proper to require that the "cross-channel unitarity" be true only for closed loops. There are two ways to close the strings in eq. (9), so instead we require

"cross-channel unitarity 1":  (10)

"cross-channel unitarity 2": 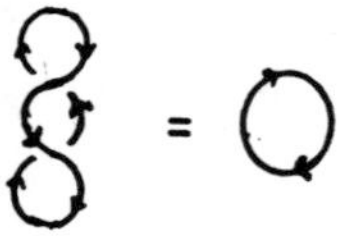 (10′)

With the invariance under both oriented type-II Reidemeister moves, one can derive from eq. (5) the invariance under type-III moves involving arbitrarily oriented strings.

To obtain a topological invariant of regular isotopy of knots or links, i.e. invariant under Reidermeister moves of both type II and type III, Kauffman has constructed a polynomial as follows:

$$[K] = \sum_{S} \langle S|K\rangle \, t^{-|S|} \tag{12}$$

Here S is a state of the knot or link K, i.e. a choice of decomposition for all the crossing of K according to the diagrammatic ansatz (6) and (7). Each (4-particle or 4-leg) vertex in the decomposition like (6) contributes a factor which is just the coefficient associated with it, and $\langle S|K\rangle$ is a product of such factors from all vertices of the state S. Note that any state S consists of closed loops, each of which carries a charge (or a label) and is called a component of S. $|S|$ in eq. (12) is the norm of a state S, defined by

$$|S| = \sum_{\ell \in \mathrm{comp}(S)} \mathrm{rot}(\ell) \cdot \mathrm{label}(\ell) \tag{13}$$

where comp(S) denotes the components of S, label(ℓ) is the charge assigned to the component ℓ and rot(ℓ) = ± 1 according to its orientation:

$$\mathrm{rot}(\circlearrowright) = -1, \qquad \mathrm{rot}(\circlearrowleft) = +1 \tag{14}$$

Finally, in eq. (12) we have to sum over all states S of K. In the language of S-matrix, a state is actually a vacuum bubble diagram consisting of only 4-leg vertices. The above procedure is quite similar to that for evaluating Feynman vacuum bubble diagrams except for the factor $t^{-|S|}$, which assigns a factor $t^{\pm a}$ for a closed loop carrying charge a according to the orientation of the loop.

To give an example, let us consider the knot in Fig. 1c. One of its states is given by Fig. 2 and it gives a term in the sum (12) as follows:

$$(\tau^{-1}-\tau)\cdot 1\cdot 1\cdot(\tau-\tau^{-1})\cdot\tau\cdot\tau^{-1}\cdot t^{a+b+2c-c} = -(\tau^{2}-2+\tau^{-2})\; t^{a+b+c} \tag{15}$$

Now the "cross-channel unitarity" condition 1 and 2, i.e. eqs.(10) and (10'), are required to hold only for the corresponding Kauffman polynomials. Namely,

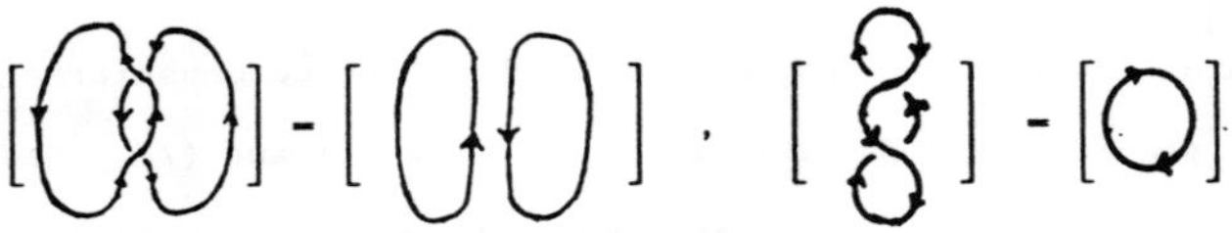

$$[\text{figure}] - [\text{figure}] \ , \ [\text{figure}] - [\text{figure}] \qquad (16)$$

Kauffman has shown that with the ansatz (6) and (7), the first condition is automatically satisfied, but the second condition requires that $t = r$ and the allowed charges must be from the ordered set consisting of integers $-n, -n+2, \cdots, n-2$ and n:

$$a, b, \cdots \in \{-n, -n+2, \cdots, n-2, n\} \qquad (17)$$

The total number of allowed charges is $N = n+1$. Therefore, the Kauffman polynomial [K] defined by eq. (12) with the vertices (6) and (7) give us a regular isotopy invariant polynomial of two variables t and N.

However, [K] is not invariant under the Reidemeister moves of type I:

$$[\text{figure}] = t^{n+1}[\bigcirc], \qquad [\text{figure}] = t^{-n-1}[\bigcirc] \qquad (18)$$

So an appropriate normalization of [K] via

$$P_K = (t^{n+1})^{-\omega(K)} \ [K]/[0] \qquad (19)$$

gives us a topological polynomial invariant under all three types of Reidemeister moves. Here [0] is the polynomial (12) for $K =$ a simple circle and $\omega(K)$ is the writhe of K defined by

$$\omega(K) = \sum_{p \in C(K)} \epsilon(p) \tag{20}$$

where C(K) is the set of crossings of K and $\epsilon[\nwarrow\!\!\!\!\nearrow] = 1$, $\epsilon[\nwarrow\!\!\!\!\nearrow] = -1$. Kauffman has shown that with the vertices (6) and (7), P_K gives us the two-variable HOMFLY polynomial[21], which satisfy

$$t^{-n-1}\, P[\nwarrow\!\!\!\!\nearrow] - t^{n+1}\, P[\nwarrow\!\!\!\!\nearrow] = (t - t^{-1})\, P[\uparrow\uparrow] \tag{21}$$

If we set N=2 (or n=1), we get the well-known Jones polynomial[20].

III. Symmetry Ansatz and Inclusion of Annihilation-Creation Vertices

To derive new link polynomials from the above-described diagrammatic approach, it is crucial to extend the ansatz (6) or (6′) for the S-matrix (or the four-leg vertices). We notice that eq. (6) or (6′) only contain scattering processes. From a particle physicist's point of view, it is natural to include annihilation-creation processes too. Moreover, the ansatz (6) looks quite ad hoc. The following experience of particle physicists may be helpful: i.e. one may start with a classification of possible symmetries in the S-matrix and, then, try to examine a general S-matrix satisfying certain symmetries. In this way, we may obtain a more complicated ansatz with the hope that by adjusting the parameters in the ansatz, all the desired conditions can be satisfied.

For definiteness, let us assume that the charge of the particles can take value only from the set (17). This set is symmetric with respect to zero. So it is tempted to think that a particle with charge (−b) is the anti-particle of the particle with charge b; such a pair can annihilate and then create another pair. In particle physics it is a common practice to reverse the arrow of a line for an anti-particle of charge (−b) and then to think of it as a particle of charge b. Doing so in a state for a knot or link is perfectly consistent with eq. (13). So we are confident that it is possible to incorporate the annihilation-creation processes into Kauffman's diagrammatic approach.

Before doing so, let us start with another common practice of particle physicists, i.e. classify and impose symmetries of S-matrix. Following Akutzu and Watadi[7], we introduce the following terminology:

charge conservation $S_{ab}^{cd} = 0 \quad \text{if } a+b \neq c+d$ (22)

C-invariance $S_{ab}^{cd} = S_{-a,-b}^{-c,-d}$ (23)

P-invariance $S_{ab}^{cd} = S_{ba}^{dc}$ (24)

T-invariance $S_{ab}^{cd} = S_{cd}^{ab}$ (25)

We notice that the ansatz (6) or (6′) satisfies (22) and (25), but not (23) and (24), since $S_{ab}^{ab} = (t - t^{-1}) \neq 0$ for $a < b$, but $S_{ba}^{ba} = 0$. It seems to us that no known solution to the special Yang-Baxter equations (3) satisfies either P- or C-invariance. However, the ansatz (6) or (6′) satisfies the combined CP-invariance:

CP-invariance $S_{ab}^{cd} = S_{-b,-a}^{-d,-c}$ (26)

Therefore, it is also CPT-invariant:

CPT-invariance $$S_{ab}^{cd} = S_{-d,-c}^{-b,-a} \tag{27}$$

Here we note that CPT-invariance has a natural diagrammatic interpretation in knot/link theory:

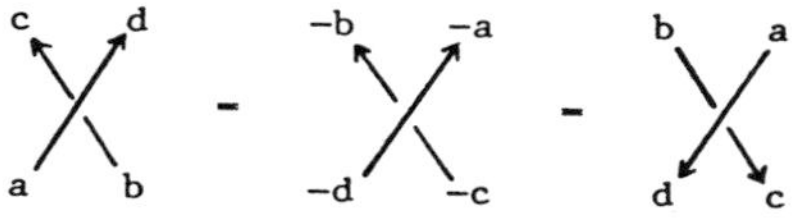

(28)

(In the last step we have changed both the arrow and the sign of charge.) As diagrams, the two opposite sides are exactly the same except the change in orientation (or rotation in the plane by 180°). This tells us that reversing the orientation of a knot or link, i.e. reversing the arrows of all lines simultaneously, will not affect the topological polynomial associated with it, as should be. Because of this knot-theory interpretation, we think CPT-invariance is an absolutely necessary requirement.

To narrow our choices, let us impose both CP- and T-invariances so that CPT-invariance is automatically true. Also the charge conservation condition seems to be necessary. So in the following when we incorporate the annihilation-creation vertices, we will consider some very general ansatz consistent with eqs. (22), (24), and (25).

First consider the N=2 case. The charges a,b,c,d can take values in {-1,1}. In this case, no S-matrix element corresponding to annihilation-creation can be added, since S_{1-1}^{1-1} and $S_{-1\,1}^{-1\,1}$ already appear in the ansatz (6) and adding S_{-11}^{1-1} is equivalent to relaxing the restriction $a < b$ in the second term in the right-hand side of eq. (6), what we do not want to do. In the following, for higher N cases, the same is true, i.e. there is no need to introduce "diagonal" annihilation-creation terms.

For the N=3 case, $a,b,c,d \in \{-2,0,2\}$. A quite general ansatz, which extends eq. (6) with annihilation-creation terms and is consistent with CP- and T-invariances and charge conservation, is given by

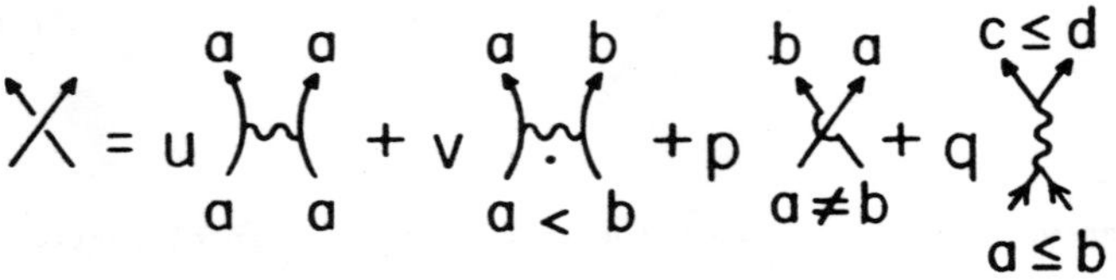

$$\chi = u \underset{a\ \ a}{\overset{a\ \ a}{\bowtie}} + v \underset{a < b}{\overset{a\ \ b}{\bowtie}} + p \underset{a \neq b}{\overset{b\ \ a}{\chi}} + q \underset{a \leq b}{\overset{c \leq d}{\chi}} \qquad (29)$$

with

$$u = u_1(\delta_{a2}+\delta_{a,-2}) + u_2\delta_{ao} \qquad (30.a)$$

$$v = v_1(\delta_{ao}\delta_{b2}+\delta_{a,-2}\delta_{bo}) + v_2\delta_{a,-2}\delta_{b2} \qquad (30.b)$$

$$p = p_1(\delta_{ao}\delta_{b,-2}+\delta_{a,-2}\delta_{bo}+\delta_{a2}\delta_{bo}+\delta_{ao}\delta_{b2})+p_2(\delta_{a2}\delta_{b,-2}+\delta_{a,-2}\delta_{b2}) \qquad (30.c)$$

$$q = q_1(\delta_{ao}\delta_{bo}\delta_{c,-2}\delta_{d2} + \delta_{a,-2}\delta_{b2}\delta_{co}\delta_{do}) \qquad (30.d)$$

The expressions (30.a-d) should be understood as accompanying corresponding diagrams in the ansatz (29). In short, this ansatz means that

$$S^{22}_{22} = S^{-2-2}_{-2-2} = u_1, \quad S^{00}_{00} = u_2, \quad S^{02}_{02} = S^{-20}_{-20} = v_1, \quad S^{-22}_{-22} = v_2 \qquad (31.a)$$

$$S^{-20}_{0-2} = S^{02}_{20} = S^{0-2}_{-20} = S^{20}_{02} = p_1, \quad S^{-22}_{2-2} = S^{2-2}_{-22} = p_2, \quad S^{-22}_{00} = S^{00}_{-22} = q_1 \qquad (31.b)$$

and other elements are zero. Here u_1, u_2, v_1, v_2, p_1, p_2 and q_1 are parameters to be determined. It is easy to see that eqs. (22), (24) and (25) are respected.

To invert the S-matrix, one may use the "unitarity" condition (4) with the diagrammatic techniques as follows. From eq. (29) it is easy to see that its inverse should be of the form

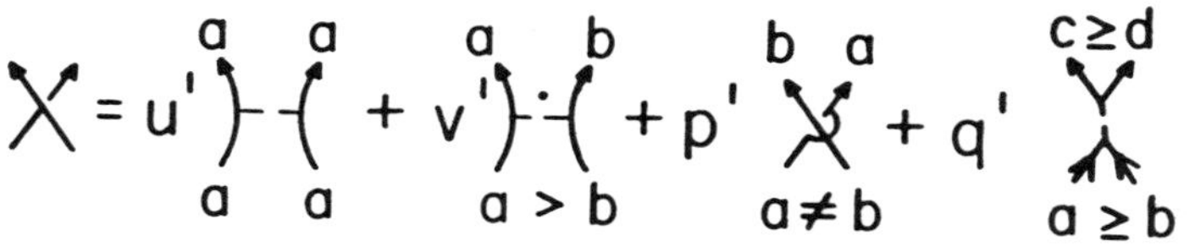

(32)

with

$$u' = u_1'(\delta_{a2}+\delta_{a,-2})+u_2'\delta_{ao} \tag{33.a}$$

$$v' = v_1'(\delta_{a2}\delta_{bo}+\delta_{ao}\delta_{b,-2})+v_2'\delta_{a2}\delta_{b,-2} \tag{33.b}$$

$$p' = p_1'(\delta_{ao}\delta_{b,-2}+\delta_{a,-2}\delta_{bo}+\delta_{a2}\delta_{bo}+\delta_{ao}\delta_{b2})+p_2'(\delta_{a2}\delta_{b,-2}+\delta_{a,-2}\delta_{b2}) \tag{33.c}$$

$$q' = q_1'(\delta_{ao}\delta_{bo}\delta_{c2}\delta_{d,-2}+\delta_{a2}\delta_{b,-2}\delta_{co}\delta_{do}) \tag{33.d}$$

Note that we have used different diagrams to distinguish between the terms in (29) and those in (30). Then the unitarity conditions (4), with (29) and (30) substituted, lead to

(i) for $a = b = c = d$ or $a = c \neq b = d$:

$$u_iu_i' = p_ip_i' = 1 \quad (i = 1,2) \tag{34.a}$$

(ii) for $a = d > b = c$ (see Fig. 3a):

$$v_1p_1' + p_1v_1' = 0 \quad (a \neq 2,\ b \neq -2) \tag{34.b}$$

$$v_2p_2' + p_2v_2' + q_1q_1' = 0 \quad (a = 2,\ b = -2) \tag{34.c}$$

(iii) for $a = -b = 2$, $c = d = 0$ (see Fig. 3b):

$$u_2q_1' + q_1p_2' = 0 \tag{34.d}$$

Therefore, the inverse matrix elements are

$$\begin{cases} u_1' = u_1^{-1},\ u_2' = u_2^{-1},\ p_1' = p_1^{-1},\ p_2' = p_2^{-1},\ q_1' = -q_1/u_2p_2 \\ v_1' = -v_1/p_1^2,\ \ v_2' = (q_1^2 - u_2v_2)/u_2p_2^2 \end{cases} \tag{35}$$

For higher N, the generalization is straightforward. See Sec. VI for the N=4 case.

IV. Diagrammatic Techniques and Regular Isotopy Invariants

Similar diagrammatic techniques as above can be used to solve the Yang-Baxter ("factorization") equations (5) and the "cross-channel unitarity" conditions (10) and (10′) or (16). As one will see later, these conditions completely determine all the parameters in the ansatz (29) and (30.a-d) in terms of the parameter t in the definition of the bracket polynomial (12), but the solution may not be unique.

First consider the Yang-Baxter equations. For a = b = d = e < c = f, we have Fig. 4a and obtain

$$\begin{cases} u_1^2v_1 = u_1v_1^2 + v_1p_1^2 & (a = -2,\ c = 0) \\ u_1^2v_2 = u_1v_2^2 + v_2p_2^2 + v_1q_1^2 & (a = -2,\ c = 2) \\ u_1^2v_1 + q_1^2u_1 = u_2v_1^2 + v_1p_1^2 & (a = 0,\ c = 2) \end{cases} \tag{36.a}$$

For a = d ≠ b = e ≠ c = f, we have Fig. 4b and obtain

$$v_1p_1^2 = v_1p_2^2 + u_2q_1^2 \quad (a = -2,\ b = 2,\ c = 0 \text{ or } a = 0,\ b = -2,\ c = 2) \tag{36.b}$$

For $a = b = 0$, $c = 2$, we have Fig. 4c and obtain

$$u_1v_2 + u_2v_1 = u_1v_1, \quad u_1p_2 = p_1^2 \tag{36.c}$$

For $a = c = 0$, $b = 2$, we have Fig. 4d and obtain

$$u_1 = p_2 + v_1 \tag{36.d}$$

In deriving eqs. (36.c) and (36.d) we have assumed that $q_1 \neq 0$. (Otherwise, $q_1 = 0$ would lead back to the cases that Kauffman has considered.) All other Yang-Baxter equations give no new restrictions. Solving eqs. (36.a-d) we obtain

$$\begin{cases} u_2 = -p_1, p_2 = p_1^2/u_1, \; v_1 = u_1 - p_1^2/u_1 \\ v_2 = (u_1 + p_1)(u_1^2 - p_1^2)/u_1^2, \; q_1^2 = -(u_1^2 - p_1^2)^2 p_1/u_1^3 \end{cases} \tag{37}$$

To check the "cross-channel unitarity" condition (16), we note that annihilation-creation terms do not contribute at all. So we have Fig. 5, which is true in the sense of the bracket polynomial (12), as emphasized in Sec. II. The first term on the right-hand side in Fig. 5 should be just equal to the term on the left-hand side and the remaining terms should make vanishing contributions to the bracket polynomial (12). This leads to

$$u_1(v'_1 + v'_2t^{-2}) + u_2v'_1t^{-2} + u_1^{-1}(v_1+v_2t^2)+u_2^{-1}v_1t^2+v_1v'_1=0 \tag{39}$$

Using eqs. (35) and (37) and setting $x = u_1/p_1$ it can be cast into

$$(x + t^2)[x^3 + (1+t^2)x^2 - x(1+t^2) - t^2] = 0 \tag{40}$$

Similarly it can be shown that the other condition in eq. (16) is automatically satisfied.

One obvious solution of eq. (4) is $x = u_1/p_1 = -t^2$. Together with eqs. (35) and (37) this gives (up to an overall sign)

$$\begin{cases} u_1 = t^2,\ u_2 = 1,\ p_1 = -1,\ p_2 = t^{-2},\ q_1 = t - t^{-3} \\ v_1 = t^2 - t^{-2},\ v_2 = t^2 - 1 - t^2 + t^{-4} \end{cases} \tag{41}$$

$$\begin{cases} u_1' = t^{-2},\ u_2' = 1,\ p_1' = -1,\ p_2' = t^2,\ q_1' = t^{-1} - t^3 \\ v_1' = t^{-2} - t^2,\ v_2' = t^4 - t^2 - 1 + t^{-2} \end{cases} \tag{41'}$$

It is amusing to notice that eqs. (41') can be obtained from (41) by the substitution $t \to t^{-1}$.

The other solutions to eq.(40) can be obtained by solving

$$x^3 + (1+t^2)x^2 - x(1+t^2) - t^2 = 0 \tag{42}$$

Once $x = u_1/p_1$ is determined, all parameters u_1, u_2, p_1, p_2, v_1, v_2 and q_1 in the ansatz (29) can be fixed through eqs. (35) and (37). Then, the bracket polynomial (12) with <S|K> determined by each set of these parameters will give us a regular isotopy invariant polynomial. In next section we will see that only the set (41) and (41') will lead to a link polynomial which is ambient isotopy invariant.

V. New Link Polynomials and Reduction Relation for $N = 3$

To obtain link polynomials one needs further invariance under Reidemeister moves of type I. First of all, we should have (for the bracket polynomials)

(43)

From eqs. (29), (30) and (12), this gives

$$2u_1 + u_2 + 2v_1 t^2 + v_2 t^4 = u_1(t^4+t^{-4}) + u_2 + v_1(t^2+t^{-2}) + v_2 \quad (44)$$

where the left-hand side is the middle term in (43) and the right-hand side those on the two sides. Therefore, upon using eq. (37), one has

$$(x + t^2)[x^2(1+t^2) - t^2(x + 1)] = 0 \quad (45)$$

It is obvious that only for $x = -t^2$ are both eqs. (40) and (45) satisfied. So we will consider only this solution, i.e. the solution (41) and (41'). It is easy to check that it also satisfies a similar equation for the other crossing:

(46)

Here we remark that the conditions (43) and (46) did not appear explicitly in Kauffman's discussion, since eq. (44), say, is automatically respected by the ansatz (6). In our case, we need eqs. (43) and (46) to distinguish between the regular isotopy and ambient isotopy invariants and uniquely fix the parameters in our ansatz for S-matrix.

However, these conditions (43) and (46) are not sufficient to guarantee the invariance under Reidemeister moves of type I. In fact, it is easy to see that

[] $= 2u_1+u_2+2v_1t^2+v_2t^4 = t^4(t^2+1+t^{-2}) = t^4\ [\bigcirc]$ (47)

[] $= 2u'_1+u'_2+2v'_1t^{-2}+v'_2t^{-4} = t^{-4}(t^2+1+t^{-2}) = t^{-4}\ [\bigcirc]$ (47′)

where we have used eqs. (41) and (41′). Recall that [K] for a simple circle is independent of the orientation, since the allowed values for the charge are symmetric in sign. Also we note that, since according to our rules only terms in (29) are allowed to appear in eq. (47), there is no contribution from the annihilation-creation terms. It is amazing that although in the present case u_1, u_2, v_1, and v_2 become very different, they still lead to eqs. (47) and (47′) which are very similar to eqs. (18) with no annihilation-creation terms included. Now it is easy to see that in order to obtain a topological polynomial also invariant under Reidemeister moves of type I, the proper normalization of [K] should be

$$\tilde{P}_K = t^{-4\omega(K)}[K]/[O] \qquad (48)$$

where $\omega(K)$ is the writhe of the knot K as defined by eq. (20).

In summary, the polynomials (48), together with eqs. (12), (41) and (41′), provide us with an ambient isotopy invariant (under Reidemeister moves of all three types). Although the annihilation-creation terms we have included in our ansatz contribute neither to the "cross-channel unitarity" condition (16) nor to the topological requirements (43), (46), (47) and (47′) under type-I moves, the fulfilment of these conditions is highly nontrivial, since the scattering vertices have undergone a significant change with the inclusion of annihilation-creation terms through the Yang-Baxter equations. So the polynomial $\tilde{P}_K$ is a new topological polynomial, different from the P_K given in Sec. II with no annihilation-creation terms in the S-matrix elements.

A few examples suffice to exhibit this difference:

K	P_K	$\tilde{P}_K$
	$t^{-2}(1+t^{-4}+t^{-6})$	$t^{-2}(1+t^{-6}+t^{-12})$
	$t^{4}(1+t^{4}-t^{8})$	$t^{4}(1+t^{6}-t^{10}+t^{12}-t^{14}t^{16}+t^{18})$
	$t^{-6}(1-t^{4}+t^{6}-t^{8}+t^{12})$	$t^{-12}(1-t^{2}-t^{4}+2t^{6}-t^{8}-t^{10}+3t^{12}$ $-t^{14}-t^{16}+2t^{18}-t^{20}-t^{22}+t^{24})$

Let us consider more examples about knots which are the closure of some braids. For example, for σ_1^{-3} we have

$$\begin{aligned}
&\tilde{P}(\sigma_1^{3}) = t^{12}[\sigma_1^{3}]/[0] \\
&= t^{12}(t^{2}+1+t^{-2})^{-1}(t^{12}-t^{8}-t^{6}-t^{4}+t^{-2}+t^{-4}+t^{-6}+t^{-8}+t^{-10}) \\
&\longrightarrow\ t^{2}(t^{9}-t^{8}-t^{7}+t^{6}-t^{5}+t^{3}+1) \qquad \text{(with } t^{2}\to t) \qquad (49)
\end{aligned}$$

For $\sigma_1^{-1}\sigma_2\sigma_1^{-1}\sigma_2$ we have, after $t^2\to t$,

$$\tilde{P}(\sigma_1^{-1}\sigma_2\sigma_1^{-1}\sigma_2) = t^{-6}(t^{12}-t^{11}-t^{10}+2t^{9}-t^{8}-t^{7}+3t^{6}-t^{5}-t^{4}+2t^{3}-t^{2}-t+1) \qquad (50)$$

Some remarks are in order. First, in deriving eq. (49), we have met diagrams like those in Fig. 6a and 6b in which we have a pair creation and annihilation. The problem is how to calculate $|S|$ for such a state or a diagram. Our rule is the following: we shrink such a pair (together with the dot lines connected to them) to a point and then calculate $|S|$ according to the rule (13). This procedure and the results are presented in Fig. 6. Thus, $|S|$ is the same as if the annihilation-creation loop is absent. As we will see later, this rule enables us to derive the generalized Alexander-Conway relation for $\tilde{P}_K$ directly form the reduction relation for the S-matrix without worrying about the factor $t^{-|S|}$ in the definition (12).

Secondly, we observe that eqs. (49) and (50) coincide with the results of Akutzu and Wadati[8]. Unlike their polynomials, which are defined only for braid-closures, ours are defined for arbitrary oriented knots or links with explicit invariance under Reidemeister moves. To show that whenever the knot or link is a closed braid, our link polynomial coincides with Akutzu-Wadati's, we are going to prove that the reduction relation for our S-matrix and the generalized Alexander-Conway (skein) relation for our link polynomial $\tilde{P}_K$ are exactly the same as theirs.

To obtain the reduction relation, let us consider the elements of $S^{-2}-(S^{-1})^2$. We are looking for a relation like

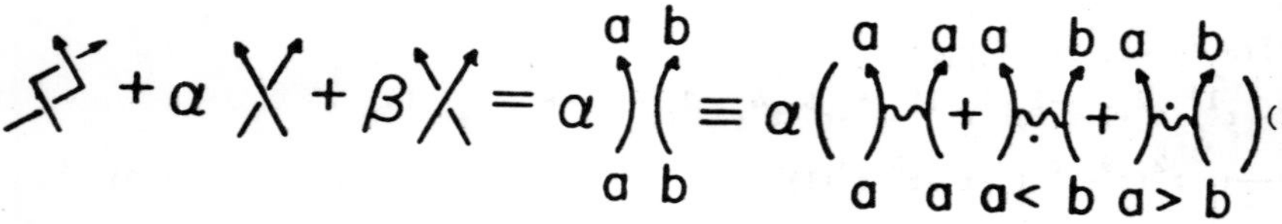

(51)

Substituting eqs. (29) and (32) into the left-hand side, one finds that

$$\alpha = -(t^4 - t^2 + t^{-2}) \quad \text{and} \quad \beta = t^4 \tag{52}$$

$$\gamma = t^6 - t^2 + 1 \tag{53}$$

Namely, such α and β make all non-diagonal elements vanish and what is left is proportional to the identity matrix with the coefficient γ. After changing $t^2 \rightarrow t$, one can rewrite eq. (51) as

$$S^{-3} = (t^2 - t + t^{-1})\, S^{-2} + (t^3 - t + 1)\, S^{-1} - t^2 \tag{54}$$

Setting $b_i = tS^{-1}$,

$$b_i^{\,3} = (t^3 - t^2 + 1)\, b_i^{\,2} + (t^5 - t^3 + t^2)\, b_i - t^5 \tag{55}$$

This is nothing but the cubic relation of Akutzu and Wadati for their braid group representation in the N=3 case[8].

For the generalized Alexander-Conway relation among the link polynomials $\tilde{P}_K$, consider four knots or link with all crossings the same except for one part which is respectively one of the four diagrams in the left-hand side and the middle of eq. (51). Now we are also looking for a relation among them which is formally similar to eq. (51). What is different here is that we have to take into account the writhe-dependent factors, which are respectively t^8, t^4, t^{-4} and t^0 for the left four diagrams in eq. (51). Therefore, with $\alpha \rightarrow t^4\alpha$, $\beta \rightarrow t^{12}\beta$, $\gamma \rightarrow t^8\gamma$ we obtain

$\tilde{P}\Big[$  $\Big] = (t^8-t^6+t^2)\ \tilde{P}[$ 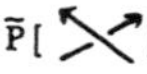 $] + t^8(t^6-t^2+1)\ \tilde{P}[$  $] - t^{16}\ \tilde{P}[$  $]$ (56)

Again, changing $t^2 \to t$, one obtains what Akutzu and Wadati has for N=3. We note that the rule for $|S|$ given in Fig. 6 for diagrams involving annihilation-creation loops is just such that the four sub-diagrams in eq. (56) all make the same contribution. This is why we are able to derive eq. (56) from eq. (51) without worrying about the factor $t^{-|S|}$ in the definition (12).

VI. The N=4 Case

For the N=4 case, the charges of particles can take values only in the set $\{-3, -1, +1, +3\}$. In this section we only write down the results without proofs which proceed in exactly the same way we described in above sections.

The diagrammatic representations of the S-matrix and its inverse are the same as eqs. (29) and (32). But eqs. (30.a-d) for the S-matrix elements are changed to

$$u = u_1(\delta_{a,-3}+\delta_{a,3}) + u_2(\delta_{a,-1}+\delta_{a1}) \tag{57.a}$$

$$v = v_1(\delta_{a,-3}\delta_{b,-1}+\delta_{a1}\delta_{b3}) + v_2(\delta_{a,-3}\delta_{b1} + \delta_{a,-1}\delta_{b3}) + v_3\delta_{a,-1}\delta_{b1} + v_4\delta_{a,-3}\delta_{b3} \tag{57.b}$$

$$p = p_1(\delta_{a,-3}\delta_{b,-1} + \delta_{a,-1}\delta_{b,-3} + \delta_{a1}\delta_{b3} + \delta_{a3}\delta_{b1}) + p_2(\delta_{a,-1}\delta_{b1} + \delta_{a1}\delta_{b,-1}) + p_3(\delta_{a,-3}\delta_{b3} + \delta_{a3}\delta_{b,-3}) \tag{57.c}$$

$$q = q_1(\delta_{a1}\delta_{b1}\delta_{c,-1}\delta_{d,3} + \delta_{a,-1}\delta_{b,-1}\delta_{c,-3}\delta_{d1} + \delta_{a,-1}\delta_{b3}\delta_{c,1}\delta_{d1} + \delta_{a,-3}\delta_{b1}\delta_{c,-1}\delta_{d,-1}) + q_2(\delta_{a1}\delta_{b,-1}\delta_{c,-3}\delta_{d3} + \delta_{a,-3}\delta_{b3}\delta_{c1}\delta_{d,-1}) + q_3(\delta_{a,-1}\delta_{b1}\delta_{c,-3}\delta_{d3} + \delta_{a,-3}\delta_{b3}\delta_{c,-1}\delta_{d1}) \tag{57.d}$$

For the inverse matrix elements we have

$$\bar{S}^{33}_{33}=u'_1,\quad \bar{S}^{11}_{11}=u'_2,\quad \bar{S}^{31}_{31}=v'_1,\quad \bar{S}^{3-1}_{3-1}=v'_2,\quad \bar{S}^{1-1}_{1-1}=v'_3,\quad \bar{S}^{3-3}_{3-3}=v'_4$$

$$\bar{S}^{13}_{13}=p'_1,\quad \bar{S}^{-11}_{1-1}=p'_2,\quad \bar{S}^{-33}_{3-3}=p'_3,\quad \bar{S}^{3-1}_{11}=q'_1,\quad \bar{S}^{-11}_{3-3}=q'_2,\quad \bar{S}^{1-1}_{3-3}=q'_3 \tag{58}$$

Here $\bar{S} = S^{-1}$. All other non-vanishing elements are obtained from the symmetry ansatz (25) and (26). In addition, our ansatz satisfies the charge conservation (22) too. Also in eq. (57.d) only the q_2-, q_3- terms correspond to annihilation-creation diagrams. The q_1-terms are a sort of generalization in which either the initial or the final states contain a pair of particles of the same charges. Later we will see that this is necessary for reproducing the Akutzu-Wadati N=4 polynomials for closed braids.

Imposing the "unitarity," "factorization," "cross-channel unitarity" conditions (4), (5), (16) and the topological conditions (43) and (44), we obtain the following solution for the parameters:

$$\begin{cases} u_1=t^2,\ u_2=t^{-2},\ v_1=t^2(1-t^{-6}),\ v_2=t^2(1-t^{-4})(1-t^{-6}) \\ v_3=(1-t^{-4})(1+t^{-2}), v_4=t^2(1-t^{-2})(1-t^{-4})(1-t^{-6}) \\ p_1= -t^{-1},\ p_2=t^{-4},\ p_3= -t^{-3},\ p_4= -t^{-7} \\ q_1=(1+t^{-2})[(1-t^{-2})(1-t^{-6})]^{1/2},\ q_2=t^{-2}(1-t^{-6}), q_3= -t(1-t^{-4})(1-t^{-6}) \end{cases} \tag{59}$$

and

$$\begin{cases} u'_1=t^{-2},\ u'_2=t^2,\ v'_1= -t^4(1-t^{-6}),\ v'_2=t^8(1-t^{-4})(1-t^{-6}) \\ v'_3= -t^6(1-t^{-4})(1+t^{-2}),\ v'_4= -t^{10}(1-t^{-2})(1-t^{-4})(1-t^{-6}) \\ p'_1= -t,\ p'_2=t^4,\ p'_3 = -t^3,\ p'_4 = -t^7, \\ q'_1=t^6(1+t^{-2})[(1-t^{-2})(1-t^{-6})]^{1/2}, q'_2=-t^8(1-t^{-6}), q'_3=-t^9(1-t^{-4})(1-t^{-6}) \end{cases} \tag{60}$$

Again, eqs. (60) can be obtained from eqs. (59) by $t \to t^{-1}$.

Similar to eqs. (47) and (47') we have, for N=4,

$[$  $] = t^{5}[\bigcirc]$, $\quad [$  $] = t^{-5}[\bigcirc]$ (61)

where

$$[\bigcirc] = t^{3} + t + t^{-1} + t^{-3} \tag{62}$$

So the corresponding topological link polynomials are

$$\tilde{P}_K = t^{-5\omega(K)}[K]/[O] \tag{63}$$

with [K] defined by eq.(12).

The reduction relation for S^{-1}, which is given by eq. (60), reads

$$S^{-4} = (t^{-2}-t^{4}+t^{8}-t^{10})\, S^{-3} + t^{2}(1-t^{4}+t^{6}+t^{10}-t^{12}+t^{16})\, S^{-2}$$
$$+ t^{10}(-1+t^{2}-t^{6}+t^{12})\, S^{-1} - t^{20} \tag{64}$$

with $b_i = tS^{-1}$ and $t^{2} \to t$, this is exactly the Akutzu-Wadati reduction relation for the N=4 case. Finally, for the generalized Alexander-Conway relation we have

$\tilde{P}[$ 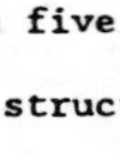 $] = t^{3}(1-t^{6}+t^{10}-t^{12})\, \tilde{P}[$ 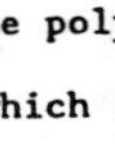 $] + t^{12}(1-t^{4}+t^{6}+t^{10}-t^{12}+t^{16})\, \tilde{P}[$ 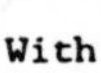 $]$

$+ t^{25}(-1+t^{2}-t^{6}+t^{12})\, \tilde{P}[$  $] - t^{40}\, \tilde{P}[$  $]$ (65)

where the five $\tilde{P}[\ \ ]$ above denote the polynomials for the five knots which have the same structure except one part which is indicated in the bracket. With $t^{2} \to t$ once again, this relation coincides with Akutzu-Wadati's for N=4. Also a rule similar to that given below eq. (50) and in Fig. 6 is needed for evaluating |S| for diagrams involving loops from anninihilation and creation such that the five subdiagrams appearing in eq. (65) all make the same contribution to the factor $t^{-|S|}$.

In principle, the generalization to higher N is possible and straightforward, but it is much more time-consuming. Perhaps one can write a computer program to systematically search for new link polynomials in the present diagrammatic approach.

VII. Conclusions and Discussions

In the above we have developed Kauffman's state model for Jones and HOMFLY polynomials, which uses Feynman diagrams for scattering processes, into a powerful and systematic diagrammatic approach for new link polynomials. In doing so we have exploited techniques which have been familiar to particle physicists.

First, we have classified the discrete symmetries for the S-matrix and found that CPT symmetry and charge conservation seems indispensable. In particular, we gave a knot-theory interpretation for CPT symmetry. C- and P-symmetries seems to be violated. And in our attempt to construct explicitly link polynomials we have imposed CP- and T-symmetries to narrow down the choices of ansatz. Violation of these symmetries, but maintaining CPT, is worthwhile to try. (At this stage we do not quite understand what symmetry except CPT is essential for obtaining link polynomials or braid-group representations. Especially why we need the charge conservation and have no separate C or P symmetry is a mystery.) Despite this, symmetry consideration is very helpful for systematizing ansatz for S-matrix, which should contain enough parameters to satisfy desired conditions when we include annihilation-creation processes and, probably, other complications.

Once an appropriate ansatz for S-matrix is set, the diagrammatic approach for finding braid group representations or link polynomials can be carried out step by step. First, one needs to check the unitarity and factorization conditions (4) and (5). This will restrict part of parameters in the ansatz and give braid-group representations containing some parameters. Then, one can impose the "cross-channel unitarity" condition (16). More parameters are determined and one will obtain a regular isotopy invariant polynomial from the bracket polynomials (12) defined by Kauffman. Then, to obtain an ambient isotopy invariant, the topological requirements (43) and (44) which are the first time explicitly formulated in this paper have to be satisfied. In all known situations, at this stage only one parameter, i.e. the parameter t appearing in the definition (12), survives. However, one still has to check the proportionality conditions (18) or (47), (47') or (61). Finally with appropriate normalization which is writhe-dependent, one can remove the proportionality factor and obtain an ambient isotopy invariant polynomial. It would be interesting to see whether one can obtain more two-variable link polynomials from this approach other than the HOMFLY polynomial.

We have explicitly carried out this diagrammatic approach for the N=3,4 cases. (N is the number of different charges.) With appropriate ansatz we have found new link polynomials which are defined for any oriented knots or links. For braid-closures our link polynomials coincide with those found previously by Akutzu and Wadati in a different approach. Since their polynomials are defined only on braid-closures, our results constitute the generalization of their polynomials to generic oriented knots or links which are obtained from braid closures by Reidemeister moves. The invariance under all three types of the Reidemeister moves is explicit by our step-by-step diagrammatic construction. From our point of view, the incorporation of annihilation-creation diagrams and other complications is essential to obtain the Akutzu-Wadati polynomials and their generalizations.

We believe more new link polynomials will come out of the diagrammatic approach by carefully choosing alternative ansatz for the S-matrix satisfying perhaps different symmetries. If N gets higher, there is more room for different ansatz. However, the diagrammatic analysis will become more time-consuming too. Perhaps a computer program would be helpful. But an analysis for a general N may still be available for certain symmetry ansatz.

We have seen that the number of conditions to be satisfied is actually bigger than the number of parameters introduced in the ansatz, so we feel that the success is a miracle: what is essential for such a success is unclear at the moment. The diagrammatic approach is used as a systematic trial-and-error procedure. Theoretically, it would be very interesting to see if there is a physical model which produces the desired S-matrix.

Acknowledgements

Two of us, M.L. Ge and Y.-S. Wu, are deeply grateful to Prof. C.N. Yang for bringing our attention to this field. We also thank Dr. I. Bakas for sending a copy of his notes on ref. [19]. This work was supported in part by Chinese National Science Foundation through Nankai University and U.S. National Science Foundation through grant No. PHY-8706501.

References

1. Y.-S. Wu, Phys. Rev. Lett. 53 (1984) 111.

2. H.C. Lee, M.L. Ge, M. Couture and Y.-S. Wu, Int. J. Mod. Phys. (to be published).

3. J. Fröhlich, 1987 Cargese Lectures, to appear in "Nonperturbative Quantum Field Theory" (Plenum Press).

4. M.L. Ge, H.C. Lee and Y.-S. Wu, talk given at the XVII Int. Collq. on Group Theoretical Methods in Physics, Saint-Adøle (Quebec), Canada; June 1988.

5. V.F.R. Jones, Notes of a talk in Atiyah's Seminar; and manuscript to be published.

6. L. Kauffman, Topology 26 (1987) 395; and in Proceeding of the Conference on Artin's Braid Group, Snata Cruz, California; July 1986.

7. Y. Akutzu and M. Wadati, J. Phys. Soc. Japan 56 (1987) 839, 3039; Commun. Math. Phys. 117 (1988) 243.

8. Y. Akutzu, T. Deguchi and M. Wadati, J. Phys. Soc. Japan 56 (1987) 3464; ibid. 57 (1988) 1173.

9. V.G. Turaev, LOMI preprint E-3-87.

10. Z.Q. Ma and B.H. Zhao, preprint BIHEP-88-31.

11. A. Tsuchiya and Y. Kanie, Lett. Math. Phys. 13 (1987) 303; and in "Conformal Field Theory and Solvable Lattice Models", Advanced Studies in Pure Mathematics 12 (1988) 297;

12. E. Verlinde, Nucl. Phys. B300 (1988) 360.

13. G. Moore and N. Sieberg, Phys. Lett. B212 (1988) 451; IAS preprints HEP-88/31 and 88/39.

14. B. Schroer, Nucl. Phys. B295 (1988) 4; K.-H. Rehren abd B. Schroer, FU preprint (1987).

15. G. Segal, "Conformal Field Theory", Oxford preprint 1987.

16. E. Witten, IAS preprint HEP-88/33.

17. C. Rovelli and L. Smolin, Phys. Rev. Lett. 61 (1988) 1155; L. Smolin, in "New Perspectives in Canonical Gravity", by A. Ashtekar, (Bibliopolis, Napoli; 1988), Part V, Chapter 6; C. Rovelli, ibid, Part V, Chapter 8.

18. C.N. Yang, Phys. Lett 19, 1312 (1967); R.J. Baxter, Ann. Phys. 70 (1972) 193; see, also, R.J. Baxter, "Exactly Solved Models in Statistical Mechanics", (Academic Press, London; 1982).

19. L. Kauffman, Lectures at Univ. Texas at Austin, March 1988.

20. V.F.R. Jones, Bull. Am. Math. Soc. 12 (1985) 103; Ann. Math. 126 (1987) 335.

21. P. Freyd, D. Yetter, J. Hoste, W.B.R. Lickorish, K. Millett and A. Ocneanu, Bull. Am. Math. Soc. 12 (1985) 239.

22. J.S. Birman, "Braids, Links, and Mapping Class Groups," Annals of Mathematics Studies No. 82, (Princeton University Press, 1974).

23. L. Kauffman, "On Knots", (Princeton University Press, 1985).

24. A.B. Zamolodchikov and A.B. Zamolodchikov, Nucl. Phys.

25. J.W. Alexander, Trans. Amer. Math. Soc. 30 (1928) 275.

Figure Captions

Fig. 1 Form a knot from the closure of a braid.

Fig. 2 A state of the knot in Fig. 1 with $a \neq b$, $a < c$, $b > c$.

Fig. 3 Diagrammatic representation of the "unitarity" condition

(3a) The $a = d > b = c$ case

(3b) The $a = -b = 2$, $c = d = 0$ case

Fig. 4 Diagrammatic representation of special Yang-Baxter equations

(4a) The $a = b = d = e < c = f$ case

(4b) The $a = d \neq b = e \neq c = f$ case

(4c) The $a = b = 0$, $c = 2$ case

(4d) The $a = 0$, $b = -2$, $c = 0$ case

Fig. 5 Diagrammatic representation of the "cross-channel unitarity" condition

Fig. 6 The rule for $|S|$ for diagrams with internal annihilation-creation loops

(6a) A diagram with $|S| = 0$

(6b) A diagram with $|S| = 4$

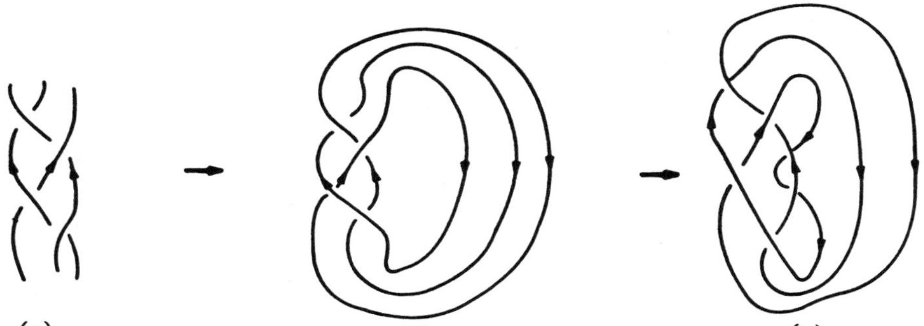

(a) (b) (c)

Fig. 1

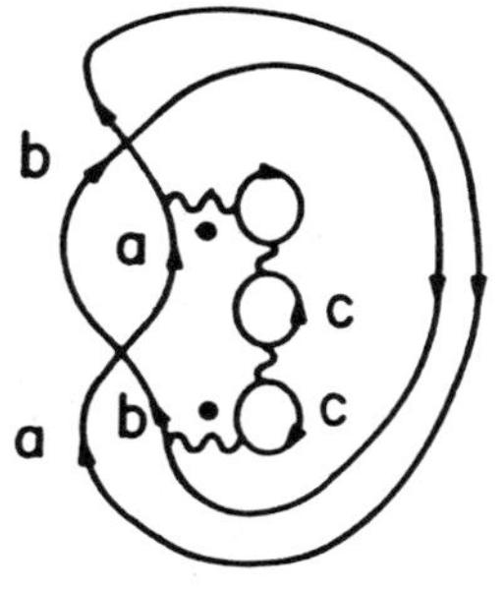

Fig. 2

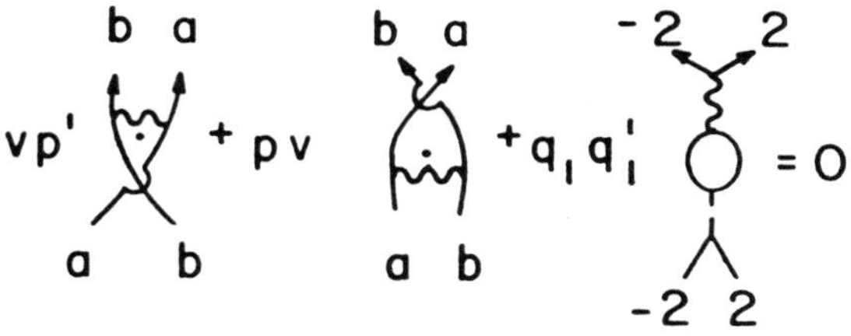

Fig. 3a

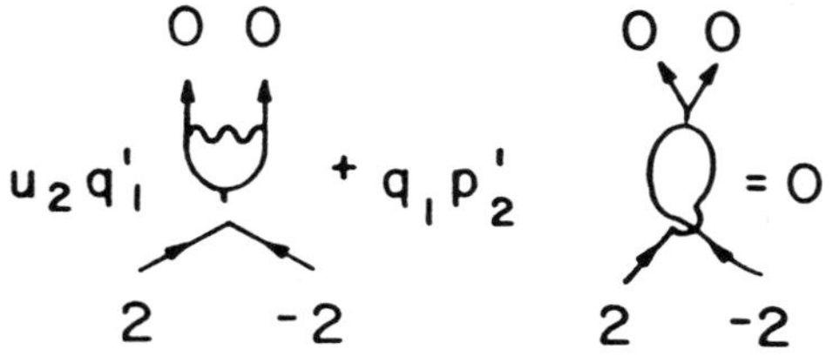

Fig. 3b

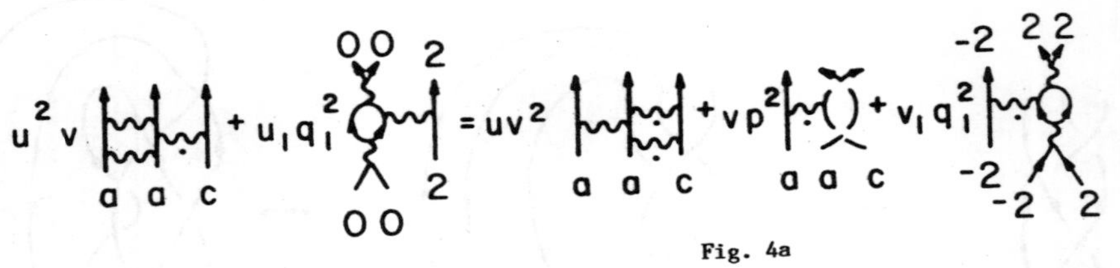

Fig. 4a

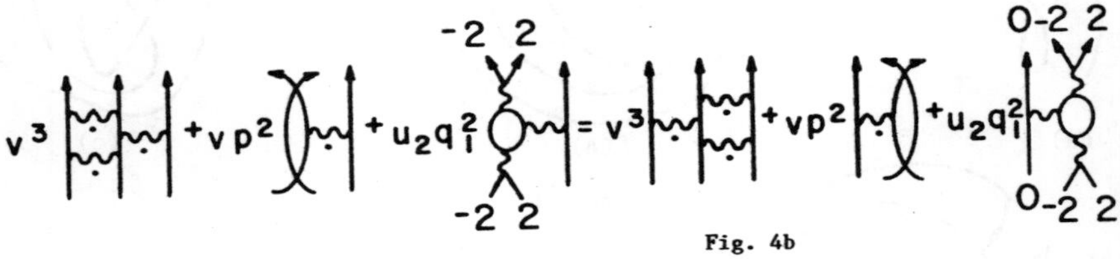

Fig. 4b

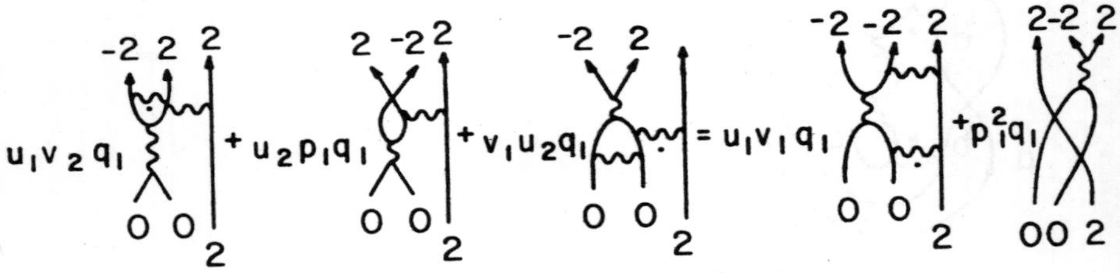

Fig. 4c

Fig. 4d

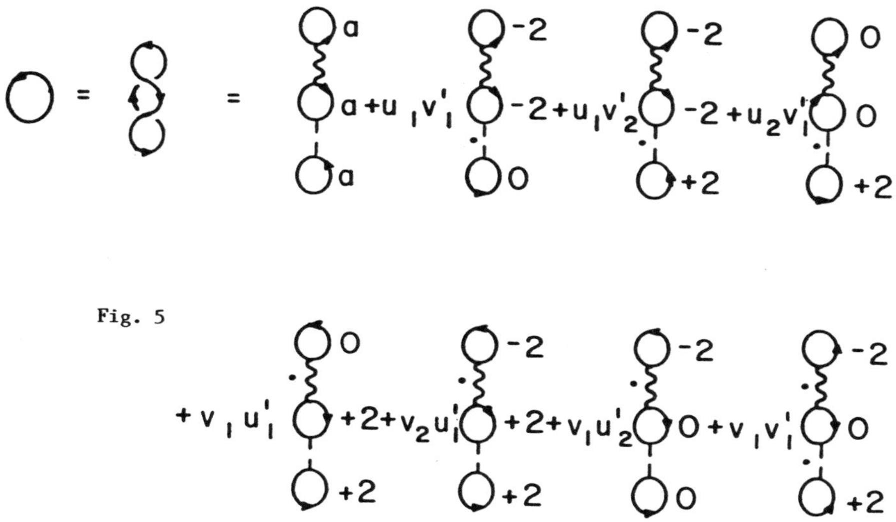

Fig. 5

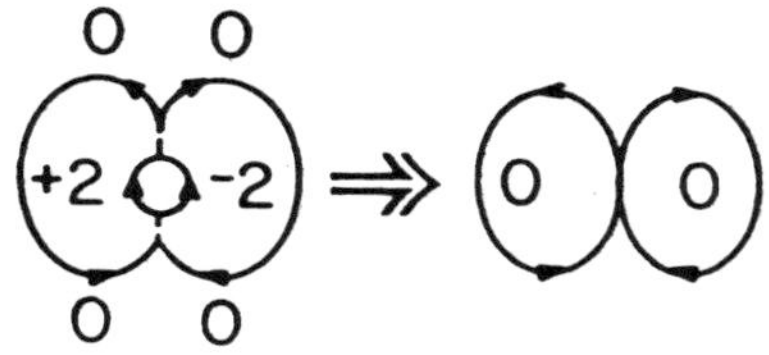

Fig. 6a

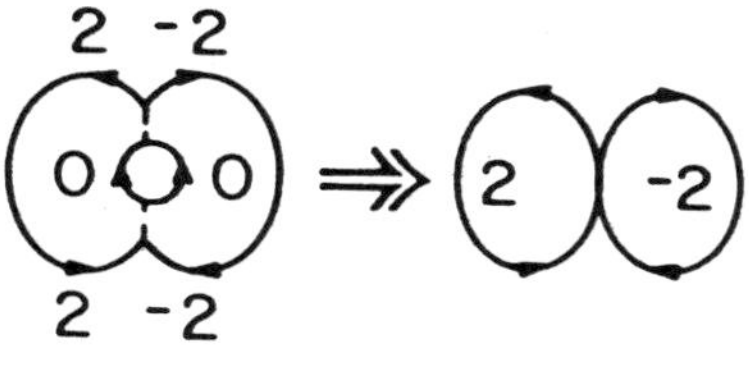

Fig. 6b

QUANTUM FIELD THEORY AND THE JONES POLYNOMIAL*

EDWARD WITTEN†

School of Natural Sciences
Institute for Advanced Study
Olden Lane
Princeton, N.J. 08540

ABSTRACT

It is shown that $2+1$ dimensional quantum Yang-Mills theory, with an action consisting purely of the Chern-Simons term, is exactly soluble and gives a natural framework for understanding the Jones polynomial of knot theory in three dimensional terms. In this version, the Jones polynomial can be generalized from S^3 to arbitrary three manifolds, giving invariants of three manifolds that are computable from a surgery presentation. These results shed a surprising new light on conformal field theory in $1+1$ dimensions.

⋆ An expanded version of a lecture at the IAMP Congress, Swansea, July, 1988.

† Research supported in part by NSF Grant No. 86-20266, and NSF Waterman Grant 88-17521.

In a lecture at the Hermann Weyl Symposium last year [1], Michael Atiyah proposed two problems for quantum field theorists. The first problem was to give a physical interpretation to Donaldson theory. The second problem was to find an intrinsically three dimensional definition of the Jones polynomial of knot theory. These two problems might roughly be described as follows.

Donaldson theory is a key to understanding geometry in four dimensions. Four is the physical dimension at least macroscopically, so one may take a slight liberty and say that Donaldson theory is a key to understanding the geometry of space-time. Geometers have long known that (via de Rham theory) the self-dual and anti-self-dual Maxwell equations are related to natural topological invariants of a four manifold, namely the second homology group and its intersection form. For a simply connected four manifold, these are essentially the only classical invariants, but they leave many basic questions out of reach. Donaldson's great insight [2] was to realize that moduli spaces of solutions of the self-dual Yang-Mills equations can be powerful tools for addressing these questions.

Donaldson theory has always been an intrinsically four dimensional theory, and it has always been clear that it was connected with mathematical physics at least at the level of classical nonlinear equations. The puzzle about Donaldson theory was whether this theory was tied to more central ideas in physics, whether it could be interpreted in terms of quantum field theory. The most important evidence for the existence of such a connection had to do with Floer's work on three manifolds [3] and the nature of the relation between Donaldson theory and Floer theory. Also, the 'Donaldson polynomials' had an interesting formal analogy with quantum field theory correlation functions. It has turned out that Donaldson theory can indeed be given a physical interpretation [4].

As for the Jones polynomial and its generalizations [5–11], these deal with the mysteries of knots in three dimensional space (figure 1). The puzzle on the mathematical side was that these objects are invariants of a three dimensional situation, but one did not have an intrinsically three dimensional definition. There were many elegant definitions of the knot polynomials, but they all involved looking in some way at a two dimensional projection or slicing of the knot, giving a two dimensional algorithm for computation, and proving that the result is independent of the chosen projection. This is analogous to studying a physical theory that is in fact relativistic but in which one does not know of a manifestly relativistic formulation – like quantum electrodynamics in the 1930's.

On the physical side, the puzzle about the knot polynomials was the following. Unlike the Donaldson theory, where a connection with quantum field theory was not obvious, the knot polynomials have been intimately connected almost from the beginning with two dimensional many body physics. In fact, constructions of the knot polynomials have related them to two dimensional (or $1+1$ dimensional) many-body physics in a bewildering variety of ways, mainly involving soluble lattice models [7], solutions of the Yang-Baxter equation [8], and monodromies of conformal field theory [11]. In the latter interpretation, the knot polynomials are related to aspects of conformal field theory that have been particularly fruitful recently [12–16]. On the statistical mechanical side, studies of the knot polynomials have related them to Temperley-Lieb algebras and their generalizations, and to other aspects of soluble statistical mechanics models in $1+1$ dimensions. For physicists the challenge of the knot polynomials has been to bring order to this diversity, find the unifying themes, and learn what it is that is three dimensional about two dimensional conformal field theory.

Now, the Donaldson and Jones (and Floer and Gromov [17]) theories deal with topological invariants, and understanding these theories as quantum field theories involves constructing theories in which all of the observeables are topological invariants. Some physicists might consider this to be a little bit strange, so let us pause to explain the physical meaning of 'topological invariance.' The physical meaning is really 'general covariance.' Something that can be computed from a manifold M as a topological space (perhaps with a smooth structure) without a choice of metric is called a 'topological invariant' (or a 'smooth invariant') by mathematicians. To a physicist, a quantum field theory defined on a manifold M without any *a priori* choice of a metric on M is said to be generally covariant. Obviously, any quantity computed in a generally covariant quantum field theory will be a topological invariant. Conversely, a quantum field theory in which all observeables are topological invariants can naturally be seen as a generally covariant quantum field theory. Indeed, the Donaldson, Floer, Jones, and Gromov theories can be seen as generally covariant quantum field theories in four, three, and two space-time dimensions. The surprise, for physicists, perhaps comes in how general covariance is achieved. General relativity gives us a prototype for how to construct a quantum field theory with no *a priori* choice of metric – we introduce a metric, and then integrate over all metrics. This example is so influential in our thinking that we tend to think of a generally covariant theory as being, by definition, a theory in which the metric is a dynamical variable. The lesson from the Donaldson, Floer, Jones, and Gromov theories is precisely that there are highly non-trivial quantum field theories in which general covariance is realized in other ways. In particular, in these notes we will describe an exactly soluble generally covariant quantum field theory in which general covariance is achieved not by integrating over metrics but because we begin with a gauge invariant Lagrangian that does not contain a metric.

This work originated with the realization that some results about conformal field theory described by G. Segal could be given a three dimensional interpretation by considering a gauge theory with Chern-Simons action. I am grateful to Segal for explaining his results, and to M. Atiyah for interesting me in and educating me about the Jones polynomial. V. F. R. Jones and L. Kauffman, and other participants at the IAMP Congress, raised many relevant questions. Finally, I must thank S. Deser and D. J. Gross for pointing out Polyakov's paper, G. Moore and N. Seiberg for explanations of their work, and the organizers of the IAMP Congress for their hospitality.

1. The Chern-Simons Action

We have been urged [1] to try to interpret the Jones polynomial in terms of three dimensional Yang-Mills theory. So we begin on an oriented three manifold M with a compact simple gauge group G. We pick a G bundle E, which may as well be trivial, and on E we place a connection $A_i{}^a$, which can be viewed as a Lie algebra valued one form (a runs over a basis of the Lie algebra, and i is tangent to M). An infinitesimal gauge transformation is

$$A_i \to A_i - D_i\epsilon, \tag{1.1}$$

where ϵ, a generator of the gauge group, is a Lie algebra valued zero form and the covariant derivative is $D_i\epsilon = \partial_i\epsilon + [A_i, \epsilon]$. The curvature is the Lie algebra valued two form $F_{ij} = [D_i, D_j] = \partial_i A_j - \partial_j A_i + [A_i, A_j]$. Now we need to choose a Lagrangian. We will *not* pick the standard Yang-Mills action*

$$\mathcal{L}_0 = \int\limits_M \sqrt{g} g^{ik} g^{jl} \operatorname{Tr}(F_{ij} F_{kl}), \tag{1.2}$$

as this depends on the choice of a metric g_{ij}. We want to formulate a generally covariant theory (in which all observeables will be topological invariants), and to this aim we want to pick a Lagrangian which does not require any choice of metric. Precisely in three dimensions there is a reasonable choice, namely the integral of the Chern-Simons three form:

$$\begin{aligned}\mathcal{L} &= \frac{k}{4\pi} \int\limits_M \operatorname{Tr}\left(A \wedge dA + \frac{2}{3} A \wedge A \wedge A\right) \\ &= \frac{k}{8\pi} \int\limits_M \epsilon^{ijk} \operatorname{Tr}\left(A_i(\partial_j A_k - \partial_k A_j) + \frac{2}{3} A_i [A_j, A_k]\right).\end{aligned} \tag{1.3}$$

The Chern-Simons term in three dimensional gauge theory has a relatively long history. The abelian gauge theory with only a Chern-Simons term was studied

---

* In what follows, the symbol 'Tr' denotes an invariant bilinear form on the Lie algebra of G, a multiple of the Cartan-Killing form; we will specify the normalization presently.

by A. Schwarz [18] and in unpublished work by I. Singer. Three dimensional gauge theories with the Chern-Simons term added to the usual action (1.2) were introduced in [19–21]. The nonabelian theory with only Chern-Simons action was studied classically by Zuckerman [22]. The abelian Chern-Simons theory has recently been studied in relation to fractional statistics by Hagen [24] and by Arovas, Schrieffer, Wilczek, and Zee [25] and recently in relation to linking numbers by Polyakov [23] and Frohlich [15]. The novelty in our present discussion is that we will consider the quantum field theory defined by the nonabelian Chern-Simons action and argue that it is exactly soluble and has important implications for three dimensional geometry and two dimensional conformal field theory.

The first fundamental property of the Chern-Simons theory is the quantization law first discussed in [21]. It arises because the group $\widehat{G}$ of continuous maps $M \to G$ is not connected. In the homotopy classification of such maps one meets at least the fact that $\pi_3(G) \simeq \mathbf{Z}$ for every compact simple group G. Though (1.3) is invariant under the component of the gauge group that contains the identity, it is not invariant under gauge transformations of non-zero 'winding number,' gauge transformations associated with non-zero elements of $\pi_3(G)$. Under a gauge transformation of winding number m, the transformation law of (1.3) is

$$\mathcal{L} \to \mathcal{L} + \text{constant} \cdot m. \tag{1.4}$$

As in Dirac's famous work on magnetic monopoles, consistency of quantum field theory does not quite require the single-valuedness of $\mathcal{L}$, but only of $\exp(i\mathcal{L})$. For this purpose, it is necessary and sufficient that the 'constant' in (1.4) should be an integral multiple of 2π. This gives a quantization condition on the parameter called k in (1.3). If G is $SU(N)$ and 'Tr' means a trace in the N dimensional representation, then the requirement is that k should be an integer. In general, for any G, we can uniquely fix the so far unspecified normalization of 'Tr' so that the quantization condition is $k \in \mathbf{Z}$.

We will see later that k is very closely related to the central charge in the theory of highest weight representations of affine Lie algebras. It is no accident that the reasoning which shows that k must be quantized in (1.3) has a $1+1$ dimensional analogue [26] which leads to quantization of the central charge in the representation theory of affine algebras.

In quantum field theory, in addition to a Lagrangian, one also wishes to pick a suitable class of gauge invariant observeables. In the present context, the usual gauge invariant local operators would not be appropriate, as they spoil general covariance. However, the 'Wilson lines' so familiar in QCD give a natural class of gauge invariant observeables that do not require a choice of metric. Let C be an oriented closed curve in M. Intrinsically C is simply a circle, but the topological classification of embeddings of a circle in M is very complicated, as we observe in figure 1. Let R be an irreducible representation of G. One then defines the 'Wilson line' $W_R(C)$ to be the following functional of the connection A_i. One computes the holonomy of A_i around C, getting an element of G that is well-defined up to conjugacy, and then one takes the trace of this element in the representation R. Thus, the definition is

$$W_R(C) = \mathrm{Tr}_R\, P \exp \int_C A_i dx^i. \tag{1.5}$$

The crucial property of this definition is that there is no need to introduce a metric, so general covariance is maintained.

We now can formulate the general problem of interest. In an oriented three manifold M, we take r oriented and non-intersecting knots C_i, $i = 1 \dots r$, whose union is what knot theorists would call a 'link' L. We assign a representation R_i to each C_i, and we propose to calculate the Feynman path integral

$$\int D\mathcal{A} \exp(i\mathcal{L}) \prod_{i=1}^{r} W_{R_i}(C_i). \tag{1.6}$$

The symbol $D\mathcal{A}$ represents Feynman's integral over all gauge orbits, that is, an integral over all equivalence classes of connections modulo gauge transformations.

Of course, (1.6) has exactly the formal structure of some familiar observeables in QCD, the difference being that we are in three dimensions instead of four and we have chosen a somewhat exotic gauge theory action. We will call (1.6) the 'partition function' of M with the given link, or the (unnormalized) 'expectation value' of the given link; we will denote it as $Z(M; C_i, R_i)$ or simply as $Z(M; L)$ for short.

For the case of links in S^3, we will claim that the invariants (1.6) are exactly those that appear in the Jones theory and its generalizations. Simply replacing S^3 with a general oriented three manifold M gives a very intriguing (and as we will see, effectively computable) generalization of the known knot polynomials. Taking $r = 0$ (no knots), (1.6) gives invariants of the oriented three manifold M which also turn out to be effectively computable. Before getting into any details, let us note a few preliminary indications of a possible connection between (1.6) and the Jones theory:

(1) In (1.6) we see the right variables, namely a compact Lie group G, a choice of representation R_i for each component C_i of the link L, and an additional variable k. (In knot theory one usually makes an analytic continuation and replaces k by a complex variable q, but it has been known since Jones' original work that there are special properties at special values of q. We claim that these properties reflect the fact that the three dimensional gauge theory with action (1.3) is well-defined only if k is an integer.) The two variable generalization of the Jones polynomial corresponds to the case that G is $SU(N)$, and the R_i are all the defining N dimensional representation of $SU(N)$. The two variables are N and k, analytically continued to complex values.

(2) As a further check on the plausibility of a relation between (1.6) and the knot polynomials, let us note first of all that (1.6) depends on a choice of the orientation of M, as this enters in fixing the sign of the Chern-Simons form. Likewise, (1.6) depends on the orientations of the C_i, since these enter in defining the Wilson lines (in computing the holonomy around C_i, one must decide in which

direction to integrate around C_i). If, however, one reverses the orientation of one of the C_i and simultaneously exchanges the representation R_i with its complex conjugate $\overline{R}_i$, then the definition of the Wilson lines is unchanged, so (1.6) is invariant under this process. And if (without changing the R_i) one reverses the orientations of *all* components C_i of the link L, then (1.6) is unchanged because of a symmetry that physicists would call 'charge conjugation.' This is an involution of the Lie algebra of G that exchanges all representations with their complex conjugates; applying this involution to all integration variables in (1.6) leaves (1.6) invariant while exchanging all R_i with their conjugates or equivalently reversing the orientation of all the C_i. These are important formal properties of the knot polynomials.

2. The Weak Coupling Limit

To begin with, since a non-abelian gauge theory with only a Chern-Simons action may seem unfamiliar, one might ask whether this Lagrangian really does lead to a sensible quantum theory, and really can be regulated to give topologically invariant results. In this section, we will briefly investigate this point by studying the theory in a weak coupling limit in which computations are comparatively straightforward. This is the limit of large k.* For large k, the path integral

$$Z = \int D\mathcal{A}\ \exp\left(\frac{ik}{4\pi}\int_M \mathrm{Tr}\left(A \wedge dA + \frac{2}{3}A \wedge A \wedge A\right)\right) \tag{2.1}$$

(for the moment we omit knots) contains an integrand which is wildly oscillatory. The large k limit of such an integral is given by a sum of contributions from the points of stationary phase. The stationary points of the Chern-Simons action are

---

* The reader may wish to bear in mind that the discussion in this section and the next contains a number of technicalities which are part of the logical story but perhaps not essential on a first reading.

precisely the 'flat connections,' that is, the gauge fields for which the curvature vanishes

$$F_{ij}{}^{a} = 0. \tag{2.2}$$

Gauge equivalence classes of such flat connections correspond to homomorphisms

$$\phi : \pi_1(M) \to G. \tag{2.3}$$

or more exactly to equivalence classes of such homomorphisms, up to conjugation. If for simplicity we suppose that the topology of M is such that there are only finitely many classes of homomorphisms (2.3), then the large k behavior of (2.1) will be a sum

$$Z = \sum_{\alpha} \mu(A^{(\alpha)}), \tag{2.4}$$

where the $A^{(\alpha)}$ are a complete set of gauge equivalence classes of flat connections, and $\mu(A^{(\alpha)})$ is to be obtained by stationary phase evaluation of (2.1), expanding around $A^{(\alpha)}$. This reduction to a stationary phase evaluation means that the nonabelian theory, for large k, is closely related to the abelian theory. This in turn has been shown [18] to lead to Ray-Singer analytic torsion [27], which is closely related to the purely topological Reidemeister torsion. The $\mu(A^{(\alpha)})$ may be evaluated as follows. We make in (2.1) the change of variables $A_i = A_i{}^{(\alpha)} + B_i$, where B_i is the new integration variable. An important invariant of the flat connection $A^{(\alpha)}$ is its Chern-Simons invariant

$$I(A^{(\alpha)}) = \frac{1}{4\pi} \int_M \mathrm{Tr}\left(A^{(\alpha)} \wedge dA^{(\alpha)} + \frac{2}{3} A^{(\alpha)} \wedge A^{(\alpha)} \wedge A^{(\alpha)}\right). \tag{2.5}$$

When the Chern-Simons action is expanded in powers of B_i, the first terms are

$$\mathcal{L} = k \cdot I(A^{(\alpha)}) + \frac{k}{4\pi} \int_M \mathrm{Tr}\left(B \wedge DB\right). \tag{2.6}$$

Here it is understood that in (2.6), the expression DB denotes the covariant exterior derivative of B with respect to the background gauge field $A^{(\alpha)}$; it does not depend on a metric on M. A salient point is that in (2.6) there is no term linear in B, since $A^{(\alpha)}$ is a critical point of the action.

To carry out the Gaussian integral in (2.6), gauge fixing is needed. There is no way to carry out this gauge fixing without picking a metric on M (or in some other way breaking the symmetry of the problem). After picking such a metric, a convenient gauge choice is $D_i B^i = 0$ (with D_i the covariant derivative constructed from the metric and the background gauge field $A^{(\alpha)}$). The standard Faddeev-Popov construction then gives rise to a gauge fixing Lagrangian

$$\mathcal{L}_{gauge} = \int_M \left(\operatorname{Tr} \phi D_i B^i + \operatorname{Tr} \bar{c} D_i D^i c\right). \tag{2.7}$$

Here ϕ is a Lagrangian multiplier that enforces the gauge condition $D_i B^i = 0$, and c, $\bar{c}$ are anticommuting 'ghosts' that are introduced to get the right measure on the space of gauge fields modulo gauge transformations. The quadratic terms in ϕ and B that can be found in (2.6), (2.7) have a natural geometric interpretation, described (in the abelian case) in [18]. Let D be the exterior derivative on M, twisted by the flat connection $A^{(\alpha)}$, and let $*$ be the Hodge operator that maps k forms to $3-k$ forms. On a three manifold one has a natural self-adjoint operator $L = *D + D*$ which maps differential forms of even order to forms of even order and forms of odd order to forms of odd order. Let L_- denote its restriction to forms of odd order. With B and ϕ regarded as a one form and a three form, respectively, the boson kinetic operator in (2.6), (2.7) is precisely this operator L_-. The kinetic operator of the ghosts is also a natural geometrical operator, the Laplacian, which we will call Δ. We can now give a formula for the stationary point contributions $\mu(A^{(\alpha)}$ that appear in (2.4). This is

$$\mu(A^{(\alpha)}) = \exp\left(ikI(A^{(\alpha)})\right) \cdot \frac{\sqrt{\det(L_-)}}{\det(\Delta)}. \tag{2.8}$$

The phase factor in (2.8) is the value of the integrand in (2.1) at the point of stationary phase, and the determinants (whose absolute values can be defined by zeta functions) result from the Gaussian integral over B, ϕ, c, and $\bar{c}$.

Now we come to the crucial point. To regularize the path integral, we have had to pick a Riemannian metric on M. Therefore, it is not obvious *a priori* that the $\mu(A^{(\alpha)})$ computed this way will really be topological invariants. Perhaps the Chern-Simons theory suffers from anomalies, and cannot be regularized in a generally covariant fashion. Happily, we can now appeal to [18], where it was shown (in the context of the abelian theory, but this aspect of [18] generalizes) that the absolute value of the ratio of determinants appearing in (2.8) is precisely the Ray-Singer analytic torsion of the flat connection $A^{(\alpha)}$, and so in particular is a topological invariant. (The phase of this ratio of determinants is more delicate, and will be discussed later.) This is the first indication that topological invariants really can be obtained from the Chern-Simons theory.

The Phase Of The Determinant

Though the absolute value of the ratio of determinants in (2.8) is the analytic torsion discussed long ago by Schwarz, the phase requires additional study. The ghost determinant $\det\Delta$ is real and positive, so the real issue is to study the phase of $\det L_-$. Because the operator L_- can be interpreted as a twisted Dirac operator, the phase of its determinant can be related to the study of the phase of odd dimensional fermion determinants, as studied by various authors [28]. However, I will here give a brief derivation of the relevant facts from the bosonic point of view, which is perhaps more natural in the present context. After an irrelevant rescaling of B and ϕ, the integral of interest is

$$\int DB\,D\phi \;\; \exp\Big(i\int_M \operatorname{Tr}(B\wedge DB + \phi D * B)\Big). \tag{2.9}$$

Upon changing variables to an orthonormal basis of eigenfunctions x_i of the operator L_-, with eigenvalues λ_i, (2.9) becomes

$$\prod_i \int_{-\infty}^{\infty} \frac{dx_i}{\sqrt{\pi}}\; e^{i\lambda_i x_i^2} \tag{2.10}$$

Therefore the crucial integral to understand is

$$I = \int_{-\infty}^{\infty} \frac{dx}{\sqrt{\pi}}\, e^{i\lambda x^2}, \tag{2.11}$$

for real λ. We consider this integral to be defined by taking the limit as $\epsilon \to 0$ of the absolutely convergent integral

$$\int_{-\infty}^{\infty} \frac{dx}{\sqrt{\pi}}\, e^{i\lambda x^2} \cdot e^{-\epsilon x^2}. \tag{2.12}$$

With this or any other physically reasonable definition, the integral (2.11) is

$$I = \frac{1}{|\sqrt{\lambda}|} \cdot \exp\left(\frac{i\pi}{4}\text{sign}\lambda\right). \tag{2.13}$$

The phase of the path integral is thus proportional to $\sum_i \text{sign}\lambda_i$, or better, to its regularized version which is the 'eta invariant' of Atiyah, Patodi, and Singer [29]:

$$\eta(A^{(\alpha)}) = \frac{1}{2}\lim_{s\to 0}\sum_i \text{sign}\lambda_i |\lambda_i|^{-s}. \tag{2.14}$$

Thus, the phase of the path integral may be expressed in the formula

$$\frac{1}{\sqrt{\det L_-}} = \frac{1}{|\sqrt{\det L_-}|} \cdot \exp\left(\frac{i\pi}{2}\eta(A^{(\alpha)})\right). \tag{2.15}$$

This can be made more explicit by using the Atiyah-Patodi-Singer theorem, which for our purposes can be regarded as a formula that expresses the dependence of η on the flat connection $A^{(\alpha)}$ about which we are expanding. In fact, in the case of the operator L_-, the formula is

$$\frac{1}{2}\left(\eta(A^{(\alpha)}) - \eta(0)\right) = \frac{c_2(G)}{2\pi} \cdot I(A^{(\alpha)}). \tag{2.16}$$

Here $I(A^{(\alpha)})$ is the Chern-Simons invariant of the flat connection $A^{(\alpha)}$, as defined in (2.5), $\eta(0)$ is the eta invariant of the trivial gauge field $A = 0$, and $c_2(G)$

is the value of the quadratic Casimir operator of the group G in the adjoint representation, normalized so that $c_2(SU(N)) = 2N$. The effect of this factor is to replace k in (2.8) by $k + c_2(G)/2$; in fact, the partition function (2.4) may now be written

$$Z = e^{i\pi\eta(0)/2} \cdot \sum_\alpha e^{i(k+c_2(G)/2)I(A^{(\alpha)})} \cdot T_\alpha \tag{2.17}$$

with T_α (the absolute value of the ratio of determinants in (2.8)) being the torsion invariant of the flat connection $A^{(\alpha)}$.

Unfortunately, although $I(A^{(\alpha)})$ and T_α are topological invariants, $\eta(0)$ is not; it depends on the choice of a metric on M in gauge fixing. Thus, to make sense of the phase of (2.17) requires further discussion, in the next subsection.

Before launching into that technical discussion, let us note that the computation just sketched actually has a very interesting spin-off. The fact that k in (2.8) has been replaced by $k + c_2(G)/2$ in (2.17) appears to be the beginning of an explanation of the fact that in many formulas of $1 + 1$ dimensional current algebra, quantum corrections have the effect of replacing k by $k + c_2(G)/2$. In turn, this is probably related to the fact that in various integrable models in $1+1$ dimensions, such as the sine-gordon model, the WKB approximation is exact if one makes suitable and seemingly *ad hoc* changes in the values of the parameters, analogous to replacing k by $k + c_2(G)/2$.

The Signature Of A Bounding Four Manifold

Now, let us discuss how the mysterious phase factor $e^{i\pi\eta(0)/2}$ in (2.17) should be interpreted.

First of all, $\eta(0)$ is the η invariant of the L_- operator coupled to (i) some metric g on M, and (ii) the trivial gauge field $A = 0$. Let $d = \dim G$ be the dimension of the gauge group G. Since the gauge field is trivial, the L_- operator consists of d copies of the purely gravitational L_- operator coupled to the metric only. Thus, as a preliminary, we write

$$\eta(0) = d \cdot \eta_{grav}, \tag{2.18}$$

where η_{grav} is the eta invariant of the purely gravitational operator. Our problematical phase factor is

$$\Lambda = \exp\left(\frac{id\pi}{2} \cdot \eta_{grav}\right). \tag{2.19}$$

Now, with a particular regularization of the Chern-Simons quantum field theory, we have obtained the formula (2.17) which contains the ambiguous phase factor Λ. The goal is to find a different regularization which will preserve general covariance. Two regularizations should differ by a local counterterm, and in this case, since the problem phase (2.19) depends on the background metric only, we want a counterterm that depends on the background metric only. It is easy to see that the counterterm with the right properties is a multiple of the gravitational Chern-Simons term, which is defined (by analogy with the Yang-Mills Chern-Simons term) as

$$I(g) = \frac{1}{4\pi} \int_M \mathrm{Tr}\left(\omega \wedge d\omega + \frac{2}{3}\omega \wedge \omega \wedge \omega\right). \tag{2.20}$$

Here ω is the Levi-Civita connection on the spin bundle of M.* $I(g)$ suffers from an ambiguity just similar to that of the Yang-Mills Chern-Simons action. To define $I(g)$ as a number, one requires a trivialization of the tangent bundle of M. Although the tangent bundle of a three manifold can be trivialized, there is no canonical way to do this. Any two trivializations differ by an invariantly

---

* (2.20) is not the integral of an intrinsic local functional, so it would not usually arise as a counterterm. Whether or not 'counterterm' is the right word, we will have to view (2.20) as a correction that must be added to the action if one wishes to work in the gauge $D_i A^i = 0$.

defined integer, which is the number of relative 'twists.' The gravitational Chern-Simons functional has the property that if the trivialization of the tangent bundle of M is twisted by s units, $I(g)$ transforms by

$$I(g) \to I(g) + 2\pi s. \tag{2.21}$$

Now, the Atiyah-Patodi-Singer theorem says that the combination

$$\frac{1}{2}\eta_{grav} + \frac{1}{12} \cdot \frac{I(g)}{2\pi} \tag{2.22}$$

is a topological invariant, depending that is on the oriented three manifold M with a choice of trivialization of the tangent bundle, but not on the metric of M.† It is clear, therefore, what we must do. We replace $\eta(0)/2$ in (2.17) by d times the combination that appears in (2.22) (the factor of d is the one that entered in (2.19)), so (2.17) is replaced by

$$Z = \exp\left(i\pi d\left(\frac{\eta_{grav}}{2} + \frac{1}{12} \cdot \frac{I(g)}{2\pi}\right)\right) \cdot \sum_{\alpha} e^{i(k+c_2(G)/2)I(A^{(\alpha)})} \cdot T_\alpha \tag{2.23}$$

So, finally, we can see that the Chern-Simons partition function, at least for large k, can be defined as a topological invariant of the oriented, framed three manifold M (a framed three manifold being one that is presented with a homotopy class of trivializations of the tangent bundle).

The fact that it is necessary to specify a framing of the three manifold may look like a nuisance, but there is no real loss of information. From (2.21) we see that if the framing is shifted by s units, the partition function is transformed by

$$Z \to Z \cdot \exp\left(2\pi i s \cdot \frac{d}{24}\right). \tag{2.24}$$

A topological invariant of framed, oriented three manifolds, together with a law for the behavior under change of framing, is more or less as good as a topological invariant of oriented three manifolds without a choice of framing.

---

† The crucial factor of 1/12 in (2.22) reflects the discrepancy between the Chern character $e^x = 1 + x^2/2 + \ldots$ that appears in gauge theory index theorems and the $\widehat{A}$ genus $(x/2)/\sinh(x/2) = 1 - x^2/24 + \ldots$ that appears in gravitational index theorems.

Of course, all of the discussion in this section, and in particular (2.24), **has** been limited to the behavior at large k. In section (4.5), we will see that the generalization of (2.24) to finite k is

$$Z \to Z \cdot \exp\left(2\pi i s \cdot \frac{c}{24}\right), \tag{2.25}$$

with c being the central charge of two dimensional current algebra with symmetry group G at level k. It is well known that the large k limit of c is exactly $d/24$.

Moduli Spaces Of Flat Connections

There is still an important gap in the above discussion of the large k behavior. The formula (2.8) is really only valid if the determinants that appear are all non-zero. In fact, the flat connection $A^{(\alpha)}$ determines a flat bundle E. The determinants in (2.8) are non-zero if and only if $A^{(\alpha)}$ is such that the de Rham cohomology of M, with values in E, is zero. If $H^1(M,E) \neq 0$, then the flat connection $A^{(\alpha)}$ is not isolated but lies on a moduli space $\mathcal{S}$ of gauge inequivalent flat connections; and the proper evaluation of the path integral (2.1) leads not to the discrete sum (2.4) but to an integral on $\mathcal{S}$. If $H^0(M,E)$ is not zero, then the fields ϕ, c and $\bar{c}$ in the above treatment have zero modes, and the gauge fixing requires more care. It is plausible that by more careful study of the path integral, the large k contribution of arbitrary flat connections can be extracted without assumptions about $H^*(M,E)$. But we will not attempt this.

Some Examples

We will later on determine the partition functions of some simple three manifolds, giving results that can be compared to large k computations. For $S^2 \times S^1$, $Z = 1$, for any G and any k. For S^3 and $G = SU(2)$, we will obtain the formula*

$$Z(S^3) = \sqrt{\frac{2}{k+2}} \sin\left(\frac{\pi}{k+2}\right). \tag{2.26}$$

* The appearance of $k+2$ in this formula is presumably an illustration of the $k + c_2(G)/2$ in (2.17).

Of course, on S^3 the only flat connection is the trivial connection, for which (2.8) is not valid, since $H^0(M, E) \neq 0$ in this case. For $G = SU(2)$, the behavior $Z \sim k^{-3/2}$ in (2.26) is probably the general behavior of the contribution of the flat connection for homology spheres (on which the flat connection is isolated); it would be interesting to know how to obtain this behavior from path integrals. In Donaldson and Floer theory, the trivial connection, which has a negative formal dimension, is the cause of many subtleties. The vanishing of (2.26) in the classical limit of large k appears to be an interesting quantitative reflection of the 'negative dimension' of the trivial connection.

2.1 INCORPORATION OF KNOTS

We now wish to consider the large k behavior in the presence of knots. For simplicity, we will limit ourselves to the case of S^3, and an abelian gauge group $G = U(1)$. Though the abelian gauge group is relatively trivial in the context of knot theory, it gives a quick and simple way to confirm the fact that the Chern-Simons action really does lead to topological invariants, and it also gives a simple context for explaining a technicality that is crucial in all that follows.

In the abelian theory, the gauge field is simply a one form A and the Lagrangian is

$$\mathcal{L} = \frac{k}{8\pi} \int_M \epsilon^{ijk} A_i \partial_j A_k. \tag{2.27}$$

We pick some circles C_a and some integers n_a (corresponding to representations of the gauge group $U(1)$). As always in this paper, we assume C_a does not intersect C_b for $a \neq b$. We wish to calculate the expectation value of the product

$$W = \prod_{a=1}^{s} \exp(i\, n_a \int_{C_a} A) \tag{2.28}$$

with respect to the Gaussian measure determined by $e^{i\mathcal{L}}$. As was recently discussed by Polyakov (in a paper [23] in which he proposed to apply the Abelian

Chern-Simons theory to high temperature superconductors), the result **can be** written in the form

$$< W >= \exp\left(\frac{i}{2k}\sum_{a,b} n_a n_b \int_{C_a} dx^i \int_{C_b} dy^j \epsilon_{ijk} \cdot \frac{(x-y)^k}{|x-y|^3}\right) \tag{2.29}$$

Here one has identified a region U of S^3 containing the knots with a region of three dimensional Euclidean space, and x^i, y^j are the Euclidean coordinates of U evaluated along the knots. For $a \neq b$, the integral in (2.29) is essentially the Gauss linking number, which can be written as

$$\Phi(C_a, C_b) = \frac{1}{4\pi}\int_{C_a} dx^i \int_{C_b} dy^j \epsilon_{ijk} \frac{(x-y)^k}{|x-y|^3}. \tag{2.30}$$

As long as C_a and C_b do not intersect, $\Phi(C_a, C_b)$ is a well defined integer; in fact, it is the most classic invariant in knot theory. Thus, if we could ignore the term $a = b$, we would have

$$< W >= \exp\left(\frac{2i\pi}{k}\sum_{a,b} n_a n_b \Phi(C_a, C_b)\right). \tag{2.31}$$

The appearance of the Gauss linking number illustrates the fact that the Chern-Simons theory does lead to topological invariants as we hope. But we have to worry about the term with $a = b$. This integral is ill-defined near $x = y$; how do we wish to interpret it?

It is well known in knot theory that there is no natural and topologically invariant way to regularize the self-linking number of a knot. Polyakov in [23] used a regularization that is not generally covariant to get an answer that is interesting geometrically but not a topological invariant. We need a different approach for our present treatment in which general covariance is a primary goal. Though there is no completely invariant substitute for Polyakov's regularization, in the sense that there is no way to get a natural topological invariant from the integral in (2.29) or (2.30) with $a = b$, we cannot simply throw away the self-linking term and its non-abelian generalizations (which are sketched in figure (3(a))), since these terms are in fact not naturally zero. There is no reason to think that one could retain general covariance by dropping these terms. In the abelian theory, on a general three manifold M, on topological grounds the self-linking number can be a non-zero fraction, well-defined only modulo one. In such a case, it cannot be correct to set the self-linking number to zero, since it is definitely not zero. (Topologically, in such a situation, the self-linking number is well defined only modulo an integer, and this precision is definitely not good enough to evaluate (2.31).) In the non-abelian theory, we will get results later which amount to assigning definite, non-zero values to the non-abelian generalizations of the self-linking integral, so it would not be on the right track to try to throw these terms away.

Topologically, it is clear what data are needed to make sense of the self-linking of a knot C. One needs to give a 'framing' of C; this is a normal vector field along C. The idea is that by displacing C slightly in the direction of this vector field one gets a new knot C', and it makes sense to calculate the linking number of C and C'. This can be defined as the self-linking number of the framed knot C. One can think of the framing as a thickening of the knot into a tiny ribbon bounded by C and C'; this is how it is drawn in figure (3(b)). It is clear that the self-linking number defined this way depends not on the actual vector field used to displace C to C' but only on the topological class of this vector field; and indeed by a 'framing' we mean only the topological class. Though a choice of framing gives a definition of the self-linking number of a knot C, it is clear that by picking a convenient framing of C one can get any desired answer for its self-linking number; as illustrated in figure (3(c)), a t-fold twist in the framing of C will change its self-linking by t.*

---

* The discussion should make is clear that the need to frame knots is analogous to the need to frame three manifolds, as found in the last section. This hopefully justifies the use of the same word 'framing' in each case.

Physically, the role of the framing is that it makes possible what **physicists** would call a point-splitting regularization. This is defined as follows: when one has to do the self-linking integral in (2.29), one lets x run on C and y on C'. This gives a well-defined integral, though of course it depends on the framing. In this paper, we will assume, without proof, that the framing gives sufficient information to make possible a consistent point-splitting regularization of all the non-abelian generalizations of the self-linking integral, without further arbitrary choices. This question is, perhaps, comparable to the question of whether the non-abelian Chern-Simons action defines a sensible quantum theory in the first place (even without introducing Wilson lines as observeables); neither of these questions will be tackled here.

Of course, if it were always possible to pick a canonical framing of knots, then we could pick this framing and hide the question. On S^3, there is a canonical framing of every knot; it is determined by asking that the self-linking number should be zero. (This makes the abelian linking integral zero, but not its non-abelian generalizations.) On general three manifolds, this cannot be done since the self-linking number may be ill-defined or may differ from an integer by a definite fraction (so that it does not vanish with any choice of framing). Even when the canonical framing does exist, it is not convenient to be restricted to using it, since natural operations (like the surgery we study in section (4)) may not preserve it.

In general, therefore, we give up on finding a natural choice, and simply pick some framing and proceed. It would be rather unpleasing if the 'physical' results depended uncontrollably on the framing of knots. What saves the day is that although we cannot in general make a natural choice of the framing, we can state a general rule for how expectation values of Wilson lines change under a change of the framing. First of all, let us note that while, in general, there is no canonical zero in the set of possible framings of a knot in a three manifold, if one compares two framings they always differ by a definite integer, which is the relative twist in going around the knot (figure(3(c)). (That is, in general there is no natural way to count how many times the ribbon in figure (3(b)) is twisted, but there is a natural local operation of adding t extra twists to this ribbon.) In the abelian theory, it is clear from (2.29) and (2.30) how the partition function transforms under a change of framing. If we shift the framing of the link C_a by t units, its self-linking number is increased by t, and the partition function is shifted by a phase

$$< W > \to \exp\left(2\pi i t \cdot ({n_a}^2/k)\right) \cdot < W > . \tag{2.32}$$

The nonabelian analog of that result will be derived in section (5.1); the transformation law in the non-abelian case is

$$< W > \to \exp(2\pi i t \cdot h) < W >, \tag{2.33}$$

where h is the conformal weight of a certain primary field in $1+1$ dimensional current algebra. This result, though it may seem rather technical, is a key ingredient enabling the Chern-Simons theory to work. It means that although we need to pick a framing for every link, because the self-linking integrals have no natural definition otherwise, there is no loss of information since we have a definite law for how the partition functions transform under change of framing.

Actually, it can be shown [13] that the structure of rational conformal field theory requires non-trivial monodromies. In the relationship that we will develop between the $2+1$ dimensional Chern-Simons theory and rational conformal field theory in $1+1$ dimensions, the need to frame all knots is the $2+1$ dimensional analog of the monodromies that arise in $1+1$ dimensions. (This will be clear in the derivation of (2.33).) Were it not for the seeming nuisance that knots must be framed to define the Wilson lines as quantum observeables, one would end up proving that the Jones knot invariants were trivial.

An alternative description may make the physical interpretation of the framing of knots more transparent. A Wilson line can be regarded as the space-time trajectory of a charged particle. In $2+1$ dimensions, it is possible for a particle to have fractional statistics, meaning that the quantum wave function changes by a phase $e^{2\pi i\delta}$ under a 2π rotation. (See [30] for a discussion of these issues.) If one wishes to compute a quantum amplitude with propagation of a particle of fractional statistics, it is not enough to specify the orbit of the particle; it is necessary to also count the number of 2π rotations that the particle undergoes in the course of its motion. Equations (2.32) and (2.33) mean that the particles represented by Wilson lines in the Chern-Simons theory have fractional statistics with $\delta = {n_a}^2/2k$ in the abelian theory or $\delta = h$ in the non-abelian theory. This fractional statistics is the phenomenon claimed by Polyakov in [23], so in essence we agree with his substantive claim, though we prefer to exhibit this phenomenon in the context of a generally covariant regularization, where it appears in the behavior of Wilson lines under change of framing.

In this section, we have obtained some important evidence that the Chern-Simons theory can be regularized to give invariants of three manifolds and knots. We have also obtained the important insight that doing so requires picking a homotopy class of trivializations of the tangent bundle, and a 'framing' of all knots. To actually solve the theory requires very different methods, to which we turn in the next section.

3. Canonical Quantization

The basic strategy for solving the Yang-Mills theory with Chern-Simons action on an arbitrary three manifold M is to develop a machinery for chopping M in pieces, solving the problem on the pieces, and gluing things back together. So to begin with we consider a three manifold M, perhaps with Wilson lines, as in figure (4(a)). We 'cut' M along a Riemann surface Σ. Near the cut, M looks like $\Sigma \times R^1$, and our first step in learning to understand the theory on an arbitrary three manifold is to solve it on $\Sigma \times R^1$.

The special case of a three manifold of the form $\Sigma \times R^1$ is tractable by means of canonical quantization. Canonical quantization on $\Sigma \times R^1$ will produce a Hilbert space $\mathcal{H}_\Sigma$, 'the physical Hilbert space of the Chern-Simons theory quantized on Σ.' * These will turn out to be finite dimensional spaces, and moreover spaces that have already played a noted role in conformal field theory. In rational conformal field theories, one encounters the 'conformal blocks' of Belavin, Polyakov, and Zamolodchikov. Segal has described these in terms of 'modular functors' that canonically associate a Hilbert space to a Riemann surface, and has described in algebra-geometric terms a particular class of modular functors, which arise in current algebra of a compact group G at level k [16]. The key observation in the present work was really the observation that precisely those functors can be obtained by quantization of a three dimensional quantum field theory, and that this three dimensional aspect of conformal field theory gives the key to understanding the Jones polynomial.

---

* It is conventional in physics to call vector spaces obtained in this fashion 'Hilbert spaces,' and we will follow this terminology. In fact, the claim that comes most naturally from path integrals and that we will actually use is only that $\mathcal{H}_\Sigma$ is a vector space canonically associated with Σ, and exchanged with its dual when the orientation of Σ is reversed. However, a Hilbert space structure is natural in the Hamiltonian viewpoint, and in the particular problem we are considering here, an inner product on $\mathcal{H}_\Sigma$ is important in more delicate aspects of conformal field theory; such an inner product gives a 'metric on the flat vector bundle' in the language of Friedan and Shenker [31]. According to Segal [16], $\mathcal{H}_\Sigma$ in fact has a canonical projective Hilbert space structure.

Actually, the general situation that must be studied is that in which possible Wilson lines on M are 'cut' by Σ, as in the figure. In this case Σ is presented with finitely many marked points $P_1, \ldots P_k$, with a G representation R_i assigned to each P_i (since each Wilson line has an associated representation). To this data – an oriented topological surface with marked points, and for each marked point a representation of G – we wish to associate a vector space. This is also the general situation that arises in conformal field theory – the marked points are points at which operators with non-vacuum quantum numbers have been inserted. If one reverses the orientation of Σ (and replaces the representations R_i associated with the marked points with their complex conjugates) the vector space $\mathcal{H}_\Sigma$ must be replaced with its dual.

The Canonical Formalism At first sight, (1.3) might look like a typically intractable nonlinear quantum field theory, but this is far from being so. Working on $\Sigma \times R^1$, it is very natural to choose the gauge $A_0 = 0$ (with A_0 being the component of the connection in the R^1 direction). In this gauge we immediately see that the Lagrangian becomes quadratic. It reduces to

$$\mathcal{L} = \frac{k}{8\pi} \int dt \int_\Sigma \epsilon^{ij} \operatorname{Tr} A_i \frac{d}{dt} A_j. \tag{3.1}$$

For the time being we will ignore extra complications due to Wilson lines that may be present on $\Sigma \times R^1$. From (3.1) we may deduce the Poisson brackets,†

$$\{A_i{}^a(x), A_j{}^b(y)\} = \frac{8\pi}{k} \cdot \epsilon_{ij} \delta^{ab} \delta^2(x-y). \tag{3.2}$$

Before rushing ahead to quantize these commutation relations, we should remember that the system is subject to a 'Gauss law' constraint, which is $\delta\mathcal{L}/\delta A_0 = 0$,

† This is a typical problem in which it is not appropriate to 'introduce canonical momenta.' The purpose of introducing such variables is to reexpress a given Lagrangian in a form which is first order in time derivatives, but (3.1) is already first order in time derivatives. The variables in (3.1) are already canonically conjugate, as indicated in the following equation.

or (ignoring the Wilson lines)

$$\epsilon^{ij} F_{ij}{}^{a} = 0. \tag{3.3}$$

This constraint equation is nonlinear (since F contains a quadratic term), and – as (3.1) is certainly a free theory – this nonlinearity is what remains of the underlying nonlinearity of (1.3).

In quantum field theory, one very often quantizes first and then imposes the constraints. The situation that we are considering here is a situation in which it is far more illuminating to first impose the constraints and then quantize. For the phase space $\mathcal{M}_0$ of connections $A_i{}^a(x)$ without the constraints is an infinite dimensional phase space; imposing the constraints will reduce us to a rather subtle but eminently finite dimensional phase space $\mathcal{M}$. The problem that faces us here, of reducing from $\mathcal{M}_0$ to $\mathcal{M}$ by imposing the constraints (3.2), has been studied before – and has proved to have extremely rich properties – in the work of Atiyah and Bott on equivariant Morse theory, two dimensional Yang-Mills theory, and the moduli space of holomorphic vector bundles [33]. In our present investigation, this familiar problem appears from a novel three dimensional vantage point.

It is necessary to recall the nature of constraint equations in classical physics. The constraints (3.2) are functions that should vanish, but they also generate gauge transformations via Poisson brackets. Imposing the constraints means two things classically: First, we restrict ourselves to values of the canonical variables for which the constraint functions vanish; and second, we identify two solutions of the constraint equations if they differ by a gauge transformation. In the case at hand, the first step means that we should consider only 'flat connections,' that is, connections for which $F_{ij}{}^{a} = 0$. The second step means that we identify two flat connections if they differ by a gauge transformation. Taking the two steps together, we see that the physical phase space, obtained by imposing the constraints (3.2), is none other than the moduli space of flat connections on Σ, modulo gauge transformations. Such flat connections are completely characterized by the 'Wilson lines,' that is, the holonomies around non-contractible loops on Σ. A simple count of parameters shows that on a Riemann surface of genus $g > 1$, the moduli space $\mathcal{M}$ of flat connections modulo gauge transformations has dimension $(2g-2) \cdot d$, where d is the dimension of the group G.

The topology of $\mathcal{M}$ is rather intricate (and this was in fact the main subject of interest in [33]). On general grounds $\mathcal{M}$ inherits a symplectic structure (that is, a structure of Poisson brackets) from the symplectic structure present on $\mathcal{M}_0$ before imposing the constraints. $\mathcal{M}$ is a compact space (with some singularities), and in particular its volume with the natural symplectic volume element is finite. Since in quantum mechanics there is one quantum state per unit volume in classical phase space, the finiteness of the volume of $\mathcal{M}$ means that the quantum Hilbert spaces will be finite dimensional. We would like to determine them.

3.1 The Holomorphic Viewpoint

Quantization of classical mechanics is usually carried out by separating the canonical variables into 'coordinates,' q^i, which are a maximal set of real commuting variables, and 'momenta,' p^j, which are conjugate to the q^i. The quantum Hilbert space is then the space $\mathcal{H}$ of square integrable functions of the q^i.

Such a scheme definitely requires a noncompact phase space of infinite volume, since – though the q^i may take values in a compact space – the p^j are definitely unbounded. Accordingly, the space $\mathcal{H}$ is infinite dimensional.

Quantizing a compact, finite volume phase space, such as the moduli space $\mathcal{M}$ of flat connections modulo gauge transformations, is quite a different kind of problem. It has no known general solution, but there is one important class of cases in which there is a natural notion of quantization. This arises in the case in which $\mathcal{M}$ is a Kahler manifold, and the symplectic structure on $\mathcal{M}$ is the curvature form that represents the first Chern class of a holomorphic line bundle L endowed with some metric. In this case, one carries out quantization not by separating the variables in phase space into 'coordinates' and 'momenta,' q's and p's, but by separating them into holomorphic and anti-holomorphic degrees of freedom, essentially $z \sim q + ip$ and $\overline{z} \sim q - ip$. The quantum Hilbert space $\mathcal{H}$ is then a suitable space of holomorphic 'functions.' More exactly, $\mathcal{H}$ is the space of holomorphic sections of the line bundle L. If $\mathcal{M}$ is compact, this latter space will be finite dimensional. In our problem, with $\mathcal{M}$ being the moduli space of flat connections modulo gauge transformations on an oriented smooth surface Σ, is there a natural Kahler structure on $\mathcal{M}$? The answer is crucial for all that follows. There is not quite a *natural* Kahler structure on $\mathcal{M}$, but there is a natural way to obtain such structures. Once one picks a complex structure J on Σ, the moduli space $\mathcal{M}$ of flat connections can be given a new interpretation – it is the moduli space of stable holomorphic $G_{\mathbf{C}}$ bundles on Σ which are topologically trivial ($G_{\mathbf{C}}$ is the complexification of the gauge group G). Let us refer to the latter space as $\mathcal{M}_J$. $\mathcal{M}_J$ is naturally a complex Kahler (and in fact projective algebraic) variety. Upon picking a linear representation of G (for our purposes it is convenient to pick a representation with the smallest value of the quadratic Casimir operator, e.g. the N dimensional representation of $SU(N)$ or the adjoint representation of E_8), and passing from a principal $G_{\mathbf{C}}$ bundle to the associated vector bundle, we can think of $\mathcal{M}_J$ as the moduli space of a certain family of holomorphic vector bundles. For $G = SU(N)$, $\mathcal{M}_J$ is simply the moduli space of all stable rank N holomorphic vector bundles of vanishing first Chern class.

The symplectic form on $\mathcal{M}$ that appears in (3.1) or (3.2) *without* picking a complex structure on Σ has a very special interpretation in holomorphic terms once we *do* pick such a complex structure. Let us recall the notion [34] of the determinant line bundle of the $\overline{\partial}$ operator. The $\overline{\partial}$ operator on Σ can be 'twisted' by any holomorphic vector bundle. $\mathcal{M}_J$ parametrizes a family of holomorphic vector bundles on Σ, and thus it can be regarded as parametrizing a family of $\overline{\partial}$ operators. Taking the determinant line gives a line bundle L over the base space $\mathcal{M}_J$ of this family. Furthermore [34], the Dirac determinant gives a natural metric on L, and the first Chern class of L, computed with this metric, is precisely the symplectic form that appears in (3.1) or (3.2), provided $k = 1$. For general k, the symplectic form that appears in (3.1) or (3.2) represents the first Chern class of the kth power of the determinant line bundle.*

Thus, all of the conditions are met for a straightforward quantization of (3.1), taking into account the constraints (3.3). The constraints mean that the classical space to be quantized is the moduli space $\mathcal{M}$ of flat connections. Picking an arbitrary complex structure J on Σ, $\mathcal{M}$ becomes a complex manifold, and the symplectic form of interest represents the first Chern class of $L^{\otimes k}$, the kth tensor power of the determinant line bundle. The quantum Hilbert space $\mathcal{H}_\Sigma$ is thus the space of global holomorphic section of $L^{\otimes k}$.

3.2 A Flat Vector Bundle On Moduli Space

This gives an answer to the problem of canonically quantizing the Chern-Simons theory on $\Sigma \times R^1$, but a crucial point now requires discussion.

Quantizing (3.1), with the constraints (3.3), is a problem that can be naturally asked whenever one is given an oriented smooth surface Σ. Beginning with a generally covariant Lagrangian in three dimensions, we were led to this problem in a context in which it was not natural to assume any metric or complex structure on Σ. However, to solve the problem and construct $\mathcal{H}_\Sigma$, it was very natural to pick a complex structure J on Σ. Thus, our description of $\mathcal{H}_\Sigma$ depends on the choice of J, and what we have called $\mathcal{H}_\Sigma$ might perhaps be better called $\mathcal{H}_\Sigma{}^{(J)}$. As J varies, the $\mathcal{H}_\Sigma{}^{(J)}$ vary holomorphically with J, and thus we could interpret this object as a holomorphic vector bundle on the moduli space of complex Riemann surfaces. But since $\mathcal{H}_\Sigma{}^{(J)}$ is the answer to a question that depends on Σ and not on J, we would like to believe that likewise the $\mathcal{H}_\Sigma{}^{(J)}$ canonically depend only on

---

* This description is valid for the gauge group $G = SU(N)$, but in general the following modification is needed. For groups other than $SU(N)$ the determinant line bundle L is not the fundamental line bundle on $\mathcal{M}$ but a tensor power thereof. For instance, for $G = E_8$, there is a line bundle L' with $(L')^{\otimes 30} \simeq L$. It is then L' whose first Chern class corresponds to (3.1) or (3.2) with $k = 1$.

Σ and not on J. The assertion that the $\mathcal{H}_\Sigma{}^{(J)}$ are canonically independent of J, and depend only on Σ, is the assertion that the vector bundle on moduli space given by the $\mathcal{H}_\Sigma^{(J)}$ has a canonical flat connection that permits one to identify the fibers. Such 'flat vector bundles on moduli space' first entered in conformal field theory somewhat implicitly in the differential equations of Belavin, Polyakov, and Zamolodchikov [32]. They were discussed much more explicitly by Friedan and Shenker [31], who proposed that they would play a pivotal role in conformal field theory, and they have been prominent in subsequent work such as [12,13]. At least in one important class of examples, we have just met a natural origin of 'flat vector bundles on moduli space.' The problem 'quantize the Chern-Simons action' can be posed without picking a complex structure, so the answer is naturally independent of complex structure and thus gives a 'flat bundle on moduli space.' The particular flat bundles on moduli space that we get this way are those that Segal has described [16] in connection with conformal field theory; Segal also rigorously proved the flatness, which is explained somewhat heuristically by the physical argument sketched above. (Because of the conformal anomaly, this bundle has only a projectively flat connection, with the projective factor being canonically odd under reversal of orientation.)

The role of these flat bundles in conformal field theory is as follows. If one considers current algebra on a Riemann surface, with a symmetry group G, at 'level' k, then one finds that in genus zero the Ward identities uniquely determine the correlation functions for descendants of the identity operator, but this is not so in genus ≥ 1. On a complex Riemann surface Σ of genus ≥ 1, the space of solutions of the Ward identities for descendants of the identity is a vector space $\widehat{\mathcal{H}}_\Sigma$, which might be called the 'space of conformal blocks.' Segal calls the association $\Sigma \to \widehat{\mathcal{H}}_\Sigma$ a 'modular functor,' and has given an algebra-geometric description of the modular functors that arise in current algebra. In quantizing the Chern-Simons theory we have exactly reproduced this description! This is then the secret of the relation between current algebra in $1+1$ dimensions and Yang-Mills theory in $2+1$ dimensions: the spaces of conformal blocks in $1+1$ dimensions are the quantum Hilbert spaces obtained by quantizing a $2+1$ dimensional theory. It would take us to far afield to explain here the algebra-geometric description of the space of conformal blocks. Suffice it to say that when one tries to use the Ward identities of current algebra to uniquely determine the correlation functions of descendants of the identity on a curve Σ of genus ≥ 1, one meets an obstruction which involves the existence of non-trivial holomorphic vector bundles on Σ; the Ward identities reduce the determination of the correlation functions to the choice of a holomorphic section of $L^{\otimes k}$ over the moduli space of bundles.

It seems appropriate to conclude this discussion with some remarks on the formal properties of the association $\Sigma \rightarrow \mathcal{H}_\Sigma$. It is good to first think of the functor $\Sigma \rightarrow H^1(\Sigma, R)$ which to a Riemann surface Σ associates its first de Rham cohomology group. This functor is defined for every smooth surface Σ, independent of complex structure. A diffeomorphism of Σ induces a linear transformation on $H^1(\Sigma, R)$, so $H^1(\Sigma, R)$ furnishes in a natural way a representation of the mapping class group. The formal properties of the functors $\Sigma \rightarrow \mathcal{H}_\Sigma$ that come by quantizing the Chern-Simons theory are quite analogous. Though a complex structure J on Σ is introduced to construct $\mathcal{H}_\Sigma$, the existence of a natural projectively flat connection on the moduli space of complex structures permits one locally to (projectively) identify the various $\mathcal{H}_\Sigma{}^{(J)}$ and forget about the complex structure. One might think that the global monodromies of the flat connection on moduli space would mean that globally one could not forget the complex structure, but this is not so; these monodromies just correspond to an action of the purely topological mapping class group, so that the formal properties of $\mathcal{H}_\Sigma$ are just like those of $H^1(\Sigma, R)$.

3.3 Inclusion Of Wilson Lines

So far, we have discussed the quantization of the Chern-Simons theory on a Riemann surface Σ *without* Wilson lines. Now we wish to include the Wilson lines, which, as in figure (4(b)), pierce Σ in some points P_i; associated with each such point is a representation R_i. Quantizing the Chern-Simons theory in the

presence of the Wilson lines should give a Hilbert space $\mathcal{H}_{\Sigma;P_i,R_i}$ that is canonically associated with the oriented surface Σ together with the choice of P_i and R_i.

It is pretty clear what problem in conformal field theory this should correspond to. Instead of simply considering correlation functions of the descendants of the identity, we should consider in the conformal field theory primary fields transforming in the R_i representations of G. With these fields (or their descendants) inserted at points P_i on Σ, one gets in conformal field theory a more elaborate space $\widehat{\mathcal{H}}_{\Sigma;P_i,R_i}$ of conformal blocks. Again, there is an algebra-geometric description of this space [16], and this is what we should expect to recover by quantizing the Chern-Simons theory in the presence of the Wilson lines.

I will now briefly sketch how this works out, deferring a fuller treatment for another occasion. First of all, the Wilson lines correspond to static non-abelian charges which show up as extra terms in the constraint equations. So (3.3) is replaced by

$$\frac{k}{8\pi}\epsilon^{ij}F_{ij}{}^a(x) = \sum_{s=1}^{r}\delta^2(x-P_s)T_{(s)}{}^a, \tag{3.4}$$

where P_s, $s = 1\ldots r$ are the points at which static external charges have been placed, and $T_{(s)}{}^a$, $a = 1\ldots \dim G$ are the group generators associated with the external charges. Now, a naive attempt to quantize (3.1) with the generalized constraints (3.4) would run into extremely unpleasant difficulties. One could try to quantize first and then impose the constraints, but this is difficult to see through even in the absence of the external charges. Alternatively, one can try to impose the constraints at the classical level and then quantize, as we did above. But it is hard to make sense of (3.4) as constraints in the classical theory; the solution $A_i{}^a$ of (3.4) cannot be an ordinary c-number connection, since non-commuting operators appear on the right hand side. It is clear that to solve (3.4), $A_i{}^a$ would have to be some sort of 'q-number connection,' whose holonomy would presumably be an element of a 'quantum group,' not an ordinary classical group. Indeed, it seems likely that the theory of quantum groups [35] can be considered to arise in this way.

However, there is a much better way to quantize the Chern-Simons theory with static charges. We certainly wish to impose (3.4) at the classical level. This cannot be done directly, since on the right hand side there appear quantum operators. A useful point of view is the following. A representation R_i of a group G should be seen as a quantum object. This representation should be obtained by quantizing a classical theory. The Borel-Weil-Bott theorem gives a canonical way to exhibit for every irreducible representation R of a compact group G a problem in classical physics, with G symmetry, such that the quantization of this classical problem gives back R as the quantum Hilbert space. One introduces the 'flag manifold' G/T, with T being a maximal torus in G, and for each representation R one introduces a symplectic structure ω_R on G/T, such that the quantization of the classical phase space G/T, with the symplectic structure ω_R, gives back the representation R. Many aspects of representation theory find natural explanations by thus regarding representations of groups as quantum objects that are obtained by quantization of classical phase spaces.

In the problem at hand, this point of view can be used to good effect. We extend the phase space $\mathcal{M}_0$ of G connections on Σ by including at each marked point P_i a copy of G/T, with the symplectic structure appropriate to the R_i representation. The quantum operators $T_{(i)}{}^a$ that appear on the right of (3.4) can then be replaced by the classical functions on G/T whose quantization would give back the $T_{(i)}{}^a$. The constraints (3.4) then make sense as classical equations, and the analysis can be carried out just as we did without marked points, though the details are a bit longer. Suffice it to say that after imposing the classical constraints, one gets a finite dimensional phase space $\mathcal{M}_{P_i,R_i}$ that incorporates the static charges; a point on this space is a flat G connection on Σ with a reduction of structure group to T at the points P_i. Upon picking an arbitrary conformal structure on Σ, this phase space can be quantized. In this way one gets exactly Segal's description of the space of conformal blocks in current algebra in a general situation with primary fields in the R_i representation inserted at the points P_i. (In current algebra at level k, one only permits certain representations, the 'integrable ones.' If one formally tries to include other representations, the Ward identities show that they decouple [36]. According to Segal, the analogous statement in algebraic geometry is that the appropriate line bundle over $\mathcal{M}_{P_i,R_i}$ has no non-zero holomorphic sections unless the R_i all correspond to integrable representations. For the Chern-Simons theory, this means that unless the representations R_i are all integrable, the zero vector is the only vector in the physical Hilbert space.)

Finally, let us note that the Borel-Weil-Bott theorem should not be used simply as a tool in quantization. It should be built into the three dimensional description. One should use the theorem to replace the Wilson lines (1.5) that appear in (1.6) with a functional integral over maps of the circle S into G/T (or actually an integral over sections of a G/T bundle, twisted by the restriction to S of the G-bundle E). This gives a much more unified formalism.

3.4 The Riemann Sphere With Marked Points

The above description may seem a little bit dense, and we will supplement it by giving a simple intuitive description of the physical Hilbert space $\mathcal{H}_{\Sigma;R_i,P_i}$ in the important case of genus zero. Let Σ be an oriented surface of genus zero, with static charges in the R_i representation at points P_i. Let us consider the case of very large k. Now, the gauge coupling in (1.3) is of order $1/k$, so for large k we are dealing with very weak coupling. Rather naively, one might believe that for extremely weak coupling the physical Hilbert space is the same as it would be if the charges were not coupled to gauge fields. If so, the physical Hilbert space would be simply the tensor product $\mathcal{H}_0 = \otimes_i R_i$ of the Hilbert spaces R_i of the individual charges. However, there is a key error here. No matter how weak the gauge coupling may be, we must remember that in a closed universe the total charge must be zero (since the electric flux has nowhere to go). The total charge being zero means in a nonabelian theory that all of the charges together must be coupled to the trivial representation of G. So the physical Hilbert space, for large k, is precisely the G-invariant subspace of $\mathcal{H}_0$, or

$$\mathcal{H} = \mathrm{Inv}(\otimes_i R_i). \tag{3.5}$$

This is a familiar answer in conformal field theory for the space of conformal blocks obtained, in the large k limit, in coupling representations R_i. Considerations of conformal field theory also show that for finite k the correct answer is always a subspace of (3.5). The most important modification of (3.5) that arises for finite k (and is explained algebra-geometrically in [16]) is that $\mathcal{H}$ is zero unless the R_i correspond to integrable representations of the loop group; in what follows a restriction to such representations is always understood.

Now we consider some important special cases.

(i) For the Riemann sphere with no marked points, the Hilbert space is one dimensional. This is well known in conformal field theory – for descendants of the identity on the Riemann sphere, there is only one conformal block.

(ii) For the Riemann sphere with one marked point in a representation R_i, the Hilbert space is one dimensional if R_i is trivial, and zero dimensional otherwise.

(iii) For the Riemann sphere with two marked points with representations R_i and R_j, the Hilbert space is one dimensional if R_j is the dual of R_i (so that there is an invariant in $R_i \otimes R_j$) and zero dimensional otherwise. Again, this is well known in conformal field theory.

(iv) For the Riemann sphere with three marked points in representations R_i, R_j, and R_k, the dimension of $\mathcal{H}$ is the number N_{ijk} for which Verlinde has proposed [12] and Moore and Seiberg have proved [13] rather striking properties. Here, N_{ijk} may in general be less than its large k limit which is the dimension of (3.5).

(v) From the results of Verlinde, the dimensions of the physical Hilbert spaces for an arbitrary collection of marked points on S^2 can be determined from a knowledge of the N_{ijk}. But let us consider a particularly important special case.

Suppose that there are four external charges, and that the representations are $R, R, \overline{R}$, and $\overline{R}$. If the decomposition of $R \otimes R$ is

$$R \otimes R = \oplus_{i=1}^{s} E_i, \tag{3.6}$$

with the E_i being distinct irreducible representations of G, then the physical Hilbert space $\mathcal{H}$ at large k will be s dimensional, since the possible invariants in $R \otimes R \otimes \overline{R} \otimes \overline{R}$ are uniquely fixed by giving the representation to which $R \otimes R$ is coupled. (For small k the dimension of $\mathcal{H}$ might be less than s.) In understanding the knot polynomials, an important special case is that in which G is $SU(N)$ and R is the defining N dimensional representation. In that case, $s = 2$ and the physical Hilbert space is two dimensional (except for $k = 1$ where it is one dimensional).

4. Calculability

Our considerations so far may have seemed somewhat abstract, and we would now like to show that in fact these considerations can actually be used to calculate things. As an introduction to the requisite ideas, we will first deduce a certain theoretical principle that is of great importance in its own right.

Consider, as in figure (5(a)), a three manifold M which is the connected sum of two three manifolds M_1 and M_2, joined along a two sphere S^2. There may be knots in M_1 or M_2, but if so they do not pass through the joining two sphere. If for every three manifold X we denote the partition function or Feynman path integral (1.6) as $Z(X)$, then we wish to deduce the formula

$$Z(M) \cdot Z(S^3) = Z(M_1) \cdot Z(M_2) \tag{4.1}$$

(it being understood that $Z(S^3)$ denotes the partition function of a three sphere that contains no knots). This can be rewritten

$$\frac{Z(M)}{Z(S^3)} = \frac{Z(M_1)}{Z(S^3)} \cdot \frac{Z(M_2)}{Z(S^3)}. \tag{4.2}$$

In some special cases, (4.2) is equivalent or closely related to known formulas. If M_1 and M_2 are copies of S^3 with knots in them, then the ratios appearing in (4.2) turn out to be the knot invariants that appear in the Jones theory, and (4.2) expresses the fact that these invariants are multiplicative when one takes the disjoint sum of knots. If M_1 and M_2 are arbitrary three manifolds without knots, then (in view of our discussion in section (2)) (4.2) is closely related to the multiplicativity of Reidemeister and Ray-Singer torsion under connected sums.

So let us study figure (5(a)) using the general ideas of quantum field theory. On the left of this figure, we see a three manifold M_1 with boundary S^2. According to the general ideas of quantum field theory, one associates a 'physical Hilbert space' $\mathcal{H}$ with this S^2; as we have seen in the last section, it is one dimensional. The Feynman path integral on M_1 determines a vector χ in $\mathcal{H}$. Likewise, on the right of figure (5(a)) we see a three manifold M_2 whose boundary is the same S^2 with opposite orientation; its Hilbert space $\mathcal{H}'$ is canonically the dual of $\mathcal{H}$. The path integral on M_2 determines a vector ψ in $\mathcal{H}'$, and according to the general ideas of quantum field theory, the partition function of the connected sum M is

$$Z(M) = (\chi, \psi). \tag{4.3}$$

The symbol (χ, ψ) denotes the natural pairing of vectors $\chi \in \mathcal{H}$, $\psi \in \mathcal{H}'$. We cannot evaluate (4.3), since we do not know χ or ψ. Instead, let us consider some variations on this theme. The two sphere S^2 that separates the two parts of figure (5(a)) could be embedded in S^3 in such a way as to separate S^3 into two three balls B_L and B_R. The path integrals on B_L and B_R would give vectors v and v' in $\mathcal{H}$ and $\mathcal{H}'$, and the same reasoning as led to (4.3) gives

$$Z(S^3) = (v, v'). \tag{4.4}$$

Again, we do not know v or v' and cannot evaluate (4.4). But we can say the following. As $\mathcal{H}$ is one dimensional, v is a multiple of χ; likewise, since $\mathcal{H}'$ is one dimensional, v' is a multiple of ψ. It is then a fact of one dimensional linear algebra that

$$(\chi, \psi) \cdot (v, v') = (\chi, v') \cdot (v, \psi). \tag{4.5}$$

The two terms on the right hand side of (4.5) are respectively $Z(M_1)$ and $Z(M_2)$, as we see in figure (5(c)). So (4.5) is equivalent to the desired result (4.1).

One may wonder what is the mysterious object $Z(S^3)$ that is so prominent in (4.1). Can it be set to one? Actually, the axioms of quantum field theory are strong enough so that the value of $Z(S^3)$ is uniquely determined and cannot be postulated arbitrarily; as we will see later it can be calculated from the theory of affine Lie algebras. For $G = SU(2)$ the formula has been given in (2.26).

As a special case of (4.1), pick s irreducible representations of G, say $R_1, \ldots R_s$, and consider a link in S^3 that consists of s unlinked and unknotted circles C_i, with one of the R_i associated with each circle. This is indicated in figure (6). Denote the partition function of S^3 with this collection of Wilson lines as $Z(S^3; C_1, \ldots C_s)$ (the representations R_i being understood). Then by cutting the figure to separate the circles, and repeatedly using (4.1), we learn that

$$\frac{Z(S^3; C_1, \ldots C_s)}{Z(S^3)} = \prod_{k=1}^{s} \frac{Z(S^3; C_k)}{Z(S^3)}. \tag{4.6}$$

If we introduce the normalized expectation value of a link L, defined by $< L >= Z(S^3; L)/Z(S^3)$, then (4.6) becomes

$$< C_1 \ldots C_s >= \prod_k < C_k > \tag{4.7}$$

for an arbitrary collection of unlinked, unknotted Wilson lines on S^3.

In knot theory there is another notion of connected sum, the 'connected sum of links.' The Jones invariants also have a simple multiplicative behavior under this operation, as we will sketch briefly at the end of section (4.5).

4.1 Knots in S^3

We will now describe the origin of the 'skein relation' which can be taken as the definition of the knot polynomials for knots on S^3. (A special case of the skein relation was first used by Conway in connection with the Alexander polynomial.)

Consider a link L on a general three manifold M, as indicated in figure (7(a)). The components of the link are associated with certain representations of G, and we wish to calculate the Feynman path integral (1.6), which we will denote as $Z(L)$ (with the representations understood). We will evaluate it by deducing an algorithm for unknotting knots. If the lines in figure (7) could pass through each other unimpeded, all knots could be unknotted. As it is, this is prevented by some unfortuitous crossings, such as the one circled in the figure. Let us draw a small sphere about this crossing, cut it out, and study it more closely. This cuts M into two pieces, which after rearrangement are shown in figure (7(b)) as a complicated piece M_L shown on the left of the figure and a simple piece M_R shown on the right. M_R consists of a three ball with boundary S^2; on this boundary there are four marked points that are connected by two lines in the interior of the ball.

To make the discussion concrete, let us suppose that the gauge group is $G = SU(N)$ and that the Wilson lines are all in the defining N dimensional representation of $SU(N)$, which we will call R. Then, as we saw at the end of the last section, the physical Hilbert spaces $\mathcal{H}_L$ and $\mathcal{H}_R$ associated with the boundaries of M_L and M_R are two dimensional.

The strategy is now the same as the strategy which led to the multiplicativity relation (4.1). The Feynman path integral on M_L determines a vector χ in $\mathcal{H}_L$. The Feynman path integral on M_R determines a vector ψ in $\mathcal{H}_R$. The vector spaces $\mathcal{H}_L$ and $\mathcal{H}_R$ (which are associated with the same Riemann surface S^2 with opposite orientation) are canonically dual, and the partition function or Feynman path integral $Z(L)$ is equal to the natural pairing

$$Z(L) = (\chi, \psi). \tag{4.8}$$

We cannot evaluate (4.8), since we know neither χ nor ψ. The one thing that we do know, at present, is that (for the groups and representations we are considering) this pairing is occurring in a two dimensional vector space. A two dimensional vector space has the marvelous property that any three vectors obey a relation of linear dependence. Thus, given any two other vectors ψ_1 and ψ_2 in $\mathcal{H}_R$, there would be a linear relation

$$\alpha\psi + \beta\psi_1 + \gamma\psi_2 = 0, \tag{4.9}$$

where α, β, and γ are complex numbers. Physically, there is a very natural way to get additional vectors in $\mathcal{H}_R$. If one replaces M_R in figure (7(b)) by any other three manifold X with the same boundary (and with suitable strings in X connecting the marked points on the boundary of M_R), then the Feynman path integral on X gives rise to a new vector in $\mathcal{H}_R$. Picking any two convenient three manifolds X_1 and X_2 for this computation gives vectors ψ_1 and ψ_2 that can be used in (4.9). We will consider the case in which X_1 and X_2 are the same manifold as M_R but with different 'braids' connecting the points on the boundary; this is indicated in figure (7(c)).

Once ψ_1 and ψ_2 are obtained in this way, (4.9) has the obvious consequence that

$$\alpha(\chi, \psi) + \beta(\chi, \psi_1) + \gamma(\chi, \psi_2) = 0. \tag{4.10}$$

The three terms in (4.10) have a 'physical' interpretation, evident in figure (7(c)). By gluing M_L back together with M_R or one of its substitutes X_1 and X_2, one gets back the original three manifold M, but with the original link L replaced by some new links L_1 and L_2. Thus, (4.10) amounts to a relation among the link expectation values of interest, namely

$$\alpha Z(L) + \beta Z(L_1) + \gamma Z(L_2) = 0. \tag{4.11}$$

This recursion relation is often drawn as in figure (8). The meaning of this figure is as follows. If one considers three links whose plane projections are identical outside a disc, and look inside this disc like the three drawings in the figure, then the expectation values of those links, weighted with coefficients α, β, and γ, add to zero.

It is well known in knot theory that (4.11) uniquely determines the expectation values of all knots in S^3. For convenience we include a brief explanation of this. One starts with a plane projection of a knot, indicated in figure (9). The number p of crossings is finite. Inductively, suppose that all knot expectation values for knots with at most $p-1$ crossings have already been computed. One wishes to study knots with p crossings. If one had $\beta = 0$ in (4.11), one could at each crossing pass the two strands through each other with a factor of $-\gamma/\alpha$ in replacing an over-crossing by an under-crossing. If this were possible, the lines would be effectively transparent, and one could untie all knots. As it is, $\beta \neq 0$, but the term proportional to β reduces the number of crossings, giving rise to a new link whose expectation value is already known by the induction hypothesis.

This process reduces the discussion to the case $p = 0$ where there are no crossings, and therefore we are dealing only with a certain number of unlinked and unknotted circles. For practice with (4.11), let us discuss this case explicitly. In figure (10), we sketch a useful special case of figure (8). The first and third links in (10) consist of a single unknotted circle, and the second consists of two unlinked and unknotted circles. If we denote the partition function for s unlinked and unknotted circles in the N dimensional representation of $SU(N)$ as $Z(S^3; C^s)$ then (4.11) amounts in this case to the assertion that

$$(\alpha + \gamma) Z(S^3; C) + \beta Z(S^3; C^2) = 0. \tag{4.12}$$

Together with (4.7),* this implies that the expectation value of an unknotted Wilson line in the N dimensional representation of $SU(N)$ is

$$< C >= -\frac{\alpha + \gamma}{\beta}. \tag{4.13}$$

Presently we will make this formula completely explicit by computing α, β, and γ in terms of the fundamental quantum field theory parameters N and k.

The induction sketched above expresses any knot expectation value as a rational function of α, β and γ (a ratio of polynomials), after finitely many steps. It is in this sense that the Jones knot invariants and their generalizations are 'polynomials.' While it is, as we have seen, comparatively elementary to prove that (4.11) uniquely determines the knot invariants, the converse is far less obvious. (4.11) can be used in many different ways to obtain the expectation value of a given link, and one must show that one does not run into any inconsistency. While this has been proved in a variety of ways, the proofs have not been intrinsically three dimensional – (4.11) has not previously been derived from a manifestly invariant three dimensional framework. This is the novelty of the present discussion.

<u>Change Of Framing</u> We want to compute α, β, and γ, but as a prelude we must discuss a certain technical point. At the end of section (2), we learned that choosing a circle C and a representation R is not enough to give a well defined quantum holonomy operator $W_R(C) = \mathrm{Tr}_R\, P \exp \int_C A dx$. It is also necessary to pick a 'framing' of the circle C, which enters when one has to calculate the self-linking number of C and its non-abelian and quantum generalizations. At the end of section (2), we promised to derive a formula (2.33) showing how any partition function with an insertion of $W_R(C)$ transforms under a change of framing. Now it is time to deliver on this promise.

---

* This is the only point at which (4.7) has to be used. The induction sketched in the previous paragraph reduces all computations for knots in S^3 to this special case without using (4.7).

As in figure (7(b)), let us cut the three manifold M on a Riemann surface Σ that intersects C in a point P (and perhaps in some other points that will not be material). In our previous argument, we used the fact that associated with the boundaries M_L or M_R are Hilbert spaces $\mathcal{H}_L$ and $\mathcal{H}_R$. Moreover, $\mathcal{H}_R$ (for example) is 'a flat bundle on moduli space' so the mapping class group of the boundary Σ acts naturally on $\mathcal{H}_R$. We wish to act on the boundary of M_R with a very particular diffeomorphism before gluing the pieces of figure (7(b)) back together again. The diffeomorphism that we want to pick is a t-fold 'Dehn twist' about the point P on Σ. Making this diffeomorphism and then gluing the pieces of figure (7(b)) back together again, one gets an identical looking picture, but the framing of the circle C has been shifted by t units. On the other hand, one knows in conformal field theory how the Dehn twist acts on $\mathcal{H}_R$. Associated with the representation R is a number h_R, the 'conformal weight of the primary field in the R representation.' The t-fold Dehn twist acts on $\mathcal{H}_R$ as multiplication by $e^{2\pi i t h_R}$. So we have obtained (2.33) with $h = h_R$.

Explicit Evaluation We will now determine the parameters α, β, and γ that appear in the crucial equation (4.10). We need to determine the explicit relation among the three vectors ψ, ψ_1 and ψ_2 that appear in figure (7(c)). This requires a further study of the two dimensional Hilbert space which arises as the space of conformal blocks for the $R, R, \overline{R}, \overline{R}$ four point function on S^2 (R being in this case the defining N dimensional representation of $SU(N)$ and $\overline{R}$ its dual). The three configurations in figure (7(c)) can be regarded as differing from each other by a certain diffeomorphism of S^2; the diffeomorphism in question is the 'half-monodromy' under which the two copies of R change places by taking a half-step around one another, as indicated in figure (11) . Moore and Seiberg call this operation B and study it extensively. The states ψ_1 and ψ_2 are none other than

$$\psi_1 = B\psi, \quad \psi_2 = B^2\psi. \tag{4.14}$$

The matrix B, since it acts in a two dimensional space, obeys a characteristic equation

$$B^2 - yB + z = 0. \tag{4.15}$$

where

$$y = \mathrm{Tr}\, B, \quad z = \det B. \tag{4.16}$$

In view of (4.14), the linear relation among ψ, ψ_1, and ψ_2 is (up to an irrelevant common just

$$z \cdot \psi - y \cdot \psi_1 + \psi_2 = 0, \tag{4.17}$$

and according to (4.16), to måke this explicit we need only to know the eigenvalues (and thus the determinant and trace) of B.

These can be obtained from [13], but before describing the formulas, I would like to point out an important subtlety. As we have discussed in the last subsection, all concrete results such as the values of α, β and γ depend on the framing of knots. The convention that is most natural in working on an arbitrary three manifold is not the convention usually used in discussing knots on S^3.

In studying figure (7(c)), to describe the relative framings, the task is to specify the relative framing of the three pictures on the right, since the picture on the left is being held fixed. If one just looks at these three pictures and ignores the fact that the lines cannot pass through each other, there is an obvious sense in which one would like to pick 'the same' framing for each picture; for instance, a unit vector coming out of the page defines a normal vector field on each link in the picture.

This is equivalent to the convention of Moore and Seiberg in defining the eigenvalues of B, so we can now quote their results. Let h_R be the conformal weight of a primary conformal field transforming as R, let E_i be the irreducible representations of $SU(N)$ appearing in the decomposition of $R \otimes R$, and let h_{E_i} be the weights of the corresponding primary fields. Then the eigenvalues of B

are

$$\lambda_i = \pm \exp\big(i\pi(2h_R - h_{E_i})\big), \tag{4.18}$$

where the $+$ or $-$ sign corresponds to whether E_i appears symmetrically or antisymmetrically in $R \otimes R$. If R is the N dimensional representation of $SU(N)$, then one finds* that the eigenvalues of B are

$$\lambda_1 = \exp\left(\frac{i\pi(-N+1)}{N(N+k)}\right), \quad \lambda_2 = -\exp\left(\frac{i\pi(N+1)}{N(N+k)}\right). \tag{4.19}$$

It is straightforward to put these formulas in (4.16), (4.17) and thus make our previous results completely explicit.

Before comparing to the knot theory literature, it is necessary to make a correction in these results. For a link in S^3, there is always a standard framing in which the self-intersection number of each component of the link is zero. Values of the knot polynomials for knots in S^3 are usually quoted without specifying a framing; these are the values for the link with standard framing. However, if on the right of figure (7(c)) we use the 'same' framing for each picture, then when the right of figure (7(c)) is glued to the left, one does not have the canonical framing for each link. If the first knot is framed in the standard fashion, then the second is in error by one unit and the third by two units. So after using (4.19) to compute α, β, γ, we must, if we wish to agree with the knot theory literature, multiply β by $\exp(-2\pi i h_R)$ and γ by $\exp(-4\pi i h_R)$. After these corrections, one gets

$$\begin{aligned} \alpha &= -\exp\left(\frac{2\pi i}{N(N+k)}\right), \\ \beta &= -\exp\left(\frac{i\pi(2-N-N^2)}{N(N+k)}\right) + \exp\left(\frac{i\pi(2+N-N^2)}{N(N+k)}\right), \\ \gamma &= \exp\left(\frac{2\pi i(1-N^2)}{N(N+k)}\right). \end{aligned} \tag{4.20}$$

* For this representation, $h_R = (N^2-1)/(2N(N+k))$. In the decomposition of $R \otimes R$, the symmetric piece is an irreducible representation with $h_{E_1} = (N^2+N-2)/N(N+k)$, and the antisymmetric piece is an irreducible representation with $h_{E_2} = (N^2-N-2)/N(N+k)$.

If one multiplies α, β, γ by an irrelevant common factor $\exp(i\pi(N^2-2)/N(N+k))$ and introduces the variable

$$q = \exp(2\pi i/(N+k)), \tag{4.21}$$

then the skein relation can be written more elegantly as

$$-q^{N/2}L_+ + (q^{1/2} - q^{-1/2})L_0 + q^{-N/2}L_- = 0. \tag{4.22}$$

Here L_+, L_0, and L_- (equivalent to L, L_1, and L_2 in (4.11)) are standard notation for overcrossing, zero crossing, and undercrossing; and for $i = +, 0, -$, we now write simply L_i, instead of $Z(L_i)$. (4.22) is correctly normalized to give the right answers for knots on S^3 with their standard framing, and if one is only interested in knots on S^3 one can use it without ever thinking about the framings. Finally, comparing (4.13) and (4.22), we see that the expectation value of an unknotted Wilson line on S^3, with its standard framing, is

$$< C >= \frac{q^{N/2} - q^{-N/2}}{q^{1/2} - q^{-1/2}}. \tag{4.23}$$

This formula can be subjected to several interesting checks. First of all, the right hand side of (4.23) is positive for all values of the positive integers N and k. This is required by reflection positivity of the Chern-Simons gauge theory in three dimensions. Second, in the weak coupling limit of $k \to \infty$, we have $< C > \to N$. This is easily interpreted; in the weak coupling limit, the fluctuations in the connection A_i on S^3 are irrelevant, and the expectation value of the Wilson line approaches its value for $A_i = 0$, which is the dimension of the representation, or in this case N. Third, (4.23) vanishes if $q^N = 1$. This is because for $k = 0$, the only integrable highest weight representation of the loop group is the trivial representation. In particular, at $k = 0$, the N dimensional representation of $SU(N)$ is not the highest weight space of an integrable representation of the loop group, and therefore the expectation value of a Wilson line in this representation of $SU(N)$ must vanish. On the other hand, for $k = 0$, one sees from (4.21) that $q^N = 1$, so (4.23) must vanish for such values of q.

4.2 SURGERY ON LINKS

We have seen that it is possible to effectively calculate the expectation value of an arbitrary link in S^3. We would now like to generalize this to computations on an arbitrary three manifold. The basic idea is that by the operation of 'surgery on links' any three manifold can be reduced to S^3, so it is enough to understand how the invariants that we are studying transform under surgery. The operation of surgery can be described as follows. One begins with a three manifold M and an arbitrarily selected embedded circle C. Note that there is, to begin with, no Wilson line associated with C; C is simply a mathematical line on which we are going to carry out 'surgery.' To do so we first thicken C to a 'tubular neighborhood,' a solid torus centered on C. Removing this solid torus, M is split into two pieces; the solid torus is called M_R in figure (12(b)), and the remainder is called M_L. One then makes a diffeomorphism on the boundary of M_R and glues M_L and M_R back together to get a new three manifold $\widetilde{M}$.

It is a not too deep result that every three manifold can be obtained from or reduced to S^3 (or any other desired three manifold) by repeated surgeries on knots. However, such a description is far from unique and it is often difficult to use a description of a three manifold in terms of surgery to compute the invariants of interest. We will now see that the invariants studied in this paper can be effectively computed from a surgery presentation.

We study figure (12(b)) by the standard arguments. Hilbert spaces $\mathcal{H}_L$ and $\mathcal{H}_R$, canonically dual to one another, are associated with the boundaries of M_L and M_R. The path integrals on M_L and M_R give vectors ψ and χ in $\mathcal{H}_L$ and $\mathcal{H}_R$, and the partition function on M is just the natural pairing (ψ, χ). If we act on the boundary of M_R with a diffeomorphism K before gluing M_L and M_R back together, then χ is replaced by $K\chi$ so (ψ, χ) is replaced by $(\psi, K\chi)$.

This potentially gives a way to determine how the partition function of the quantum field theory transforms under surgery. Upon gaining a suitable understanding of $K\chi$, we will be able to reduce calculations on $\widetilde{M}$ to calculations on

4.3 The Physical Hilbert Space In Genus One

At this point we need a description of the physical Hilbert space in genus one. A beautiful description, perfectly adapted for our needs, appears in the work of Verlinde [12].

First of all, the loop group LG has at level k finitely many integrable highest weight representations. Let t be the number of these. For each such highest weight representation of the loop group, the highest weight space is an irreducible representation of the finite dimensional group G. In this way there appear t distinguished representations of G; we label these as $R_0, R_1 \dots R_{t-1}$, with R_0 denoting the trivial representation (which is always one of those on this list). Verlinde showed that if Σ is a Riemann surface of genus one, then the dimension of the physical Hilbert space $\mathcal{H}_\Sigma$ is t. Moreover, though there is no canonical basis for $\mathcal{H}_\Sigma$, Verlinde showed that every choice of a homology basis for $H^1(\Sigma, \mathbf{Z})$, consisting of two cycles a and b, gives a canonical choice of basis in $\mathcal{H}_\Sigma$. For our purposes, this can be described as follows. Topologically, there are many inequivalent ways to identify a torus Σ as the boundary of a solid torus U. The choice of U can be fixed by requiring that the cycle a is contractible in U. This is indicated in figure (13(a)). Next, for every $i = 0 \dots t-1$, one defines a state v_i in $\mathcal{H}_\Sigma$ as follows. One places a Wilson line in the R_i representation in the interior of U, running in the b direction,* and one performs the Feynman path integral in U to define a vector v_i in $\mathcal{H}_\Sigma$. The v_i make up the Verlinde basis in $\mathcal{H}_\Sigma$. It must be understood that a Wilson line in the trivial representation is equal to 1, so the vector v_0 obtained by this definition is the same as the vector χ which in the last subsection was obtained by a path integral on U with no Wilson lines:

$$\chi = v_0. \tag{4.24}$$

---

* The b cycle on the boundary of U gives a framing of this Wilson line.

The situation that we actually wish to apply this to is the case in which X is S^2 with some marked points P_a, $a = 1 \dots s$ to which representations $R_{i(a)}$ are assigned. (For $a = 1 \dots s$, $i(a)$ is one of the values $0 \dots t-1$ corresponding to integrable level k representations of the loop group.) In this case, the simple product $X \times S^1$ is just $S^2 \times S^1$ with some Wilson lines which are unknotted, parallel circles of the form $\{P_a\} \times S^1$, as sketched in figure (14(b)). To determine the path integral on $S^2 \times S^1$ in the presence of these Wilson lines, which we will denote as $Z(S^2 \times S^1; < R >)$, one needs to study the Hilbert space of S^2 with charges in the representations R_{a_i}; we will denote this as $\mathcal{H}_{S^2;\langle R \rangle}$. The analog of (4.31) is then

$$Z(S^2 \times S^1; < R >) = \dim \mathcal{H}_{S^2;\langle R \rangle} \tag{4.33}$$

The dimensions of these spaces were discussed at the end of section (3). Thus, if the collection of representations $\langle R \rangle$ consists of a single representation R_a, we get

$$Z(S^2 \times S^1; R_a) = \delta_{a,0}, \tag{4.34}$$

since the physical Hilbert space with a single charge in the R_a representation is one dimensional if R_a is the trivial representation ($a = 0$) and zero dimensional otherwise.* For two charges in the representations R_a and R_b, we get

$$Z(S^2 \times S^1; R_a, R_b) = g_{ab}, \tag{4.35}$$

where g_{ab}, introduced earlier, is 1 if R_b is the dual of R_a and zero otherwise. The formula (4.35) follows from the result of section (4.4) for the Hilbert space on S^2 with two charges. Finally, if there are three charges in the representations

---

* Implicit in (4.34) and subsequent formulas is the use of the standard framing of the Wilson line which is invariant under rotations of S^1; it is for this choice that the path integral on $S^2 \times S^1$ computes the trace of the identity operator in the physical Hilbert space.

R_a, R_b, R_c, we get

$$Z(S^2 \times S^1; R_a, R_b, R_c) = N_{abc}, \tag{4.36}$$

with N_{abc} the trilinear 'coupling' of Verlinde, since this is the dimension of the physical Hilbert space.

4.5 Some Concrete Surgeries

Now we would like to describe some useful results that can be obtained from concrete surgeries. The first goal is compute the partition function of S^3. Since we already know the partition function of $S^2 \times S^1$, we will try to interpret S^3 as a manifold obtained by surgery on $S^2 \times S^1$. This is readily done. We consider the circle C in $S^2 \times S^1$ indicated in figure (15(a)). A tubular neighborhood of C is a torus Σ; we pick a basis of $H^1(\Sigma; \mathbf{Z})$ consisting of cycles a and b indicated in the figure. Now we wish to make a particular surgery associated with a very special diffeomorphism $S : \Sigma \to \Sigma$. We pick S to map a to b and b to $-a$.† This surgery – removing the interior of Σ from $S^2 \times S^1$ and gluing it back after acting with S – produces a three manifold that is none other than S^3 (figure (16)). Since this point is crucial in what follows, we pause to explain it. We regard S^3 as R^3 plus a point at infinity. In figure (16(a)) a torus Σ has been embedded in R^3. Obviously, Σ with its interior make up a solid torus T. It is also relatively easy to see that the figure (a) is invariant under inversion, so that the exterior of Σ (including the point at infinity) is a second solid torus T'. Thus, S^3 can be made by gluing two solid tori along their boundaries. Now in (b) we sketch two identical solid tori T and T'; T' has been obtained by simply translating T in Euclidean space. If one glues together the boundaries pointwise with the identification that is indicated by saying that 'T' is a translate of T,' one gets $S^2 \times S^1$. (In fact, the solid torus T is $D \times S^1$, with D a two dimensional disc,

† This transformation, which acts on the upper half plane as $\tau \to -1/\tau$, is indeed usually called S in the theory of the modular group $SL(2, \mathbf{Z})$ (which can be identified, via the basis a, b, with the mapping class group of Σ).

and T' is $D' \times S^1$ with D' a second disc. Just as two discs D and D' glued on their boundary make S^2, $D \times S^1$ naturally glues to $D' \times S^1$ to make $S^2 \times S^1$). On the other hand, we know from part (a) that S^3 can be obtained by gluing two solid tori. A little mental gymnastics, comparing the argument we gave in connection with (a) to that in (b), shows that to make S^3 we must glue together T and T' after making the modular transformation S on the boundary of T'.

Now we can use (4.27), with S^3 playing the role of $\widetilde{M}$, $S^2 \times S^1$ playing the role of M, and the arbitrary diffeomorphism K replaced by S. So we learn

$$Z(S^3) = \sum_j S_0{}^j \, Z(S^2 \times S^1; R_j). \tag{4.37}$$

We have learned in the last section that $Z(S^2 \times S^1; R_j)$ is 1 for $j = 0$ and 0 otherwise, so

$$Z(S^3) = S_{0,0}. \tag{4.38}$$

Here $S_{0,0}$ can be determined from the theory of affine Lie algebras; for $G = SU(2)$ one gets the formula stated earlier in (2.26). In fact, the whole matrix S_{ij} can be written very explicitly for $G = SU(2)$. The integrable representations of level k are those of spin $n/2$ for $n = 0 \dots k$, and the matrix elements of S are

$$S_{mn} = \sqrt{\frac{2}{k+2}} \sin\left(\frac{(m+1)(n+1)\pi}{k+2}\right). \tag{4.39}$$

<u>The Phase of The Partition Function</u>

Now let us re-examine in the light of these methods a thorny question that appeared in section (2) – the framing of three manifolds, and the phase of the partition function.

We have obtained S^3 from $S^2 \times S^1$, by performing surgery on a certain circle C, using the modular transformation $S : \tau \to -1/\tau$. Apart from S, there are other modular transformations that could be used to build S^3 by surgery on the same knot C in $S^2 \times S^1$. The general choice would be $T^n S T^m$, with n and m being arbitrary integers, and T being the modular transformation $T : \tau \to \tau + 1$. (S and T are the standard generators of the modular group, obeying $S^2 = (ST)^3 = 1$.) Had we used $T^n S T^m$, we would have gotten not (4.38) but

$$Z(S^3) = (T^n S T^m)_{0,0}. \tag{4.40}$$

This may readily be evaluated. In the Verlinde basis, T is a diagonal matrix with $T \cdot v_i = e^{2\pi i(h_i - c/24)} \cdot v_i$; h_i is the conformal weight of the primary field in the representation R_i and c is the central charge for current algebra with symmetry group G at level k. Since $h_0 = 0$, if we replace (4.38) by (4.40) the partition function transforms as

$$Z \to Z \cdot \exp\left(2\pi i(n - m) \cdot \frac{c}{24}\right) \tag{4.41}$$

Though we have obtained this formula in the example of a particular surgery (giving S^3 from $S^2 \times S^1$), the same ambiguity arises in any process of surgery. Whenever one makes surgery on a circle C, in a three manifold M, with the surgery being determined by an $SL(2, Z)$ element u, one could instead consider surgery on the same circle C, using the $SL(2, Z)$ element $u \cdot T^m$. This would have the same effect topologically, but our surgery law would give a partition function containing an extra phase $\exp(-2\pi i m \cdot c/24)$.

This phase ambiguity was already encountered, in the large k limit, in formula (2.24). What is more, from the discussion in section (2), we know what topological structure on three manifolds must be considered in order to keep track of the factors of $\exp(2\pi i \cdot c/24)$. One must consider 'framed' three manifolds. Two surgeries that have the same effect on the topology of a three manifold may have different effects on the framing. I will discuss elsewhere how to systematically keep track of the factors of $\exp(2\pi i \cdot c/24)$ under surgery. In the simple applications in this paper, this will not be necessary. All of our applications will involve considering the standard surgery (by the modular transformation S) that was used in the last subsection to obtain S^3 from $S^2 \times S^1$.

Some Expectation Values

Now let us see if we can go farther and determine the path integral $Z(S^3; R_j)$ on S^3 with an unknotted Wilson line on S^3 in an arbitrary representation R_j.*

To do this, we start on $S^2 \times S^1$ with a Wilson line in the R_j representation running parallel to the circle C on which we are doing surgery, as in figure (15(b)). Carrying out the same surgery as before turns $S^2 \times S^1$ into S^3, with a Wilson line in the R_j representation on S^3. Application of (4.27) now gives

$$Z(S^3; R_j) = \sum_i S_0{}^i \, Z(S^2 \times S^1; R_i, R_j) \tag{4.42}$$

Using (4.35), we can evaluate this and determine the partition function for a Wilson line in an arbitrary representation R_j; it is

$$Z(S^3; R_j) = \sum_i S_0{}^i \, g_{ij} = S_{0,j}. \tag{4.43}$$

Let us compare this to our previous evaluation (4.23) of the expectation value of an unknotted Wilson line in S^3. We must recall that the symbol $< C >$ in (4.23) represented a ratio $< C > = Z(S^3; R)/Z(S^3)$. Let us take $G = SU(2)$, so that we can use the explicit formulas (4.39), and take R to be the two dimensional representation of $SU(2)$, so that we can compare to (4.23). Using (4.38), (4.43), and (4.39), we get

$$< C > = \frac{S_{0,1}}{S_{0,0}} = \frac{\sin\big(2\pi/(k+2)\big)}{\sin\big(\pi/(k+2)\big)}. \tag{4.44}$$

It is easy to see that setting $N = 2$ in (4.23) gives the same formula. Let us take this one step further and try to calculate by these methods the partition function $Z(S^3; R_j, R_k)$ for S^3 with two unknotted, unlinked Wilson lines in representations R_j and R_k. In figure (15(c)), we start on $S^2 \times S^1$ with *two* Wilson lines, in

* We give this Wilson line the framing described in the footnote after (4.34); after surgery this turns into the standard framing on S^3.

representations R_j and R_k, parallel to the circle C on which surgery is to be performed. Carrying out the surgery, we get to S^3 with the desired unlinked, unknotted circles. In this case, the surgery formula (4.27) tells us that

$$Z(S^3; R_j, R_k) = \sum_i S_0{}^i Z(S^2 \times S^1; R_i, R_j, R_k) \tag{4.45}$$

The right hand side can be evaluated with (4.36), while the left hand side can be reduced to (4.43) using (4.6). We get

$$\frac{S_{0,j} S_{0,k}}{S_{0,0}} = \sum_i S_0{}^i N_{ijk}. \tag{4.46}$$

Proof Of Verlinde's Conjecture The last equation is a special case of a celebrated conjecture by Verlinde, which has been proved by Moore and Seiberg [13]. We can use these methods to give a new proof of Verlinde's conjecture, in the case of current algebra. We will have to use the generalized surgery relation (4.28). We return to figure (15(b)) but now instead of treating C as a purely imaginary contour on which surgery is to be performed, we suppose that there is a Wilson line on C in the R_i representation. In this case, the standard surgery on C will still turn $S^2 \times S^1$ into S^3, but now on S^3 we will have two Wilson lines, in the R_i and R_j representations. Some mental gymnastics shows that they are linked, as in figure (17(a)); schematically, we refer to this linked pair of Wilson lines as $L(R_i; R_j)$. The use of (4.28) therefore determines the partition function of S^3 with a pair of linked Wilson lines:

$$Z(S^3; L(R_i; R_j)) = \sum_k S_i{}^k Z(S^2 \times S^1; R_k, R_j) = S_{ij}. \tag{4.47}$$

In the second step, we have used the fact that the partition function of $S^2 \times S^1$ with static charges R_k and R_j is the metric that we have called g_{kj}– one if R_k is dual to R_j and otherwise zero.

Now let us go back to figure (15(c)), and again on what was previously the purely imaginary circle C we put a Wilson line in the R_i representation. The standard surgery on this link will now produce a picture sketched in figure (17(b)), with a Wilson line on S^3 in the R_i representation that links a pair of Wilson lines R_j, R_k that are themselves unlinked and unknotted. We call this configuration $L(R_i; R_j, R_k)$. The evaluation of (4.28) now gives

$$Z(S^3; L(R_i; R_j, R_k)) = \sum_m S_i{}^m Z(S^2 \times S^1; R_m, R_j, R_k) = \sum_m S_i{}^m N_{mjk} \quad (4.48)$$

To obtain Verlinde's formula, it is now necessary to find an independent way to evaluate the left hand side.

Such a method is provided by the following generalization of the multiplicativity formula (4.1). The key point in the derivation of (4.1) was that the physical Hilbert space for S^2 with no charges was one dimensional. It is likewise true that the physical Hilbert space $\mathcal{H}$ for S^2 with a pair of charges in the dual representations R_i and $R_{\bar{i}}$ is one dimensional. Using this and otherwise repeating the derivation of (4.1) gives the following formula:

$$Z(S^3; L(R_i; R_j, R_k)) \cdot Z(S^3; R_i) = Z(S^3; L(R_i; R_j)) \cdot Z(S^3; L(R_i; R_k)). \quad (4.49)$$

The idea in (4.49) is that, as in figure (17(c)), the evaluation of $Z(S^3; L(R_i; R_j, R_k))$ can be expressed as a pairing (ψ, χ) where ψ and χ are certain vectors in $\mathcal{H}$ and its dual. Likewise the evaluation of $Z(S^3; R_i)$ is a pairing (v', v), where v' and v are vectors in $\mathcal{H}$ and its dual. Using the wonderful fact of one dimensional linear algebra $(\psi, \chi) \cdot (v', v) = (\psi, v) \cdot (v', \chi)$, we arrive at (4.49). Since all factors in (4.49) are known except the first, we arrive at the result

$$Z(S^3; L(R_i; R_j, R_k)) = S_{ij} S_{ik} / S_{0,i}. \quad (4.50)$$

Combining this with (4.48), we have

$$S_{ij} S_{ik} / S_{0,i} = \sum_m S_i{}^m N_{mjk}. \quad (4.51)$$

This is equivalent to Verlinde's statement that 'the matrix S diagonalizes the fusion rules.' In other words, in the basis v_i indicated in figure (13), the structure constants of the Verlinde algebra are by definition $v_i v_j = \sum_k N_{ij}{}^k v_k$, where $N_{ij}{}^k = \sum_r N_{ijr} g^{rk}$. If we introduce a new basis $w_i = S_{0,i} \cdot \sum S_i{}^m v_m$, then the Verlinde algebra reduces to $w_i w_j = \delta_{ij} w_j$. To verify this, we compute

$$w_i w_j = \sum_{k,l} S_i{}^k S_j{}^l v_k v_l \cdot S_{0,i} S_{0,j}. \tag{4.52}$$

Using $v_k v_l = \sum_m N_{kl}{}^m v_m$ and (4.51), this becomes

$$w_i w_j = S_j{}^l v^m \cdot S_{il} S_i{}^m \cdot S_{0,j} \tag{4.53}$$

Using the unitarity of S, in the form $S_j{}^l S_{il} = \delta_{ij}$, we see that

$$w_i w_j = \delta_{ij} \sum_m S_j{}^m v_m \cdot S_{0,j} = \delta_{ij} \cdot w_j \tag{4.54}$$

showing that the Verlinde algebra has been diagonalized and that the w_i are idempotents.

Connected Sum Of Links

At the beginning of section four, we have seen that the quantum partition function has a multiplicative behavior under connected sum of three manifolds. From (4.1), if $M = M_1 + M_2$, then $Z(M) \cdot Z(S^3) = Z(M_1) \cdot Z(M_2)$. In the special case that M_1 and M_2 are copies of S^3 with links in them, the connected sum of M_1 and M_2 is a copy of S^3 containing the *disconnected* sum of the two links.

In knot theory there is also an operation of taking the *connected* sum of two *links*. This operation has appeared in the above discussion. The link that we have called $L(R_i; R_j, R_k)$ is the connected sum of the two links that we have called $L(R_i, R_j)$ and $L(R_i, R_k)$. In fact, in figure (17(c)), $L(R_i; R_j, R_k)$ is 'cut' into two pieces, with are respectively $L(R_i, R_j)$ and $L(R_i, R_k)$ with in each case a connected segment removed. This is the defining configuration for the 'connected sum of links.' Accordingly, the reasoning that led to (4.49) has the following more general consequence. If L_1 and L_2 are two links, and $L_1 + L_2$ is their connected sum, then

$$Z(S^3; L_1 + L_2) \cdot Z(S^3; C) = Z(S^3; L_1) \cdot Z(S^3; L_2). \tag{4.55}$$

Here it is understood that (as in (4.49)) representations have been assigned to the connected components of L_1, L_2, and $L_1 + L_2$ in a compatible fashion; C is an unknot placed in whatever representation is carried by the strand 'cut' in the generalization of figure (17(c)). (4.55) has a generalization in which L_1 and L_2 are links in arbitrary three manifolds M_1 and M_2; then the connected sum of links $L_1 + L_2$ is a link in the connected sum of manifolds $M = M_1 + M_2$, and (4.55) is replaced by

$$Z(M_1 + M_2; L_1 + L_2) \cdot Z(S^3; C) = Z(M_1; L_1) \cdot Z(M_2; L_2). \tag{4.56}$$

4.6 THE KNOT POLYNOMIALS AND THE BRAID GROUP

The results in the last subsection are nice enough so that one may wonder if the partition function for an arbitrary link on S^3 can be evaluated in this way. This can indeed be done, and in a way that is closely related to the original route by which the Jones polynomial was discovered, though we cannot expect such explicit formulas as in the simple cases treated above. An arbitrary link L on S^3, whose partition function we will call $Z(S^3; L)$, can be arranged in the form of a braid, as indicated in figure (18(a)). One can imagine 'lifting' this braid B out of S^3 and putting it on $S^2 \times S^1$. To get back to S^3 one would have to do surgery on a circle running parallel to the braid, as suggested in figure (18(b)). The general surgery formula (4.27) then tells us

$$Z(S^3; L) = \sum_j S_0{}^j \, Z(S^2 \times S^1; R_j, B), \tag{4.57}$$

where $Z(S^2 \times S^1; R_j, B)$ is the partition function on $S^2 \times S^1$ in the presence of both the braid B and a parallel Wilson line in the R_j representation. We want to rewrite this in the spirit of (4.32). Suppose that the braid B contains n strands making up a collection of representations $\langle R \rangle$. Then B can be regarded as defining an element of the Artin braid group on n letters. The braid group is closely related to the mapping class group for S^2 with marked points. The reason for that is that if in figure (18(b)) we 'cut' $S^2 \times S^1$, to get back to $S^2 \times I$ (this amounts to undoing what was done in figure (14)), then the braid can be unbraided. Thus, the complete information about the braid is in the choice of a diffeomorphism of S^2 (constrained to preserve the marking of the points) by which the top and bottom of figure (18(c)) are to be identified. This, however, does not quite mean that the braid group is the same as the mapping class group. In figure (18(b)), there are $n+1$ strands, one of which arose from the surgery and does not participate in the braid, while the other n strands make up the braid. The braid group on n letters is the subgroup of the mapping class group on $n+1$ letters which fixes one of the (framed) strands. There are a number of invariant traces on the braid group that can be naturally defined with the data at our disposal. They are

$$\tau_i(B) = Z(S^2 \times S^1; R_i, B) \tag{4.58}$$

This has the key property of a trace,

$$\tau_i(B_1 B_2) = \tau_i(B_2 B_1) \tag{4.59}$$

since the two sides of (4.59) have the same path integral representation, in which the two braids are glued end to end in $S^2 \times S^1$, as in figure (18(d)). Not only does (4.58) obey (4.59); it is actually equal to the trace of the operator B in a certain representation of the braid group, namely the representation furnished by the physical Hilbert space $\mathcal{H}$ for S^2 with the $n+1$ charges, it being understood that the braid group is acting on the first n charges, and the $(n+1)^{st}$ is fixed in the R_i representation. That $\tau_i(B)$ is the trace of B in this Hilbert space is a statement just along the lines of (4.32).

So we can rewrite (4.57) in the form

$$Z(S^3; L) = \sum_j S_0{}^j \tau_j(B). \tag{4.60}$$

This shows that the link invariants on S^3 may be written as linear combinations of braid traces. This is very close to how the knot polynomials were originally discovered. It is clear from the work of Tsuchiya and Kanie [11] and Segal [16] along with what has been said above that the braid traces that arise from the Chern-Simons theory are precisely those that first appeared in the work of Jones.

5. Applications To Physics

Finally, I would like to comment on the likely implications of these results for physics. We have been exploring a three dimensional viewpoint about conformal field theory, at least for the important special case of current algebra on Riemann surfaces. Many aspects of rational conformal field theory have emerged as natural consequences of general covariance in three dimensions. It seems likely that the marvelous hexagons and pentagons of [13], and the other consistency conditions of rational conformal field theories, can be synthesized by saying that such theories come from generally covariant theories in three dimensions. (It is fairly obvious that the hexagons and pentagons and the requirements of factorization are consequences of general covariance in three dimensions; so the real issue here is whether general covariance implies unwanted conditions that are not necessarily true in conformal field theory.) If so, general covariance in three dimensions dimensions may well emerge as one of the main unifying themes governing two dimensional conformal field theory. Such considerations have motivated a study of $2+1$ dimensional gravity which will appear elsewhere [38].

The basic connection that we have so far stated between general covariance in $2+1$ dimensions and conformally invariant theories in $1+1$ dimensions is that the physical Hilbert spaces obtained by quantization in $2+1$ dimensions can be interpreted as the spaces of conformal blocks in 1+1 dimensions. This connection may seem rather abstract, and I will now make a few remarks aimed at making a more concrete connection. Starting from three dimensions we were led to the problem of quantizing the Lagrangian (3.1), which we repeat for convenience:

$$\mathcal{L} = \frac{k}{8\pi} \int dt \int_{\Sigma} \epsilon^{ij} \operatorname{Tr} A_i \frac{d}{dt} A_j. \tag{5.1}$$

This is to be quantized with constraints $\epsilon^{ij} F_{ij}{}^a = 0$, these constraints being the generators of the infinitesimal gauge transformations

$$A_i \to A_i - D_i \epsilon. \tag{5.2}$$

So far we have quantized (5.1) on Riemann surfaces without boundary, but now let us relax this requirement. Quantization of (5.2) in this more general case amounts to studying the three dimensional Chern-Simons theory on a three manifold with boundary, namely $\Sigma \times R^1$ where Σ is a Riemann surface with boundary. For instance, let Σ be a disc D. To quantize (5.1) on the disc, we must impose the constraints, which generate the gauge transformations (5.2). As D has a boundary, we must choose boundary conditions on A_i and ϵ (for closed surfaces this question did not arise). We will adopt free boundary conditions for A_i, but require $\epsilon = 0$ on the boundary of S. (A rationale for this choice is that the Chern-Simons action is not invariant under gauge transformations that do not vanish on the boundary.) With this condition, ϵ generates in (5.2) the group $\widehat{G}_1$ of gauge transformations which are the identity on the boundary of the disc. An element of this group is an arbitrary continuous map $V : D \to G$ whose restriction to S is is the identity. Now we impose the constraints. The first step is to require that $\epsilon^{ij} F_{ij}{}^a = 0$. Since the disc is simply connected, this implies that $A_i = -\partial_i U \cdot U^{-1}$ for some map $U : D \to G$. U is uniquely defined up to

$$U \to U \cdot W, \tag{5.3}$$

for a constant element $W \in G$. Then we must identify two U's that differ by an element of the restricted gauge group $\widehat{G}_1$. This means that we must impose the equivalence relation $U \simeq VU$ for any V such that $V = 1$ on the boundary S. The equivalence relation means that only the restriction of U to S is relevant. This restriction defines an element of the loop group LG, but because of the freedom (5.3), we should actually regard U as an element of LG/G. Geometrically, we have learned that the homogeneous space LG/G can be regarded as the symplectic quotient of the space of G connections on the disc D, by the group $\widehat{G}_1$ of symplectic diffeomorphisms. Now we wish to quantize the theory, which means doing quantum mechanics on LG/G, which inherits a natural symplectic structure from (5.1). Clearly, the group LG of gauge transformations on the boundary of Σ acts on LG/G, so the quantum Hilbert space will be at least a projective representation of the loop group. In fact, according to Segal and Pressley [39], the quantization of LG/G, with this symplectic structure, gives rise to the basic irreducible highest weight representation of the loop group. This makes the connection between $2+1$ dimensions and $1+1$ dimensions far more direct, since the irreducible representations of the loop group are a basic ingredient in the $1+1$ dimensional theory. It is obvious at this point that by considering more complicated Riemann surfaces with various boundary components, we can generate the whole $1+1$ dimensional conformal field theory, essentially by studying the generally covariant $2+1$ dimensional theory on various three manifolds with boundary.

REFERENCES

1. M. F. Atiyah, 'New Invariants Of Three And Four Dimensional Manifolds,' in *The Mathematical Heritage of Hermann Weyl*, Proc. Symp. Pure Math. **48**, ed. R. Wells (American Mathematical Society, 1988).

2. S. Donaldson, 'An Application Of Gauge Theory To The Topology Of Four Manifolds,' J. Diff. Geom. **18** (1983) 269, 'Polynomial Invariants For Smooth Four-Manifolds,' Oxford preprint.

3. A. Floer, 'An Instanton Invariant For Three Manifolds,' Courant Institute preprint (1987), 'Morse Theory For Fixed Points Of Symplectic Diffeomorphisms,' Bull. AMS **16** (1987) 279.

4. E. Witten, 'Topological Quantum Field Theory,' Comm. Math. Phys.**117** (1988) 353.

5. V. F. R. Jones, 'Index For Subfactors,' Inv. Math. **72** (1983) 1, 'A Polynomial Invariant For Links Via Von Neumann Algebras,' Bull. AMS **12** (1985) 103, 'Hecke Algebra Representations of Braid Groups and Link Polynomials,' Ann. Math. **126** (1987) 335.

6. P. Freyd, D. Yetter, J. Hoste, W.B.R. Lickorish, K. Millett, and A. Ocneanu, 'A New Polynomial Invariant of Knots And Links,' Bull. AMS **12** (1985) 239.

7. L. Kauffman, 'State Models And The Jones Polynomial,' *Topology* **26** (1987) 395; 'Statistical Mechanics And The Jones Polynomial,' to appear in the Proceedings of the July, 1986, conference on Artin's braid group, Santa Cruz, California; 'An Invariant Of Regular Isotopy,' preprint.

8. V. G. Turaev, 'The Yang-Baxter Equation And Invariants Of Links,' LOMI preprint E-3-87.

9. J. H. Przytycki and P. Traczyk, 'Invariants Of Links Of Conway Type,' *Kobe J. Math.*

10. J. Birman, 'On The Jones Polynomial Of Closed 3-Braids,' Inv. **Math.** **81** (1985) 287; J. Birman and H. Wenzel, 'Link Polynomials And A New Algebra,' preprint.

11. A. Tsuchiya and Y. Kanie, in *Conformal Field Theory And Solvable Lattice Models,* Advanced Studies In Pure Mathematics **16** (1988) 297, Lett. Math. Phys. **13** (1987) 303.

12. E. Verlinde, 'Fusion Rules And Modular Transformations In 2d Conformal Field Theory,' Nucl. Phys. **B300** (1988) 360.

13. G. Moore and N. Seiberg, 'Polynomial Equations For Rational Conformal Field Theories,' to appear in Phys. Lett. B, 'Naturality In Conformal Field Theory,' to appear in Nucl. Phys. B, 'Classical And Quantum Conformal Field Theory,' IAS preprint HEP-88/35.

14. B. Schroer, Nucl. Phys. **295** (1988) 4; K.-H. Rehren and B. Schroer, 'Einstein Causality and Artin Braids,' FU preprint (1987).

15. J. Frohlich, 'Statistics Of Fields, The Yang-Baxter Equation, And The Theory Of Links And Knots,' 1987 Cargese lectures, to appear in 'Nonperturbative Quantum Field Theory' (Plenum Press).

16. G. Segal, 'Conformal Field Theory,' Oxford preprint; and lecture at the IAMP Congress, Swansea, July, 1988.

17. M. Gromov, 'Pseudo-Holomorphic Curves In Symplectic Manifolds,' Invent. Math. **82** (1985) 307.

18. A. Schwarz, 'The Partition Function Of Degenerate Quadratic Functional And Ray-Singer Invariants,' Lett. Math. Phys. **2** (1978) 247.

19. J. Schonfeld, 'A Mass Term For Three Dimensional Gauge Fields,' Nucl. Phys. **B185** (1981) 157.

20. R. Jackiw and S. Templeton, 'How Superrenormalizable Theories Cure Their Infrared Divergences,' Phys. Rev. **D23** (1981) 2291.

21. S. Deser, R. Jackiw, and S. Templeton, 'Three Dimensional Massive Gauge Theories,' Phys. Rev. Lett. **48** (1983) 975, 'Topologically Massive Gauge Theory,' Ann. Phys. NY **140** (1984) 372.

22. G. Zuckerman, 'Action Principles And Global Geometry,' in the proceedings of the 1986 San Diego Summer Workshop, ed. S.-T. Yau

23. A. M. Polyakov, 'Fermi-Bose Transmutations Induced By Gauge Fields,' Mod. Phys. Lett. **A 3** (1988) 325.

24. C. R. Hagen, Ann. Phys. **157** (1984) 342.

25. D. Arovas, R. Schrieffer, F. Wilczek, and A. Zee, 'Statistical Mechanics Of Anyons,' Nucl. Phys. **B251**[**FS13**] (1985) 117.

26. E. Witten, 'Non-Abelian Bosonization In Two Dimensions,' *Comm. Math. Phys.* **92** (1984) 455.

27. D. Ray and I. Singer, Adv. Math. **7** (1971) 145, Ann. Math. **98** (1973) 154.

28. S. Deser, R. Jackiw, and S. Templeton, in ref. (21); I. Affleck, J. Harvey, and E. Witten, Nucl. Phys. **B206** (1982) 413; A. N. Redlich, 'Gauge Invariance And Parity Conservation Of Three-Dimensional Fermions,' Phys. Rev. Lett. **52** (1984) 18; L. Alvarez-Gaumé and E. Witten, 'Gravitational Anomalies,' Nucl. Phys. **B 234** (1983) 269; L. Alvarez-Gaumé, S. Della Pietra, and G. Moore, 'Anomalies And Odd Dimensions,' Ann. Phys. (NY) **163** (1985) 288; M. F. Atiyah, 'A Note On The Eta Invariant,' unpublished.

29. M. F. Atiyah, V. Patodi, and I. Singer, Math. Proc. Cambridge Philos. Soc. **77** (1975) 43, **78** (1975) 405, **79** (1976) 71.

30. F. Wilczek and A. Zee, 'Linking Numbers, Spin, And Statistics Of Solitons,' Phys. Rev. Lett. **51** (1983) 2250.

31. D. Friedan and S. Shenker, Nucl. Phys.**B281** (1987) 509.

32. A. Belavin, A. M. Polyakov, and A. Zamolodchikov, Nucl. Phys. **B** (1984).

33. M. F. Atiyah and R. Bott, 'The Yang-Mills Equations Over Riemann Surfaces,' Phil. Trans. Roy. Soc. London **A308** (1982) 523.

34. D. Quillen, 'Determinants Of Cauchy-Riemann Operators Over A Riemann Surface,' Funct. Anal. Appl. **19** (1986) 31.

35. V. Drinfeld, 'Quantum Groups,' in the Proceedings of the International Congress of Mathematicians, Berkeley, 1986.

36. D. Gepner and E. Witten, 'String Theory On Group Manifolds,' Nucl. Phys. B

37. V. G. Kac, 'Infinite Dimensional Lie Algebras,' Cambridge University Press (1985); V. G. Kac and D. H. Peterson, *Adv. Math.* **53** (1984) 125; V. G. Kac and M. Wakimoto, *Adv. Math.* **70** (1988) 156.

38. E. Witten, '2 + 1 Dimensional Gravity As An Exactly Soluble System,' IAS preprint HEP-88/32.

39. G. Segal and A. Pressley, *Loop Groups* (Cambridge University Press, 1986).

FIGURE CAPTIONS

1. A knot in three dimensional space.

2. Several linked but non-intersecting oriented knots in a three manifold M. This collection of knots is called a 'link.'

3. The self-linking integral is, in a non-abelian theory, the first in an infinite series of Feynman diagrams, with gauge fields emitted and absorbed by the same knot, as in (a); these all pose similar problems. A topologically invariant but not uniquely determined regularization can be obtained by supposing that each knot is 'framed,' as in (b). In (c), the framing is shifted by 2 units by making a 2-fold twist.

4. Cutting a three manifold M on an intermediate Riemann surface Σ is indicated in part (a). Wilson lines W on M may pierce Σ and if so Σ comes with certain 'marked points,' with representations attached. Locally, near Σ, M looks like $\Sigma \times R^1$, indicated in part (b).

5. In (a) is sketched a three manifold M which is the connected sum of two pieces M_1 and M_2, joined along a sphere S^2. Similarly, a three sphere S^3 can be cut along its equator, as in (b). Cutting both M and S^3 as indicated in (a) and (b), the pieces can be rearranged into the *disconnected* sum of M_1 and M_2, as in (c).

6. A three sphere with 3 unlinked and unknotted circles C_i, associated with representations $R_1 \ldots R_3$. The figure can be cut in various ways to separate the circles.

7. A link C on a general three manifold M is sketched in (a). A small sphere S has been drawn about an inconvenient crossing; it cuts M into a simple piece (the interior of S) and a complicated piece. In (b), the picture is rearranged to exhibit the cutting of M more explicitly; the two pieces now appear on the left and right as M_L (the complicated piece whose details

are not drawn) and M_R (the interior of S). The key to the skein relation is to consider replacing M_R with some substitutes, as shown in (c).

8. A recursion relation for links.

9. A plane projection of a knot, with four crossings.

10. A special case of the use of figure (8). The idea is that the three pictures are identical outside of the dotted lines, and look like figure ((4.11)) inside them.

11. The half-monodromy operation exchanging two equivalent points on S^2 is sketched in (a); the arrows are meant to suggest a process in which the first two points change places by executing a half-twist about one another. The idea in (b) is that if the two points on the left in the first picture undergo a half-twist about one another, the first picture becomes the second, and if this is done again, the second picture becomes the third. In this way the three pictures on the right of figure (7(c)) differ by a succession of half-monodromies.

12. Surgery on a circle C in a three manifold M is carried out by removing a tubular neighborhood of C, depicted in (a). At this point M has been separated into two pieces, M_L and M_R, with a torus Σ for their boundaries, as sketched in (b). M_R is simply a solid torus. Surgery is completed by gluing the two pieces back together after making a diffeomorphism of the boundary of M_R. At the end of this process, M has been replaced by a new three manifold $\widetilde{M}$. As we will eventually see, computations on $\widetilde{M}$ are equivalent to computations on M with a physical Wilson line where the surgery was made, as in (c). The difference between (a) and (c) is that in (a) the circle C is just a locale for surgery, but in (c) it is a Wilson line.

13. A Riemann surface Σ of genus one is shown in (a) as the boundary of a solid torus U; the indicated a-cycle is contractible in U. In (b), a basis of the physical Hilbert space is indicated consisting of states obtained by

placing a Wilson line, in the R_i representation, in the interior of U, parallel to the cycle b, and performing the path integral to get a vector v_i in $\mathcal{H}_\Sigma$.

14. Beginning with $X \times I$, shown in (a), one makes $X \times S^1$ by identifying $X \times \{0\}$ with $X \times \{1\}$. If X is S^2 with some marked points P_i, then this construction gives the picture of (b). If $S^2 \times \{0\}$ is joined to $S^2 \times \{1\}$ via a non-trivial diffeomorphism B, one makes in this way a braid, as in (c).

15. In part (a), we consider surgery on a circle C in $S^2 \times S^1$. A tubular neighborhood of this circle is a torus Σ; a useful basis of $H^1(\Sigma, \mathbf{Z})$ is indicated. In (b), in addition to the circle C on which we perform surgery, there is a parallel circle C' on which we place a Wilson line in the R_j representation. In (c), there are two parallel circles C' and C'' with Wilson lines in representations R_j and R_k.

16. The purpose of this figure is to indicate how S^3 can be made by surgery starting with $S^2 \times S^1$. In (a) we show a torus Σ, sitting in R^3 and in (b) a pair of identical solid tori.

17. In (a), we sketch linked but unknotted Wilson lines on S^3, in the R_i and R_j representations. In (b), a Wilson line R_i is linked with two Wilson lines R_j and R_k on S^3. In (c), we sketch how two crucial amplitudes can be factored through the same one dimensional space.

18. Any link on S^3 can be shaped into the form of a braid, as in (a); putting the same braid on $S^2 \times S^1$, and doing surgery on the circle C indicated with the dotted line in part (b), one gets back to the original link on S^3. If one 'cuts' $S^2 \times S^1$ to get to $S^2 \times I$, the braid can be 'unbraided'; the braid is recovered by prescribing a diffeomorphism of S^2, via which the top and bottom of part (c) are to be identified. In (d), we sketch the origin of the key property of the braid traces. In the presence of an arbitrary Wilson line on the dotted contour (reflecting the results of surgery), two braids B_1 and B_2 are joined end to end in $S^2 \times S^1$. There is no way to tell which comes first, so the partition function is invariant under exchange of B_1 and B_2.

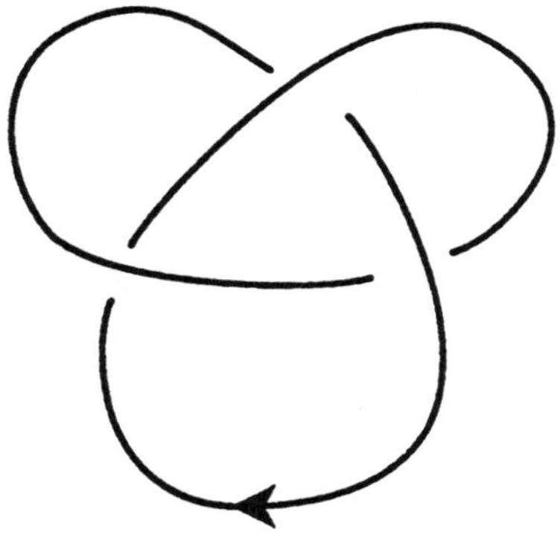

FIGURE 1

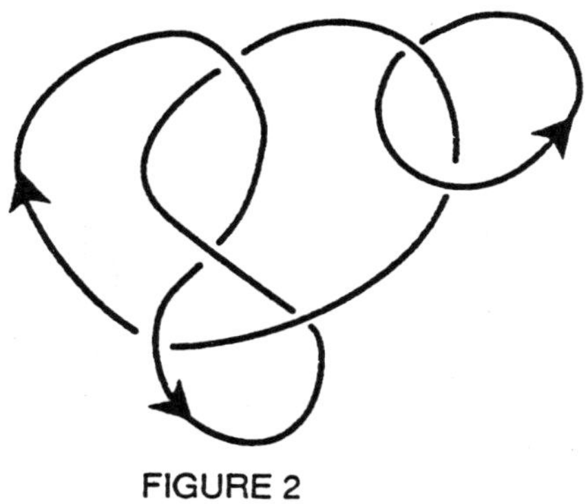

FIGURE 2

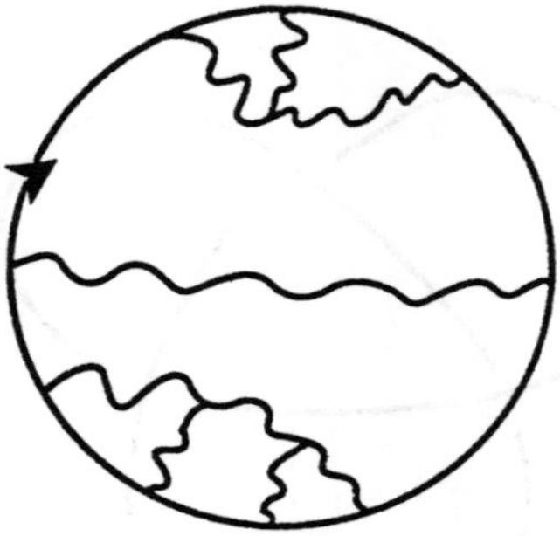

FIGURE 3-a

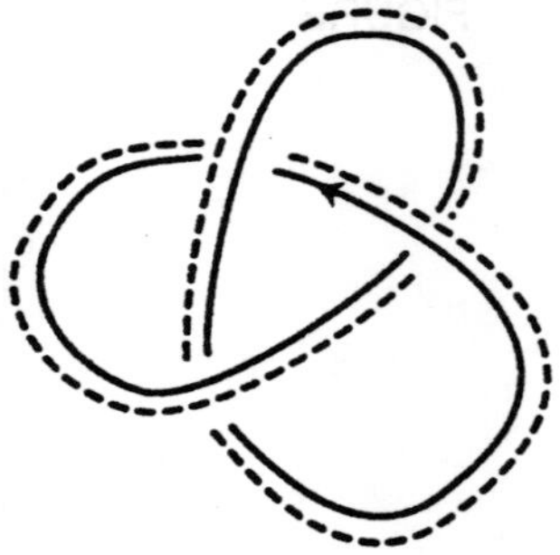

FIGURE 3-b

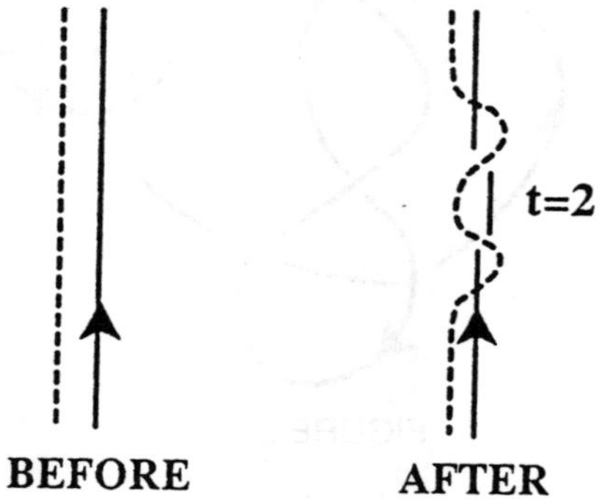

FIGURE 3-c

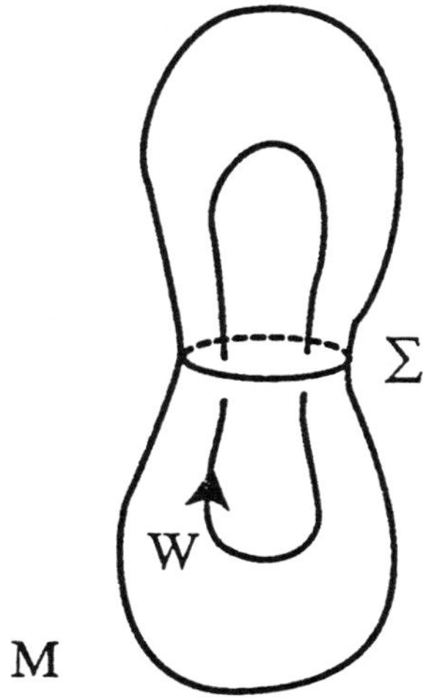

FIGURE 4-a

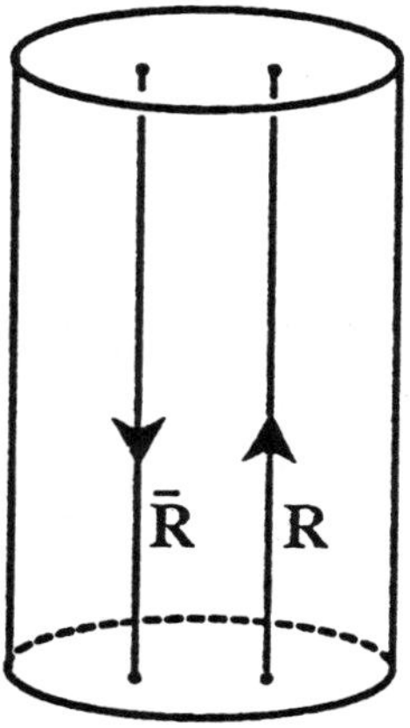

FIGURE 4-b

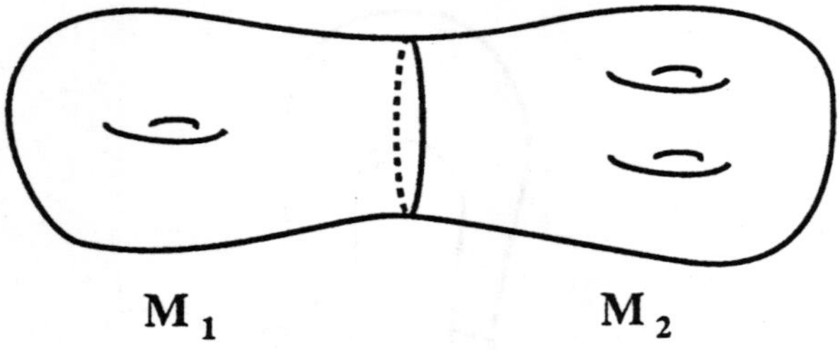

FIGURE 5-a

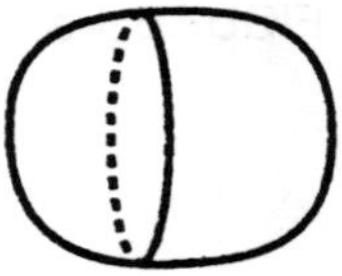

FIGURE 5-b

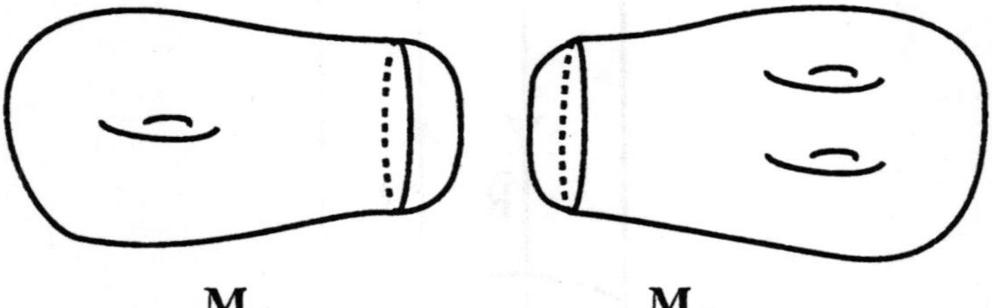

FIGURE 5-c

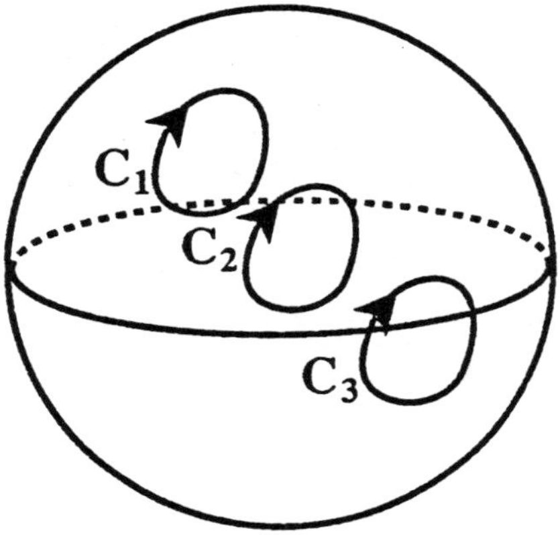

FIGURE 6

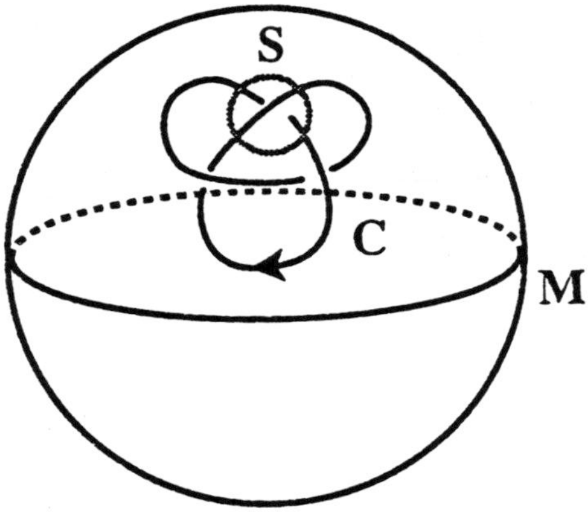

FIGURE 7-a

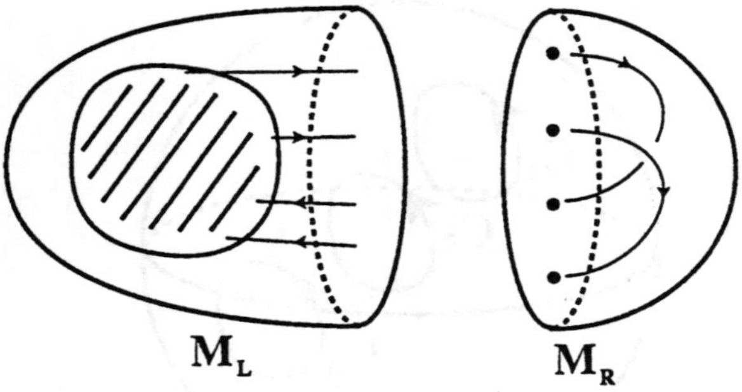

FIGURE 7-b

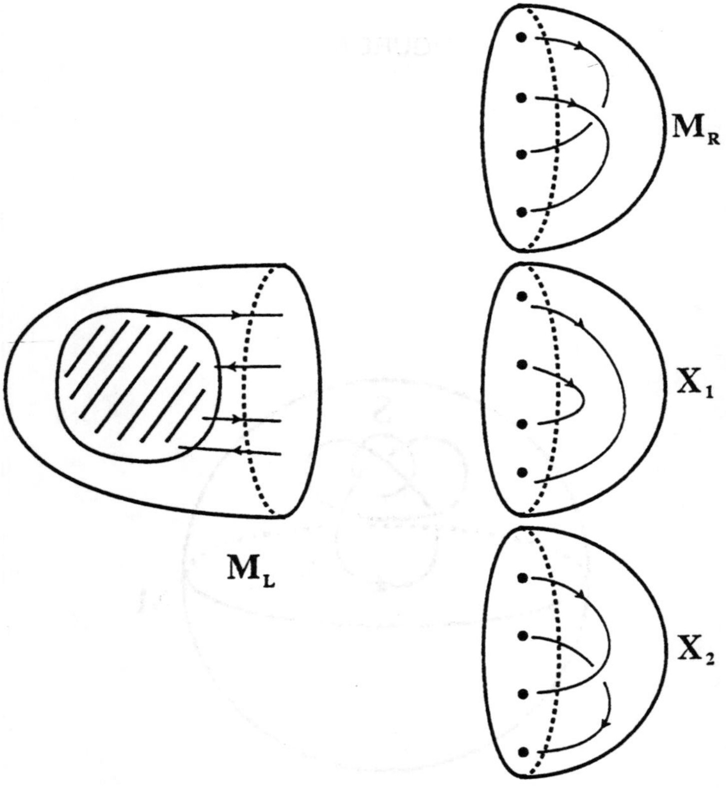

FIGURE 7-c

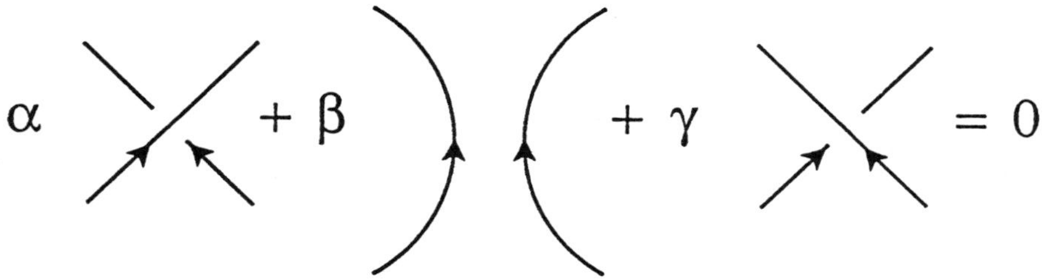

FIGURE 8

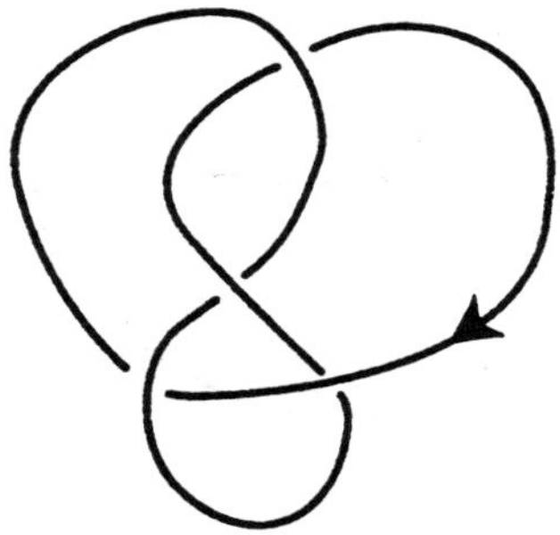

FIGURE 9

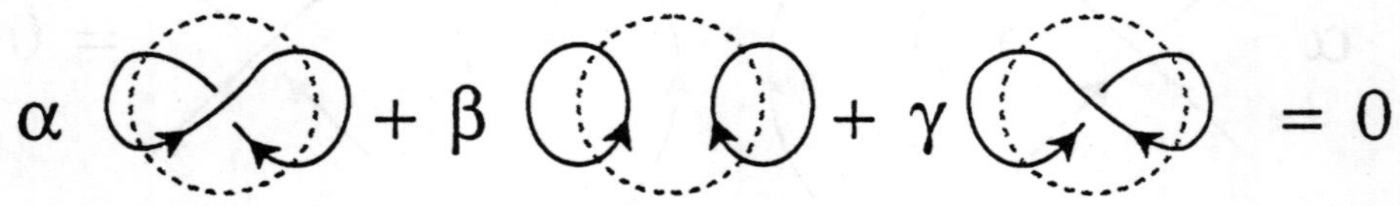

FIGURE 10

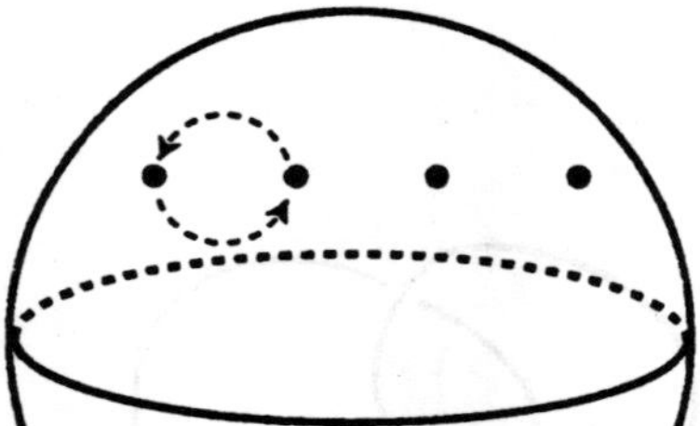

FIGURE 11-a

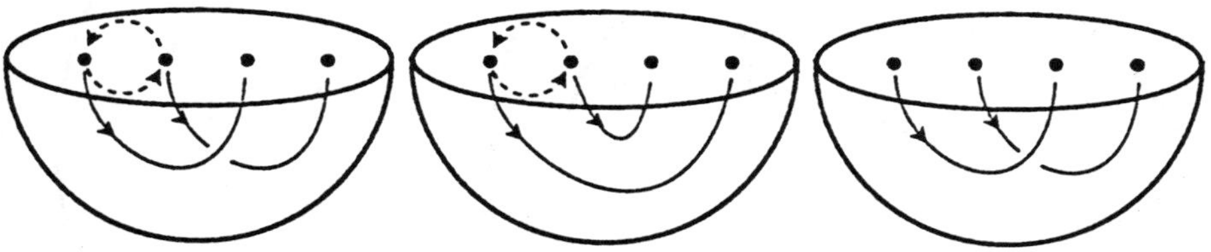

FIGURE 11-b

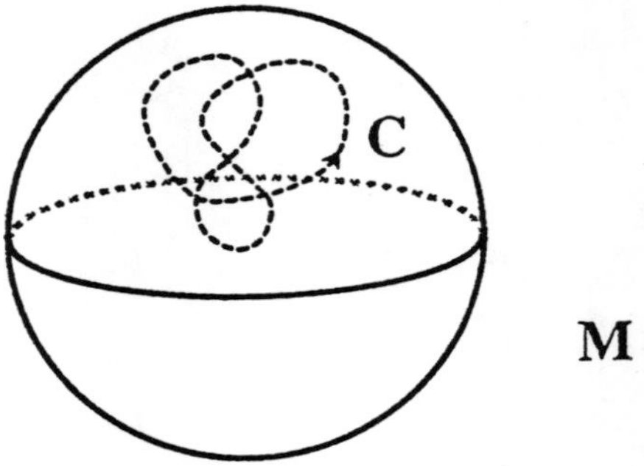

FIGURE 12-a

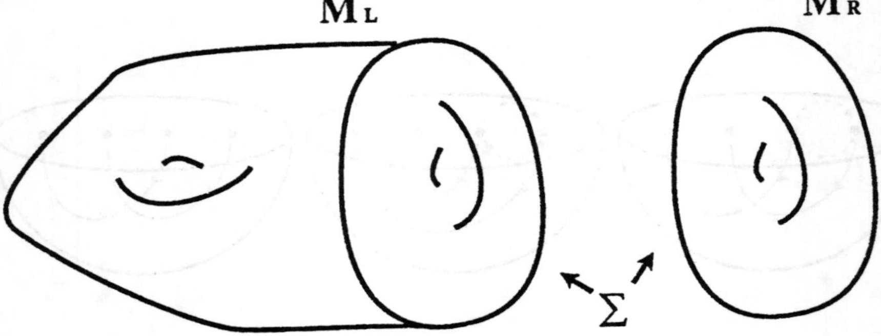

FIGURE 12-b

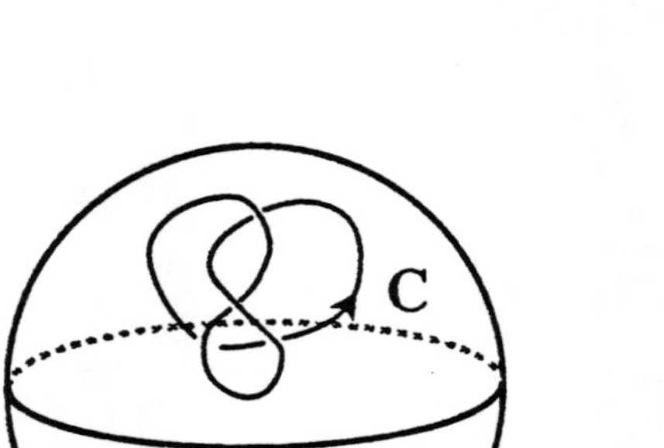

FIGURE 12-c

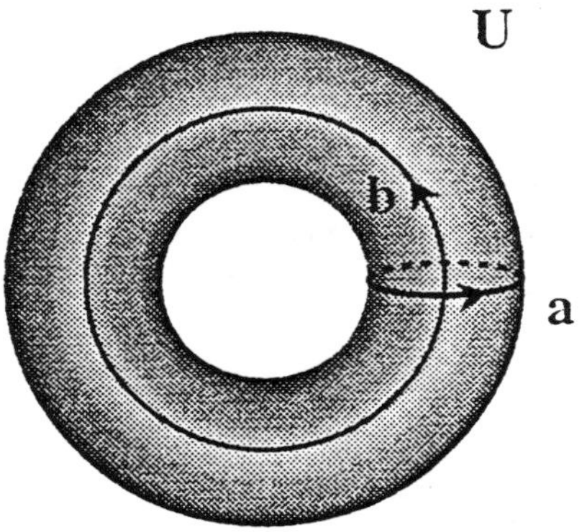

FIGURE 13-a

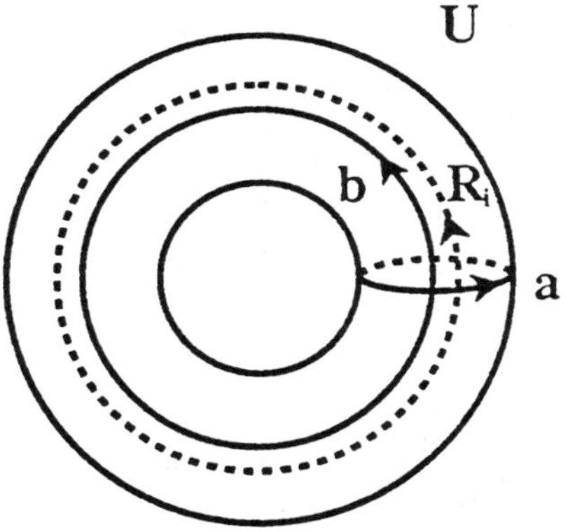

FIGURE 13-b

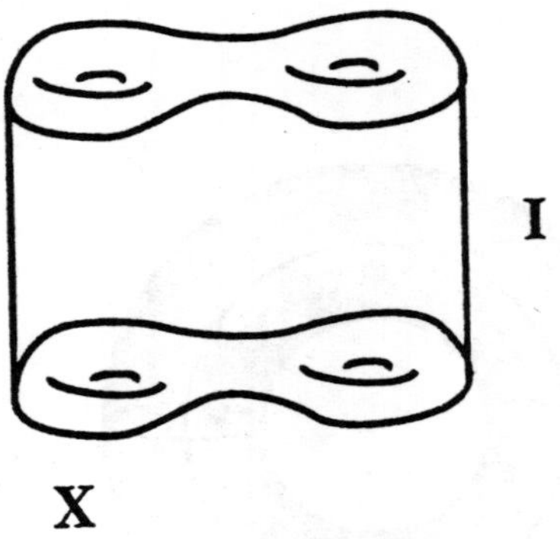

FIGURE 14-a

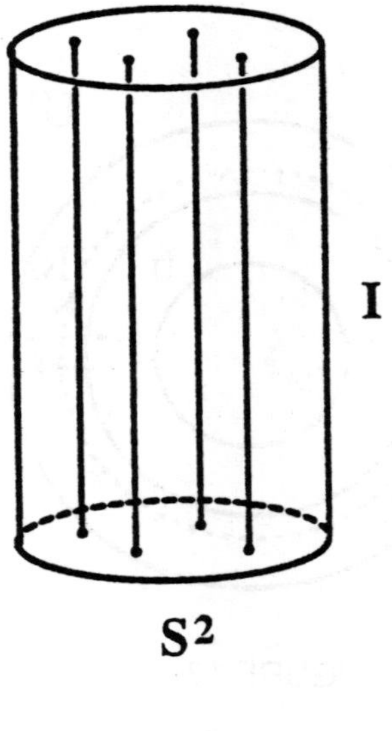

FIGURE 14-b

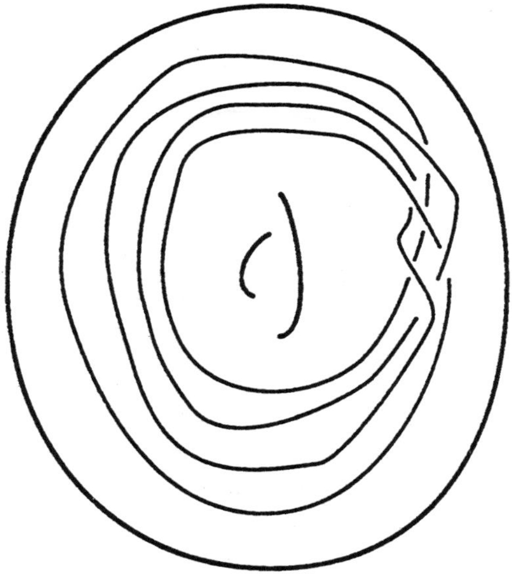

FIGURE 14-c

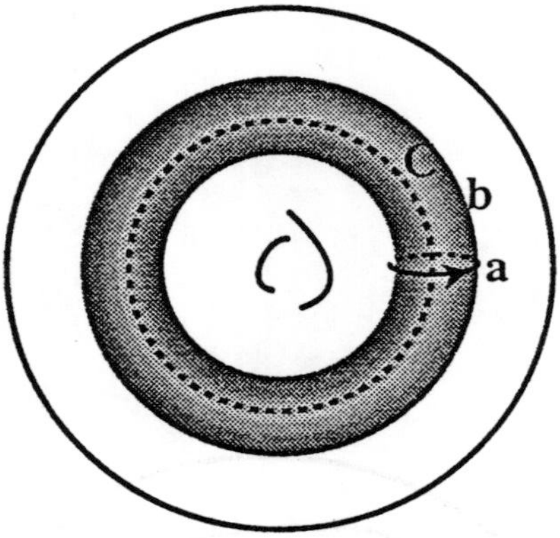

FIGURE 15-a

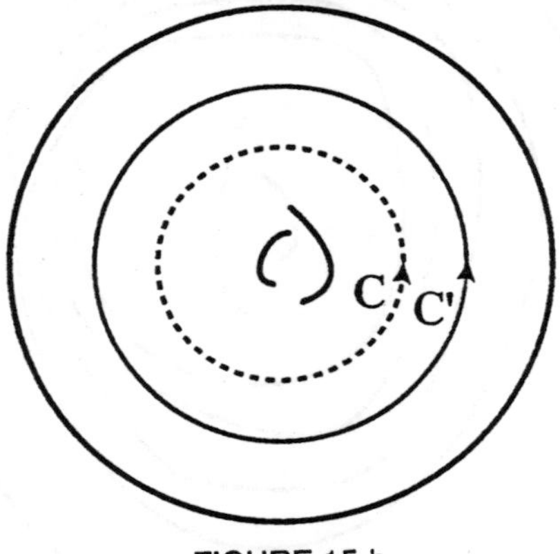

FIGURE 15-b

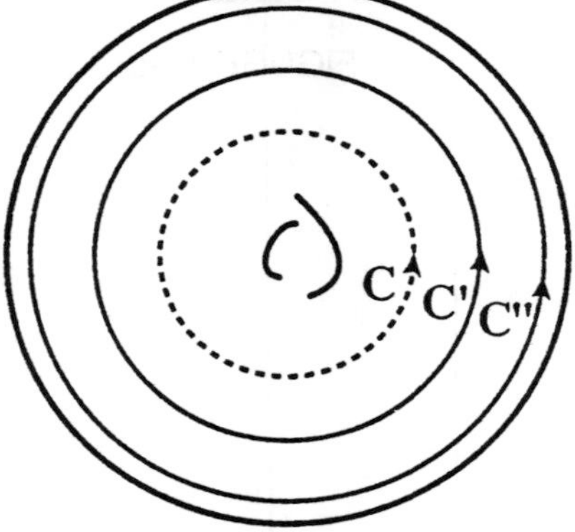

FIGURE 15-c

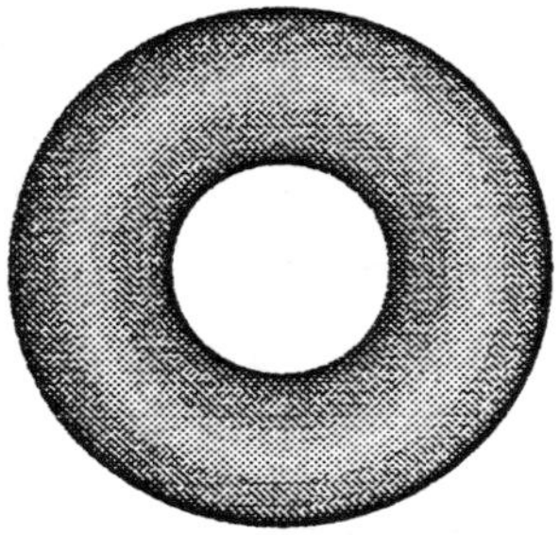

FIGURE 16-a

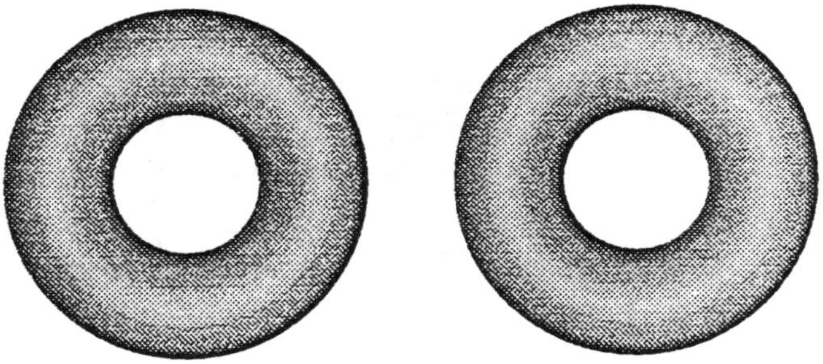

FIGURE 16-b

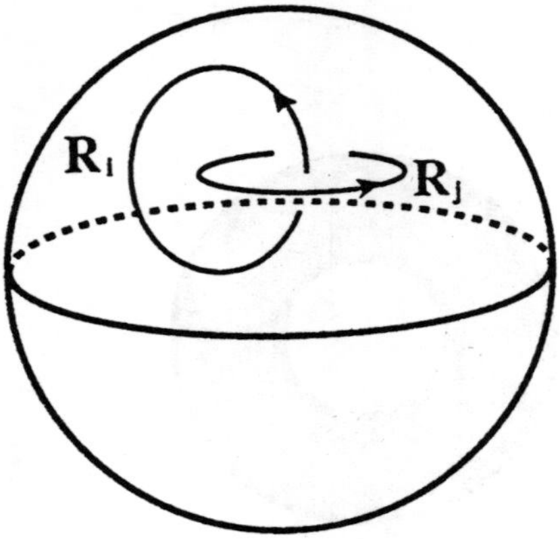

FIGURE 17-a

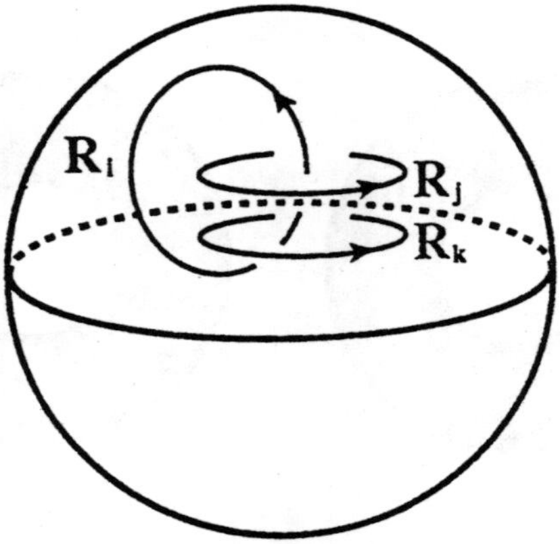

FIGURE 17-b

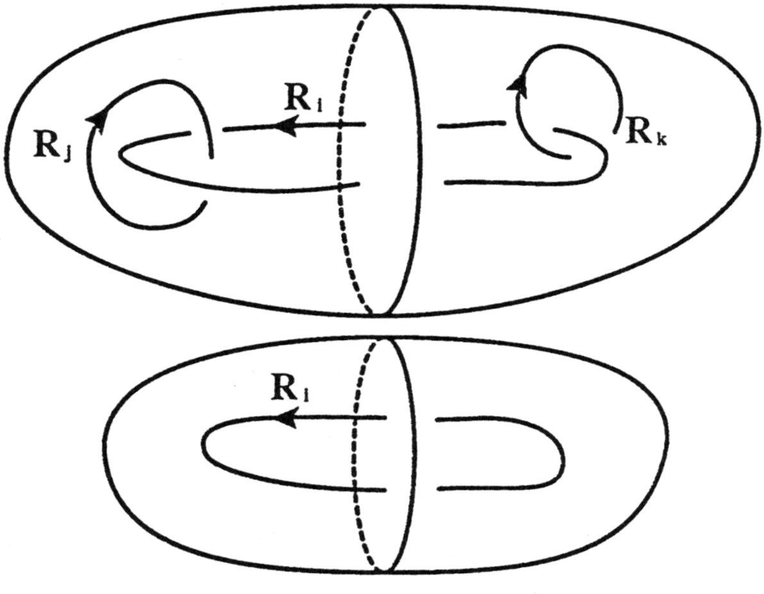

FIGURE 17-c

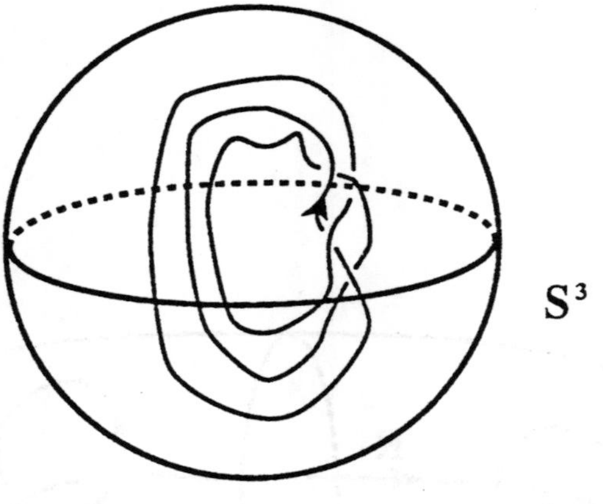

FIGURE 18-a

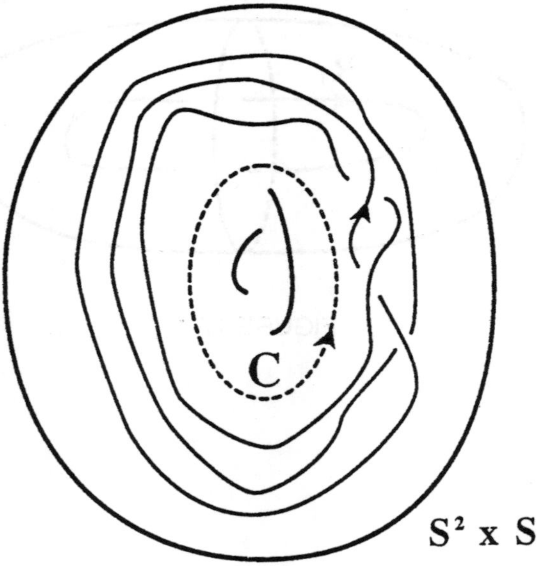

FIGURE 18-b

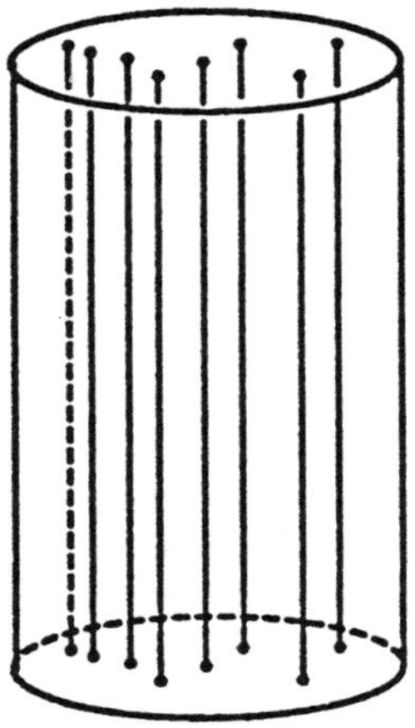

FIGURE 18-c

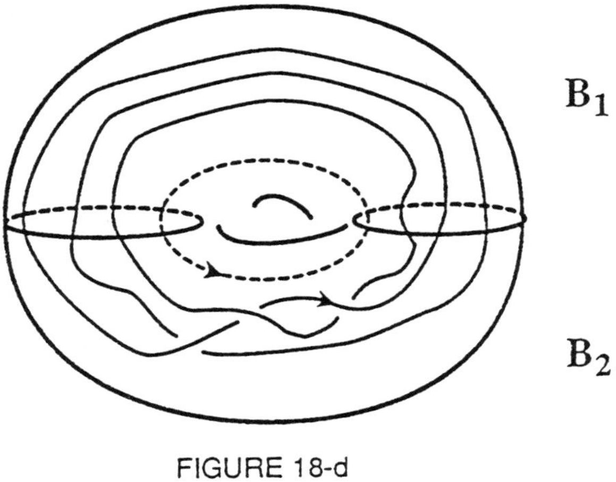

FIGURE 18-d

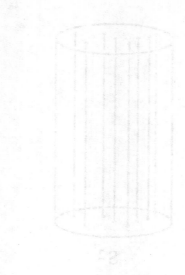